ENGINEERED POLYMER NANOCOMPOSITES FOR ENERGY HARVESTING APPLICATIONS

ENGINEERED POLYMER NANOCOMPOSITES FOR ENERGY HARVESTING APPLICATIONS

Edited by

M. T. Rahul
Department of Physics, Catholicate College, Pathanamthitta, India

Nandakumar Kalarikkal
School of Pure and Applied Physics, Mahatma Gandhi University, Kottayam, India

Sabu Thomas
International and Inter University Centre for Nanoscience and Nanotechnology, Mahatma Gandhi University, Kottayam, India; School of Energy Materials, Mahatma Gandhi University, Kottayam, India

Bruno Ameduri
ICGM, University of Montpellier, CNRS, ENSCM, Montpellier, France

Didier Rouxel
Université de Lorraine, CNRS, Institut Jean Lamour, Nancy, France

Raneesh Balakrishnan
Department of Physics, Catholicate College, Pathanamthitta, India

ELSEVIER

Elsevier
Radarweg 29, PO Box 211, 1000 AE Amsterdam, Netherlands
The Boulevard, Langford Lane, Kidlington, Oxford OX5 1GB, United Kingdom
50 Hampshire Street, 5th Floor, Cambridge, MA 02139, United States

ISBN: 978-0-12-824155-4

For Information on all Elsevier publications
visit our website at https://www.elsevier.com/books-and-journals

Publisher: Matthew Deans
Acquisitions Editor: Christina Gifford
Editorial Project Manager: Joshua Mearns
Production Project Manager: Kamesh Ramajogi
Cover Designer: Mark Rogers

Typeset by MPS Limited, Chennai, India

Contents

List of contributors

Bruno Ameduri ICGM, University of Montpellier, CNRS, ENSCM, Montpellier, France

T. Avanish Babu Ceramic Composite Laboratory, Centre for Functional Materials, Vellore Institute of Technology, Vellore, India

Raneesh Balakrishnan Department of Physics, Catholicate College, Pathanamthitta, Kerala, India

Tarak Ben Zineb Université de Lorraine, CNRS, Arts et Métiers ParisTech, LEM3, Nancy, France

Fatima Ezzahra Bouharras IMED-Lab, Faculty of Sciences and Techniques, Cadi Ayyad University (UCA), Marrakesh, Morocco; ICGM, University of Montpellier, CNRS, ENSCM, Montpellier, France

Sobi K. Chacko Department of Physics, Catholicate College, Pathanamthitta, Kerala, India

Arunkumar Chandrasekhar Nanosensors and Nanoenergy Lab, Department of Sensors and Biomedical Technology, School of Electronics Engineering, Vellore Institute of Technology, Vellore, India

Samir Chatbouri Université de Lorraine, CNRS, Institut Jean Lamour, Nancy, France

Sujoy Kumar Ghosh Laboratorio NEST, Istituto Nanoscienze-CNR and Scuola Normale Superiore, Pisa, Italy

Honey John Inter University Centre for Nanomaterials and Devices, Cochin University of Science and Technology, Kochi, India; Department of Polymer Science and Rubber Technology, Cochin University of Science and Technology, Kochi, India

Sherin Joseph Inter University Centre for Nanomaterials and Devices, Cochin University of Science and Technology, Kochi, India

W. Madhuri Ceramic Composite Laboratory, Centre for Functional Materials, Vellore Institute of Technology, Vellore, India

Biswajit Mahanty Department of Electronics & Communication Engineering, Ramgarh Engineering College, Ramgarh, India

Pralay Maiti School of Materials Science and Technology, Indian Institute of Technology (BHU), Varanasi, India

Ammu Anna Mathew School of Electrical Engineering, Vellore Institute of Technology, Vellore, India

Anshida Mayeen Inter University Centre for Nanomaterials and Devices, Cochin University of Science and Technology, Kochi, India

Guven Muslum Université de Lorraine, CNRS, Arts et Métiers ParisTech, LEM3, Nancy, France

Avinash R. Pai International and Inter University Centre for Nanoscience and Nanotechnology, Mahatma Gandhi University, Kottayam, India

Luana Persano Laboratorio NEST, Istituto Nanoscienze-CNR and Scuola Normale Superiore, Pisa, Italy

M.T. Rahul Department of Physics, Catholicate College, Pathanamthitta, Kerala, India

Mustapha Raihane IMED-Lab, Faculty of Sciences and Techniques, Cadi Ayyad University (UCA), Marrakesh, Morocco

Farsa Ram Polymer Science and Engineering Division, CSIR-National Chemical Laboratory, Pune, India; Academy of Scientific and Innovative Research (AcSIR), Ghaziabad, India

Didier Rouxel Université de Lorraine, CNRS, Institut Jean Lamour, Nancy, France

Kadhiravan Shanmuganathan Polymer Science and Engineering Division, CSIR-National Chemical Laboratory, Pune, India; Academy of Scientific and Innovative Research (AcSIR), Ghaziabad, India

Sunija Sukumaran Université de Lorraine, CNRS, Arts et Métiers ParisTech, LEM3, Nancy, France; Université de Lorraine, CNRS, Institut Jean Lamour, Nancy, France

K. Thiyagarajan School of Electrical Engineering, Vellore Institute of Technology, Vellore, India

Sabu Thomas International and Inter University Centre for Nanoscience and Nanotechnology, Mahatma Gandhi University, Kottayam, India; School of Energy Materials, Mahatma Gandhi University, Kottayam, India

Sharin Maria Thomas Centre for Nanotechnology Research, Vellore Institute of Technology, Vellore, India

Shivam Tiwari School of Materials Science and Technology, Indian Institute of Technology (BHU), Varanasi, India

S. Vivekanandan School of Electrical Engineering, Vellore Institute of Technology, Vellore, India

Preface

The ambition of researchers in the present scenario of an energy crisis is to guarantee access for the world population to reliable, durable, and modern cost-effective energy services. Hence, the development of energy-harvesting devices (EHDs) is receiving great attention within the research community. The performances of such devices greatly depend on the material's properties. As a strategy for developing promising EHDs, polymer-based composites have been widely explored. The main goal of this book is to provide a deeper understanding of various polymer systems and their composites that are suitable for energy-harvesting purposes. A detailed survey on the materials aspects of piezoelectric, triboelectric, pyroelectric, and magnetomechanoelectric EHDs are presented in this book. As far as energy harvesting is concerned, fluorinated polymers, such as polyvinylidene fluoride (PVDF) and VDF copolymers, stand out due to their remarkable electroactive properties. Better electroactive properties in PVDF and VDF copolymers can be attained by crystallizing them in the electroactive phase conformations. Electroactive modification strategies and suitable characterization methods to probe the chain conformations of the fluorinated polymers can also be understood from this book. In addition to the fluorinated polymers, nonfluorinated polymers (natural and synthetic) have been used for developing EHDs. The interesting energy-harvesting performances of the nonfluorinated polymer-based EHDs were also discussed in a detailed manner. Hybrid composites based on shape memory alloys and piezoelectric materials also have emerged as suitable material systems for converting the widely available ambient thermal energy into electric energy. Design, fabrication, energy-harvesting studies, and finite element analysis of the shape memory alloy/piezo composites are also presented. As a further step to improve the performance of the EHDs, devices which simultaneously make use of two or more energy conversion mechanisms are getting much attention. Discussions on coupled EHDs based on piezo- and pyroelectric effects are supplied. Other than the fascinating attributes of polymer nanocomposites and the performances of the devices based on such materials, attention has been given to provide the fundamental aspects of various polymeric systems.

The chapters of this book are arranged in such a way as to get a clear idea of the fundamentals, fascinating attributes, and the energy-harvesting performances of the fluorinated and nonfluorinated electro-active polymer-based composites. We believe that this book will be a very valuable reference material for those who have an interest in the research and development of EHDs. Also we would like to thank all of the contributors, each an expert in these fascinating fields of research.

CHAPTER

1

Recent advances in vinylidene fluoride copolymers and their applications as nanomaterials

Fatima Ezzahra Bouharras[1,2], *Mustapha Raihane*[1] *and Bruno Ameduri*[2]

[1]IMED-Lab, Faculty of Sciences and Techniques, Cadi Ayyad University (UCA), Marrakesh, Morocco, [2]ICGM, University of Montpellier, CNRS, ENSCM, Montpellier, France

1.1 Introduction

Piezoelectric and ferroelectric polymers are materials that display many advantages, such as light weight, low cost, flexibility, ease of processing, and high electric breakdown strength, which make them promising candidates in the dielectric field [1]. Moreover, and due to their unique features, they can be exploited in many technologies, such as in sensors and actuators [2,3], and in biomedical [4] and energy harvesting applications (Fig. 1.1) [5,6].

Poly(vinylidene fluoride), PVDF, and P(VDF-*co*-TrFE) copolymers (where TrFE stands for trifluoroethylene, and particularly when TrFE mol % is 20−50) are among the polymers with the highest ferroelectric and piezoelectric activity, making them suitable to be integrated into a variety of flexible electrical devices [7]. PVDF is a semicrystalline polymer that has a molecular repeating unit of (CH_2-CF_2), and can exist in five different crystalline polymorphs, namely, α, β, γ, δ, and ε (the α and ε being nonpolar phases, while β (all trans) is a polar one) [8−10]. In fact, the major interest lies in obtaining the electroactive polar β phase of PVDF that exhibits a piezoelectric behavior [10,11]. Such a structure is thermodynamically favored while the α phase is kinetically produced. This phase can be

Engineered Polymer Nanocomposites for Energy Harvesting Applications
DOI: https://doi.org/10.1016/B978-0-12-824155-4.00010-7

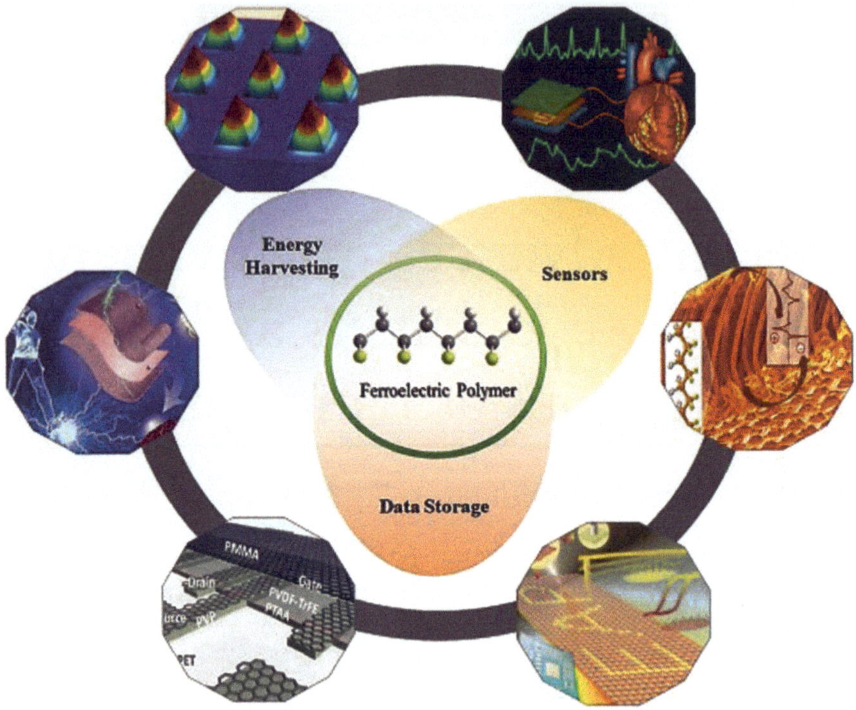

FIGURE 1.1 Applications of ferroelectric polymer in flexible electronics, including data storage, energy harvesting, and sensors. Source: *Reproduced with permission of X. Chen, X. Han, Q.D. Shen, PVDF-based ferroelectric polymers in modern flexible electronics, Adv. Electron. Mater. 3 (2017) 1600460. https://doi.org/10.1002/aelm.201600460 [7].*

obtained from various methods such as mechanical stretching of the α phase, casting process from polar solution, an ultrafast quenching process [12], drawing at low temperature, annealing at high temperature, or under high pressure or with a strong electric field, or from the addition of a second polymer to PVDF in a blend [13] or in a block [14].

In addition to the abovementioned properties, the molecular structure of PVDF can be modified in order to obtain PVDF-based co/terpolymers, such as P(VDF-*co*-CTFE), P(VDF-*co*-TrFE) and P(VDF-*ter*-TrFE-*ter*-CTFE) or P(VDF-*ter*-TrFE-*ter*-CFE) terpolymers (where CTFE and CFE stand for chlorotrifluoroethylene and chlorofluoroethylene, respectively), which will be discussed in Section 1.3. These modifications enable the tailoring of intermolecular interactions, and therefore enhance the overall features of PVDF. Thus the combination of these monomers leads to polymers with particularly promising performances acting in different fields of applications.

On the other hand, the approach of using a polymer composite could be considered. In fact, the incorporation of a ceramic filler (with a high dielectric constant) into a polymer with good flexibility and high breakdown strength could lead to a material with enhanced properties [15,16]. Mayeen et al. [17,18] studied the effect of the introduction of barium titanate ($BaTiO_3$) and nickel ferrite ($NiFe_2O_4$) nanoparticles into P(VDF-*co*-TrFE) copolymer matrix. They found that the resulting nanocomposites exhibited enhanced dielectric, piezoelectric, ferroelectric, and magnetoelectric properties, making these nanocomposites promising candidates in different fields, such as smart magnetic sensors and energy storage devices. Loos' team comprehensively summarized these magnetoelectric features in a book chapter [19].

1.2 Different classes of ferroelectric polymers

Ferroelectrics are materials belonging to a special class of dielectrics, which have a switchable spontaneous polarization that can be switched by the action of an external high electric field (0.1–0.5 MV/cm) [20–24]. A typical polarization versus electric field loop of a ferroelectric material is presented in Fig. 1.2, which can afford ferroelectric information, including remanent polarization (P_r), spontaneous polarization (P_s), and coercive field (E_c). Remanent polarization is present where no electric field is applied, saturation polarization is the maximum polarization that can be reached, while the coercive field corresponds to the field required to turn the polarization to zero [19,25,26].

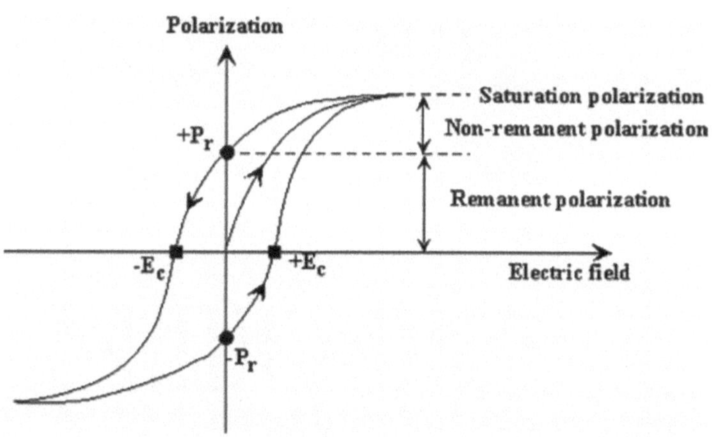

FIGURE 1.2 Typical polarization versus electric field (P-E) hysteresis loop of ferroelectrics. Source: *Reproduced with permission of N. Izyumskaya, Y.I. Alivov, S.J. Cho, H. Morkoç, H. Lee, Y.S. Kang, Processing, structure, properties, and applications of PZT thin films, Crit. Rev. Solid. State Mater. Sci. 32 (2007) 111–202. https://doi.org/10.1080/10408430701707347 [25].*

When an electric field is applied, polarization increases linearly with the amplitude of the electric field until the direction of the polarization of all domains aligns with the direction of the applied electric field. Then, when this latter becomes sufficiently high to switch dipoles, polarization changes sharply and reaches saturation at the higher field. In addition, when the field starts to decrease, the direction of some domains will switch back. However, the polarization will not become zero when the applied field decreases to zero; instead, it will become zero at the opposite coercive field ($-E_c$). Further increasing the applied field in the reverse direction will induce a new alignment in the direction of all domains and by decreasing it in reverse it completes the cycle of the hysteresis loop [27–30].

When the positive and negative parts of coercive field are equal, a symmetrical hysteresis loop is obtained. Actually, the shape of the hysteresis loop, that is, coercive field (E_c in Fig. 1.2), spontaneous polarization, and remanent polarization can be also affected by many factors, such as the device factor, including thickness, mechanical stresses, film fabrication conditions, applied electric field, etc. [31].

Ferroelectric polymers can be divided into two main classes (Fig. 1.3): the first one consists of liquid crystalline polymers, and is associated with uniaxial drawing which causes the orientation of polymer chains along the

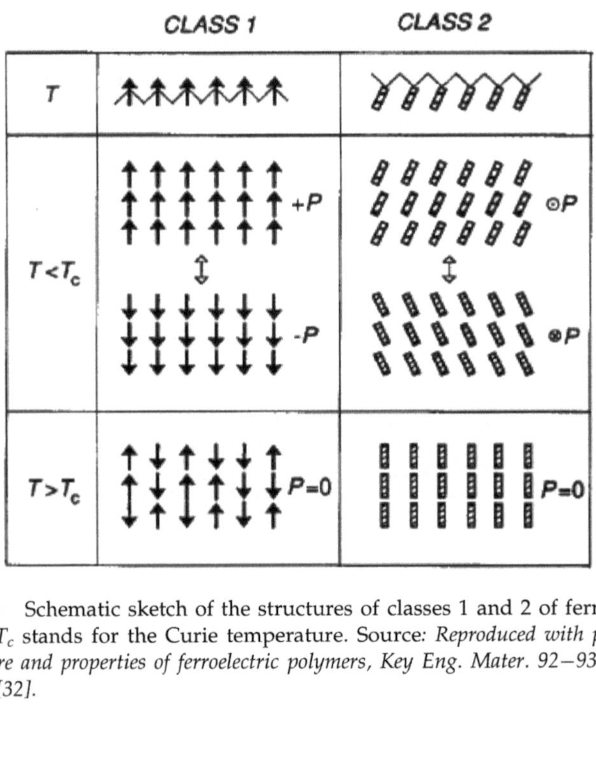

FIGURE 1.3 Schematic sketch of the structures of classes 1 and 2 of ferroelectric polymers, where T_c stands for the Curie temperature. Source: *Reproduced with permission of F. Takeo, Structure and properties of ferroelectric polymers, Key Eng. Mater. 92–93 (1994) 15–30, Scientific.Net. [32].*

draw direction. The second class is composed of polar polymers having permanent dipoles in the direction perpendicular to the chain axis and is concerned with the pooling treatment that uses the application of a high electric field to induce dipole orientation in polar polymers [32]. Examples of polar ferroelectric polymers include polyvinyl fluoride (PVF), PVDF, and copolymers of VDF with trifluoroethylene or etc. [20].

1.3 PVDF and VDF copolymers and terpolymers

VDF is a gas at room temperature, and is colorless and odorless. It boils at −84°C (Table 1.1) and can form an explosive mixture with air. Polymerization of this gas is highly exothermic and takes place above its critical temperature and pressure [33–35].

1.3.1 PVDF homopolymer

PVDF is a semicrystalline polymer containing 59.4 wt.% of fluorine atoms and 3 wt.% of hydrogen atoms [8,16,35]. Its chemical formula $(\text{-}CH_2\text{-}CF_2\text{-})_n$ is an intermediate between that of polyethylene $(\text{-}CH_2\text{-}CH_2\text{-})_n$ and polytetrafluoroethylene $(\text{-}CF_2\text{-}CF_2\text{-})_n$, providing both high flexibility (as

TABLE 1.1 Properties of vinylidene fluoride.

Property	Value
Molecular weight	64.038
Boiling point (°C)	−84
Freezing point (°C)	−144
Vapor pressure at 21°C (kPa)	3.683
Critical pressure (MPa)	4.434
Critical temperature (°C)	30.1
Critical density (kg/m^3)	417
Heat of formation at 25°C (kJ/mole)	−345.2
Heat of polymerization at 25°C (kJ/mole)	−474.2
Explosive limits (vol.% in air)	5.8–20.3
Solubility in water, cm^3/100 g at 25°C at 10 kPa	6.3
Activation energy of polymerization, E_0 (kJ/mole)	161

Reproduced with permission of A.L. Moore, Fluoroelastomer monomers. Fluoroelastomers Handbook, Elsevier, 2006, pp. 25–36. https://doi.org/10.1016/b978-081551517-3.50006-x [33].

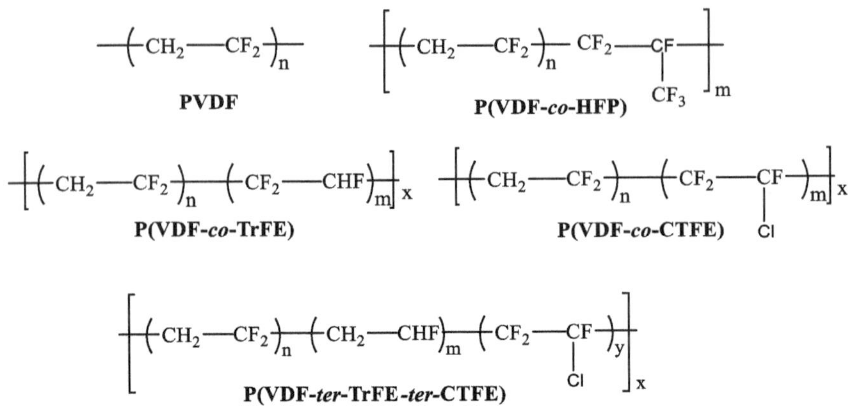

FIGURE 1.4 Schematic representation of the molecular structures of PVDF and its corresponding co/terpolymers.

much as PE) and some chemical constraints (similar to PTFE) to the main structure of PVDF [36,37]. The product characteristics and quality will be affected by variables such as pressure, temperature, additives, as well as monomer feeding. For example, the temperature of polymerization has a significant effect on the incident of "head-to-head" ($-CF_2-CF_2$) or "tail-to-tail" ($-CH_2-CH_2$) defects of chaining within the molecular backbones (Fig. 1.4) [34,38].

PVDF is commonly synthesized by a radical initiated aqueous polymerization, involving pressures of 10–300 atm at temperatures of 10°C–130°C depending on the process (emulsion, suspension, or solution) [34].

Controlled radical polymerization (or RDRP) of VDF can be achieved by iodine transfer polymerization (ITP) [39–44], cobalt mediated radical polymerization [45,46], and reversible addition fragmentation chain transfer polymerization (RAFT) [47,48].

Guerre et al. [48,49] investigated in detail the preparation of PVDF by RAFT polymerization of VDF initiated by 3-*tert*-amylperoxy-2-ethylhexanoate and controlled by *O*-ethyl-*S*-(1-methoxycarbonyl) ethyldithiocarbonate as the chain transfer agent. The resulting PVDF-Xanthate was mainly characterized by ^1H and ^{19}F NMR spectroscopy MALDI tof and DSC.

1.3.2 VDF co-/terpolymers

PVDF can feature several types of macromolecular and crystal structures, which change depending on the synthesis conditions. Indeed, its overall properties could be enhanced as well by the preparation of a copolymer or terpolymer containing VDF and other fluoromonomers, the reactivities of which must be taken into account to tune their contents [34] (Fig. 1.4).

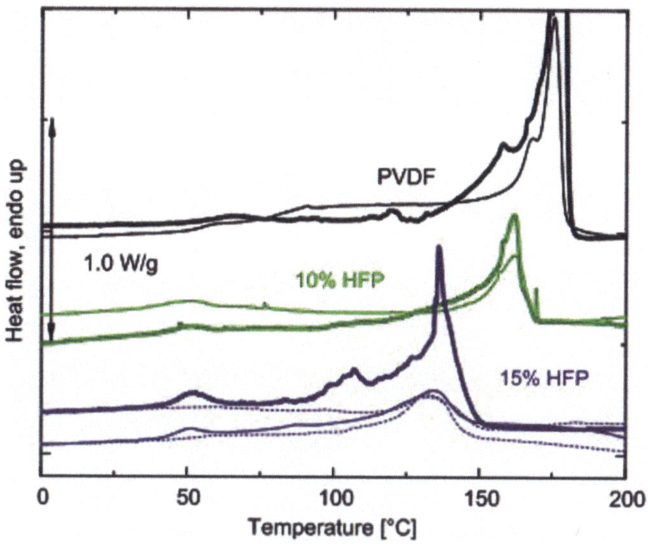

FIGURE 1.5 DSC thermograms of PVDF (toppermost) and P(VDF-*co*-HFP) copolymers with 10 wt.% and 15 wt.% HFP (first scans after preparation). Source: *Reproduced with permission of P. Frübing, F. Wang, M. Wegener, Relaxation processes and structural transitions in stretched films of polyvinylidene fluoride and its copolymer with hexafluoropropylene, Appl. Phys. A 107 (2012) 603–611. https://doi.org/10.1007/s00339-012-6838-1 [58].*

1.3.2.1 P(VDF-co-HFP) copolymers

Poly(vinylidene fluoride-*co*-hexafluoropropylene) copolymer can be either thermoplastic or elastomer (depending on the HFP content) and has attracted great attention as a desirable membrane material or component for Lithium ion batteries [50–55]. In fact, the incorporation of HFP groups into the main backbones of VDF enhances the properties of prepared copolymer resulting in lower crystallinity [56], higher solubility [57], higher free volume [53], and better mechanical strength [51]. Moreover, the addition of HFP leads to an increase in fluorine content making the P(VDF-*co*-HFP) copolymer more hydrophobic [52]. Frübing et al. [58] prepared films from PVDF and P(VDF-*co*-HFP) copolymer with 10 and 15 wt.% of HFP and further studied their thermal properties. Fig. 1.5 exhibits DSC thermograms taken from the first scans after the preparation of the different samples, showing clearly the influence of HFP content on the melting of the crystalline phase of PVDF. Indeed, the addition of HFP to the copolymer results in a decrease of thermal stability and a decreased crystallinity.

1.3.2.2 P(VDF-co-CTFE) copolymers

Poly(VDF-*co*-chlorotrifluoroethylene) copolymer shows a chemical structure of $[(CH_2CF_2)_x(CF_2CFCl)_y]_n$, where the overall properties

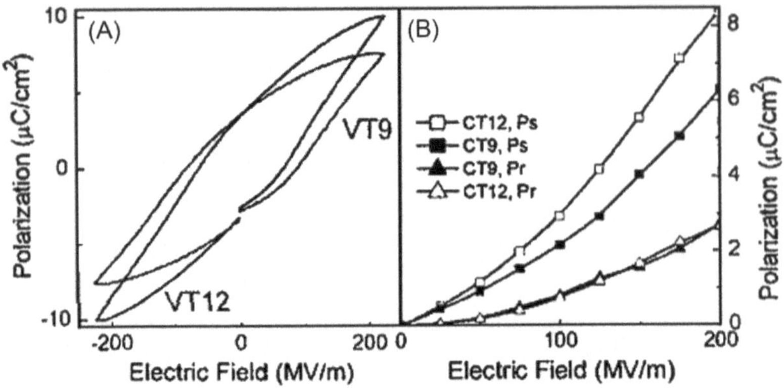

FIGURE 1.6 (A) Polarization hysteresis loop of CT9 (9 mol% of CTFE) and CT12 (12 mol% of CTFE) samples at room temperature, and (B) the maximum polarization P_s and remanent polarization P_r versus the maximum electric field applied in the experiments. Source: *Reproduced with permission of Z. Li, Y. Wang, Z.Y. Cheng, Electromechanical properties of poly(vinylidene-fluoride-chlorotrifluoroethylene) copolymer, Appl. Phys. Lett. 88 (2006) 062904. https://doi.org/10.1063/1.2170425 [61].*

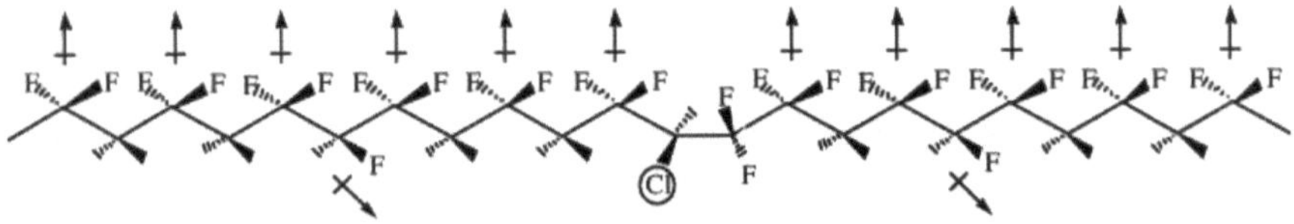

SCHEME 1.1 Sketch of the dipole moments in P(CTFE-*co*-VDF) copolymers and the kink related to the presence of bulky chlorine atoms. Source: *Reproduced with permission of Z. Zhang, T.C.M. Chung, Study of VDF/TrFE/CTFE terpolymers for high pulsed capacitor with high energy density and low energy loss, Macromolecules 40 (2007) 783–785. https://doi.org/10.1021/ma0627119 [62].*

strongly depend on CTFE content [34]. In fact, at low VDF contents (<25 mol%), P(VDF-*co*-CTFE) copolymer is semicrystalline with a hexagonal structure, thus for VDF content ranging between 25 and 70 mol %, the copolymer becomes amorphous. However, at high VDF content (>75 mol%), the copolymer has a semicrystalline structure with a monoclinic phase [59,60]. Li et al. [61] investigated the effect of CTFE content on the electromechanical response of P(VDF-*co*-CTFE) copolymers. Two films were prepared by varying the CTFE content from 9 mol% (CT9) to 12 mol% (CT12). Fig. 1.6 exhibits the hysteresis loops obtained for both copolymers. Actually, the introduction of bulky CTFE into PVDF modifies the structure and properties of the resulting copolymer, especially its ferroelectricity. As Zhang and Chung [62] reported, a "kink" was induced by the chlorine atoms of CTFE (Scheme 1.1).

TABLE 1.2 DSC and XRD parameters of the first heating, piezoelectric coefficient (d_{33}), and Elastic constant (C_{11}) of P(VDF-co-TrFE) films (VDF/TrFE content: 70/30).

P(VDF-*co*-TrFE)	T_c[a] (°C)	T_m[b] (°C)	Crystallinity (%)		d_{33} (μC/cm²)	C_{11} (GPa)
			DSC	DRX		
Spin coating[c]	106	150	30	43	26	11–12
Solvent casting[d]	101	150	30	43	26	11–12

[a]*Curie temperature.*
[b]*Melting temperature.*
[c]*Thickness 3–15 μm.*
[d]*Thickness 15–300 μm.*
Reproduced with permission of V.S. Nguyen, D. Rouxel, M. Meier, B. Vincent, A. Dahoun, S. Thomas, et al., Effect of ultrasonication and other processing conditions on the morphology, thermomechanical, and piezoelectric properties of poly(vinylidene difluoride-trifluoroethylene) copolymer films, Polym. Eng. Sci. 54 (2014) 1280–1288. https://doi.org/10.1002/pen.23670 [69].

Moreover, the higher polarization level observed in CT12 indicates that the introduction of such a bulky CTFE makes the structure loose, resulting in an easier dipole orientation when an extra electric field is applied [61].

1.3.2.3 *P(VDF-co-trifluoroethylene) copolymers*

Poly(VDF-*co*-TrFE) is considered to be the most studied and used electroactive copolymer of VDF, that exists in a β-crystal phase (for a TrFE mol% of 20–50) or ferroelectric phase, depending on the TrFE content [63,64]. Actually, the bulky TrFE induces an increase of the interchain distances. Table 1.2 shows the relevant characteristics of P(VDF-*co*-TrFE) films. In fact, the presence of TrFE units in the copolymer can be viewed as defects, and in all cases, when the defect concentration increases, the ferroelectric β phase becomes less stable with respect to the paraelectric phase [65,66]. Moreover, when VDF content ranges from 50 to 80 mol%, these copolymers can crystallize readily from the melt or by solvent casting into an electroactive phase similar to that of ferroelectric phase of PVDF without any need for stretching [67]. More recently, Zhu's team reported the key role of the amorphous zone in such a copolymer to enhance the piezoelectric properties [68].

P(VDF-*co*-TrFE) copolymer can be obtained from the copolymerization of VDF and TrFE in emulsion, suspension, and solution processes [67,69,70] (Fig. 1.7). Another strategy to get this copolymer consists of the reduction of chlorine atoms of CTFE units from P(VDF-*co*-CTFE) copolymer [67,71] (Fig. 1.8).

1.3.2.4 *P(VDF-co-2,3,3,3-tetrafluoropropene) copolymers*

In contrast to the above copolymers, quite a few studies involving P (VDF-*co*-2,3,3,3-tetrafluoropropene) copolymers have been reported [72–75].

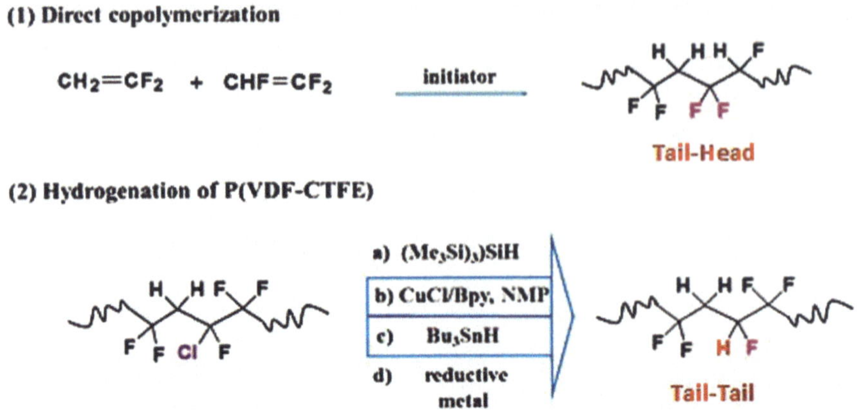

FIGURE 1.7 Different strategies to obtain P(VDF-*co*-TrFE) copolymers: (1) traditional radical copolymerization and (2) hydrogenation of P(VDF-*co*-TrFE) from P(VDF-*co*-CTFE). Source: *Reproduced with permission of W. Xia, Z. Zhang, PVDF-based dielectric polymers and their applications in electronic materials, IET Nanodielectrics 1 (2018) 17–31. https://doi.org/10.1049/iet-nde.2018.0001 [71].*

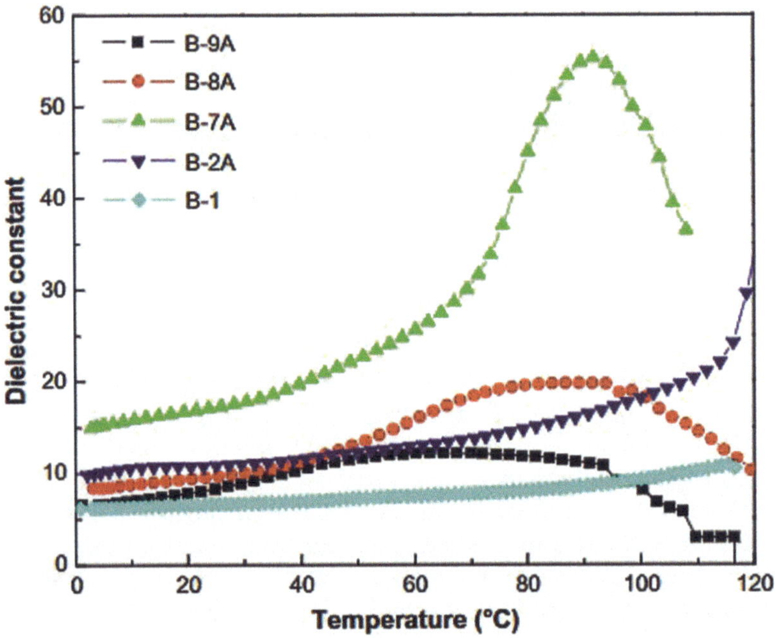

FIGURE 1.8 Effect of TrFE content on the dielectric constant of P(VDF-*ter*-TrFE-*ter*-CTFE) terpolymers, where VDF/TrFE/CTFE compositions are B1 (93.0/0.0/8.0), B2A (88.0/5.2/6.8), B7A (80.4/11.8/7.8), B8A (66/26.6/7.4), and B9A (62.7/29.3/8.0). Source: *Reproduced with permission of Z. Zhang, Q. Meng, T.C.M. Chung, Energy storage study of ferroelectric poly(vinylidene fluoride-trifluoroethylene-chlorotrifluoroethylene) terpolymers, Polymer (Guildf.) 50 (2009) 707–715. https://doi.org/10.1016/j.polymer.2008.11.005 [77].*

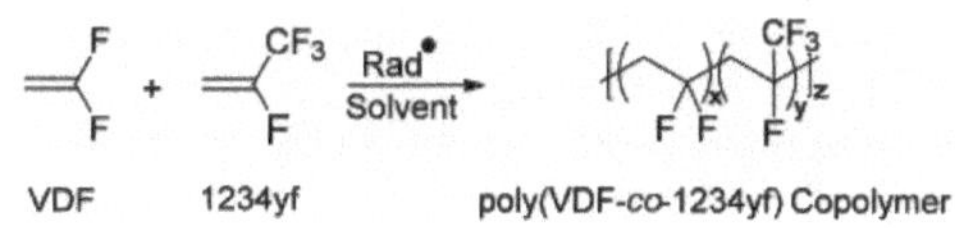

SCHEME 1.2 Radical copolymerization of vinylidene fluoride (VDF) with 2,3,3,3-tetrafluoro-1-propene (1234yf) initiated by tert-butyl peroxypivalate (TBPPi). Source: *Reproduced with permission of S. Banerjee, S. Zaghloul, A. Alaaeddine, B. Ameduri, Kinetic and mechanistic aspects of the iodine transfer copolymerization of vinylidene fluoride with 2,3,3,3-tetrafluoro-1-propene and functionalization into ω-hydroxy fluorinated copolymers, Polym. Chem. 7 (2016) 6099–6109. https://doi.org/10.1039/c6py01152g [73].*

Actually, iodine transfer copolymerization (ITP) [73] and RAFT [72] polymerization can lead to block copolymers. Banerjee et al. [73] were the first to report a detailed study on the kinetics and mechanistic aspects of ITP of VDF with 2,3,3,3-tetrafluoropropene (1234yf) (Scheme 1.2). The reactivity ratios obtained for both monomers were $r_{VDF}=0.76\pm0.34$ and $r_{1234yf}=1.23\pm0.17$ at 74°C, respectively, indicating that 1234yf is more reactive than VDF. Moreover, the authors showed that the 1234yf content in the copolymer controls the crystalline or amorphous state of the resulting copolymer, where a content higher than 15–18 mol% leads to amorphous copolymers.

1.3.2.5 P(VDF-ter-TrFE-ter-CTFE) terpolymers

A new generation of electroactive fluorinated copolymers deals with P(VDF-*ter*-TrFE-*ter*-M) terpolymers in which two products are nowadays commercially available involving either CTFE or 1-chloro-1-fluoroethylene (CFE) [65,67].

The introduction of CTFE units into P(VDF-*co*-TrFE) copolymer tends to create defects in the ferroelectric structure of this copolymer, which makes the P(VDF-*ter*-TrFE-*ter*-CTFE) terpolymers multiply functional due to their unique properties in comparison to other ferroelectric polymers [10].

Chung and Petchsuk [76] synthesized P(VDF-*ter*-TrFE-*ter*-CTFE) terpolymer by a combination of the bulk terpolymerization process boran/oxygen system at ambient temperature. The reaction consists of the homogeneous incorporation of a small amount of bulky CTFE into VDF/TrFE copolymer chain. It was found that the insertion of CTFE units induced a significant decrease in the melting and phase transition temperatures, while the degree of crystallinity was kept high (Table 1.3) [76].

In contrast to such a strategy, Zhang et al. [77] prepared P(VDF-*ter*-TrFE-*ter*-CTFE) terpolymers by a first copolymerization of VDF with CTFE followed by a reduction of P(VDF-*co*-CTFE) to convert CTFE units into TrFE ones (Fig. 1.8). These authors studied the effect of TrFE content on the dielectric constant at 1 kHz (where $\varepsilon=57$ at 90°C) (Fig. 1.9). It can be seen that as the TrFE content decreases, the

TABLE 1.3 P(VDF-*ter*-TrFE-*ter*-CTFE) terpolymers prepared by borane/oxygen initiator in bulk process (MEK and [η] stand for methylethylketone and intrinsic viscosity, respectively, while none means nonobservable).

Sample No.	Terpolymer mole ratio			Melting temperature		Phase transition temperature		[η] (MEK) (35°C)
	VDF	TrFE	CTFE	T_m (°C)	ΔH_m (J/g)	T_c (°C)	ΔH_c (J/g)	
1	72.2	17.8	10.0	108	17.8	25.0	0.4	0.38
2	66.0	22.5	11.5	117	21.6	25.2	2.5	0.42
3	66.1	21.4	12.5	116	20.5	25.8	1.8	0.42
4	63.1	25.4	11.5	114	18.5	None	None	0.49
5	61.4	25.3	13.3	111	17.8	15.6	1.5	0.69
6	58.0	33.1	8.9	130	21.3	31.6	2.5	0.63
7	60.0	36.0	4.0	141	25.7	43.8	6.5	0.58
8	55.6	36.1	8.3	125	21.2	None	None	0.54

Reproduced with permission of T.C. Chung, A. Petchsuk, Ferroelectric polymers with large electrostriction; Based on semicrystalline VDF/TrFE/CTFE terpolymers, Ferroelectr. Lett. Sect. 28 (2001) 135–143. https://doi.org/10.1080/07315170108202957 [76].

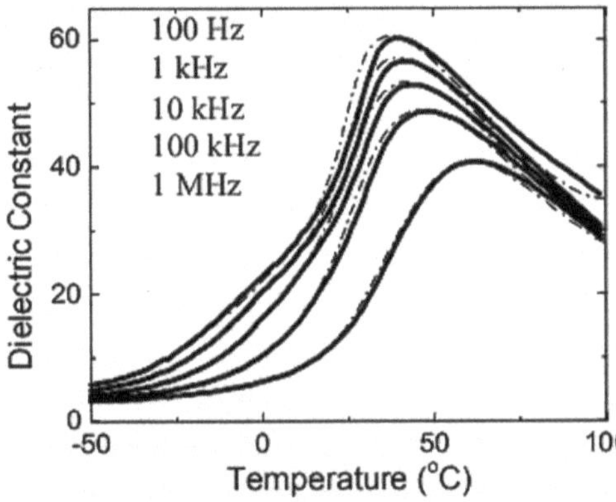

FIGURE 1.9 Dielectric constant at different frequencies versus temperature for the P (VDF-*ter*-TrFE-*ter*-CFE) (59.5/36.5/4 mol%) terpolymer. The curves from top to bottom correspond to the frequencies of 100 Hz, 1 kHz, 10 kHz, 100 kHz, and 1 MHz, respectively. The solid lines represent the data collected during the heating process, while the dashed ones represent the data obtained during cooling process. Source: *Reproduced with permission of Z.-Y. Cheng, H. Li, F. Xia, H. Xu, D. Olson, C. Huang, et al., Electromechanical properties and molecular conformation in P(VDF-TrFE)-based terpolymer, in: Y. Bar-Cohen (Ed.), Smart Structures and Materials 2002: Electroactive Polymer Actuators and Devices (EAPAD), vol. 4695, SPIE, 2002, pp. 167–175. https://doi.org/10.1117/12.475162 [80].*

TABLE 1.4 Comparison of the electromechanical properties of P(VDF-TrFE) based co- and terpolymers.

Polymer	S_M (%)	Y (GPa)	$YS_M^2/2$ (J/cm³)	k_{33}
Piezo P(VDF-TrFE)	0.2	3.3	0.01	0.27
High-Energy Electron Irradiated P(VDF-TrFE)	−5.0	0.4	0.50	0.30
P(VDF-TrFE-CTFE)	−4.0	0.4	0.32	0.28
P(VDF-TrFE-CFE)	−4.5	1.1	1.10	0.55

*S_M is the maximum strain observed in the material. Y is the Young's modulus. $YS_M^2/2$ is the elastic energy density, a parameter normally used to compare different electroactive materials. k_{33} is the thickness electromechanical coupling factor. For electrostrictive polymers, the value of k_{33} presented here is the value at high field.

Reproduced with permission of Z.-Y. Cheng, H. Li, F. Xia, H. Xu, D. Olson, C. Huang, et al., Electromechanical properties and molecular conformation in P(VDF-TrFE)-based terpolymer, in: Y. Bar-Cohen (Ed.), Smart Structures and Materials 2002: Electroactive Polymer Actuators and Devices (EAPAD), vol. 4695, SPIE, 2002, pp. 167–175. https://doi.org/10.1117/12.475162 [80].

dielectric constant peak shifts from about 40°C to 100°C, while the peak shape evolves from a broad flat to a narrow peak at about 90°C, then broadens and fades steadily [77]. In 2006, Lu et al. [78] revisited that terpolymerization initiated by AIBN, that was well-characterized by NMR spectroscopy.

1.3.2.6 P(VDF-ter-TrFE-ter-CFE) terpolymers

In addition to the abovementioned class of VDF-based ferroelectric terpolymers, P(VDF-*ter*-TrFE-*ter*-CFE) is considered to have the highest room temperature permittivity value (i.e., ~50 at 1 kHz) [79]. Cheng et al. [80] reported the synthesis of P(VDF-*ter*-TrFE-*ter*-CFE) terpolymer with the composition 59.5/36.5/4 mol%. Fig. 1.9 shows the dielectric constants of such terpolymers at different frequencies versus temperature. It can be seen that a high permittivity (c. 60) could be reached at room temperature. Moreover, the obtained terpolymers resulted in better electromechanical performances (the elastic energy density ~1.1 J/cm³ and the electromechanical coupling factor ~0.55) compared to those of P(VDF-*co*-TrFE) copolymer (Table 1.4).

Despite having high permittivity and good electroactive properties, P(VDF-*ter*-TrFE-*ter*-CFE) terpolymer suffers from low breakdown strength, which restricts its further improvement in energy storage properties [81]. To overcome this problem, Mao et al. [81] developed PVDF/P(VDF-*ter*-TrFE-*ter*-CFE) terpolymer blend achieved by the casting method at high temperature followed by quenching of the film (Fig. 1.10). The resulting polymer blends filled with 20 wt.% of

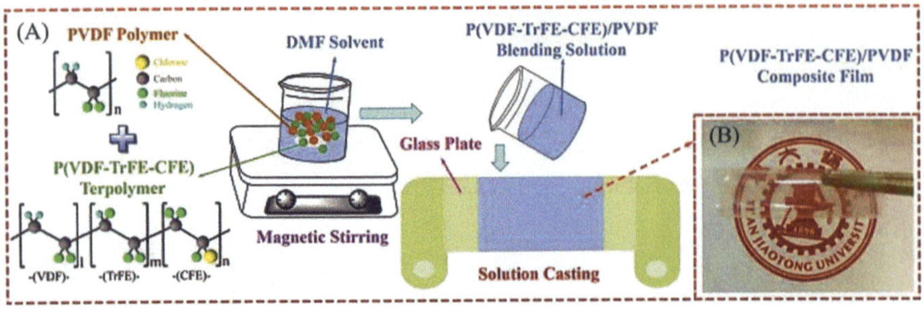

FIGURE 1.10 Schematic illustration of (A) the fabrication process for the P(VDF-*ter*-TrFE-*ter*-CFE) /PVDF polymer composite films and (B) photograph of the PVDF polymer composite films blended with 20 wt.% P(VDF-*ter*-TrFE-*ter*-CFE) terpolymer. *Source: Reproduced with permission of P. Mao, J. Wang, L. Zhang, Q. Sun, X. Liu, L. He, et al., Tunable dielectric polarization and breakdown behavior for high energy storage capability in P (VDF−TrFE−CFE)/PVDF polymer blended composite films, Phys. Chem. Chem. Phys. 22 (2020) 13143−13153. https://doi.org/10.1039/d0cp01071e [81].*

terpolymer exhibited high discharge energy density of 13.63 J/cm^3 at an enhanced breakdown strength of 480 MV/cm, making them suitable for energy storage applications.

Another example to enhance the electroactive behavior of fluoropolymers deals with P(VDF-*ter*-TrFE-*ter*-CFE) terpolymers/polyM polymer blends. Casar et al. [82] studied the dielectric properties of P(VDF-*ter*-TrFE-*ter*-CFE) terpolymer blended with P(VDF-*co*-TrFE) copolymer.

Fig. 1.11 exhibits the temperature dependences of the real (ε') and imaginary (ε'') parts of the complex dielectric constant, detected at various frequencies in the pure P(VDF-*ter*-TrFE-*ter*-CFE) terpolymer (100:0), its blends with different wt.% (5−50) of P(VDF-*co*-TrFE) copolymer, and in the pure copolymer (0:100) [82]. The terpolymer exhibits a typical relaxor broad dispersive maximum, while the copolymer is found to display a ferroelectric behavior. At low P (VDF-*co*-TrFE) copolymer content (<15 wt.%), the polymer blends show a relaxor-like dielectric response, since the interfacial couplings to the bulky defects in the terpolymer convert the normal ferroelectric copolymer into a relaxor. However, in samples where the copolymer amount is higher than 20 wt.%, both ferroelectric and relaxor behaviors are present [82].

Another PVDF/P(VDF-*ter*-TrFE-*ter*-CFE) terpolymer blend was prepared by Zhang et al. [83] where PVDF volume fractions increased from 0% to 100%. Much enhanced dielectric constant as well as electric displacement at high electric fields were achieved in the blends compared to the neat polymers. At 100 Hz, the terpolymer exhibits a dielectric constant of 53 that is larger than that of PVDF (12), while for the polymer blends, the permittivity increased with a higher terpolymer amount,

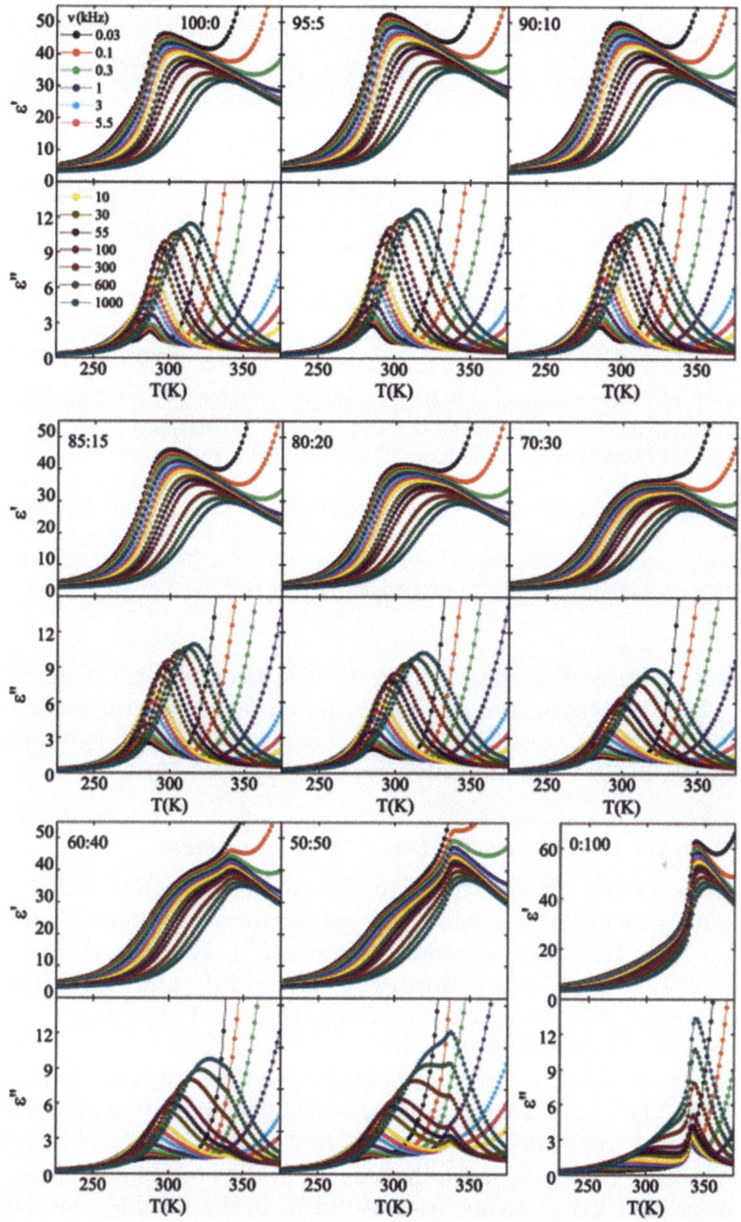

FIGURE 1.11 Temperature dependences of the real, ε', and imaginary, ε'', parts of the complex dielectric constant, detected at various frequencies in the pure P(VDF-*ter*-TrFE-*ter*-CFE) terpolymer (100:0), its blends with different P(VDF-*co*-TrFE) copolymer contents (5–50 wt.%) and in the pure copolymer (0:100). Source: *Reproduced with permission of G. Casar, X. Li, B. Malič, Q.M. Zhang, V. Bobnar, Impact of structural changes on dielectric and thermal properties of vinylidene fluoride-trifluoroethylene-based terpolymer/copolymer blends, Phys. B Condens. Matter 461 (2015) 5–9. https://doi.org/10.1016/j.physb.2014.12.010 [82].*

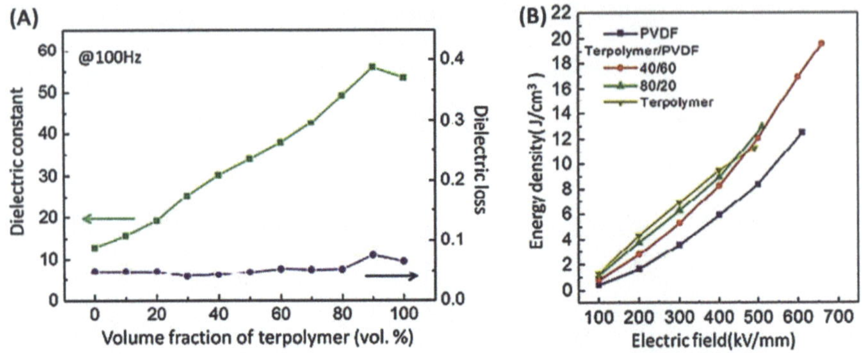

FIGURE 1.12 (A) Variations of dielectric constant (square) and dielectric loss (circle) with the volume fraction of terpolymer in the blends; and (B) maximum discharged energy density of neat PVDF, P(VDF-*ter*-TrFE-*ter*-CFE) terpolymer, and blends with composition of 40/60 vol.% and 80/20 vol.% (Terpolymer/PVDF) at their characteristic breakdown strength. Source: *Reproduced with permission of X. Zhang, Y. Shen, Z. Shen, J. Jiang, L. Chen, C.W. Nan, Achieving high energy density in PVDF-based polymer blends: suppression of early polarization saturation and enhancement of breakdown strength, ACS Appl. Mater. Interfaces 8 (2016) 27236–27242. https://doi.org/10.1021/acsami.6b10016 [83].*

and reached 56 with a composition of 90/10 vol.% (terpolymer/PVDF) (Fig. 1.12A). Moreover, the dielectric breakdown strength of the blends was also improved, resulting in high energy density of 19.6 J/cm^3 for blends with 40/60 vol.% (terpolymer/PVDF) (Fig. 1.12B).

1.3.2.7 *Other P(VDF-ter-TrFE-ter-M) terpolymers*

Another family of electroactive fluorinated P(VDF-*ter*-TrFE-*ter*-M) terpolymers, where M stands for 3,3,3-trifluoropropene (TFP) [84], *trans*-1,3,3,3-tetrafluoropropene (1234ze) [85], 2,3,3,3-tetrafluoropropene (1234yf) [74], hexafluoropropylene [86], and 2-trifluoromethacrylic acid [87], has also been reported and the resulting films led to ferroelectric materials.

Soulestin et al. [84] prepared P(VDF-*ter*-TrFE-*ter*-TFP) terpolymers, where the VDF/TrFE molar ratio was 65/35 and TFP was introduced using different compositions ranging from 0 to 10 mol%. The incorporated TFP units were mostly found to be located in the amorphous phase of the terpolymer, and led to an increase of the T_g in the resulting terpolymers. In a search of new electroactive terpolymers, the same team synthesized P (VDF-*ter*-TrFP-*ter*-1234ze) terpolymer [85]. Despite the differences in their microstructure, P(VDF-*ter*-TrFP-*ter*-1234ze) terpolymers are compositionally very homogeneous, in contrast to P(VDF-*ter*-TrFE-*ter*-TFP). However, these two terpolymers shared identical electroactive properties (Fig. 1.13). For instance, both terpolymers exhibited comparably high E_c (50–60 MV under

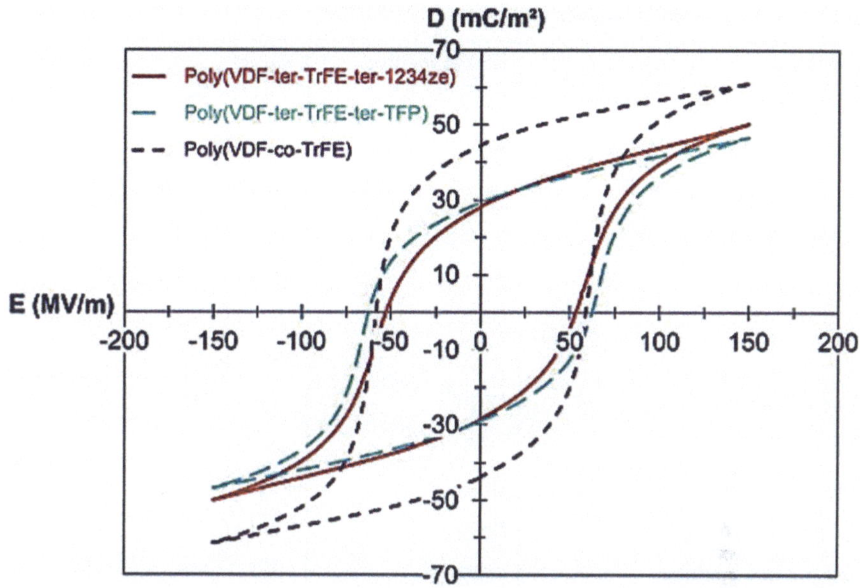

FIGURE 1.13 Comparison of continuous bipolar D − E loops for annealed poly(VDF_{65}-*co*-$TrFE_{35}$) reference copolymer (purple dashed line), poly(VDF_{67}-ter-$TrFE_{28}$-ter-TFP_5) (blue dashed line), and poly(VDF_{69}-ter-$TrFE_{25}$-ter-$1234ze_6$) (red full line), all synthesized under the same conditions (solution copolymerization). Source: *Reproduced with permission of T. Soulestin, P. Marcelino Dos Santos Filho, V. Ladmiral, C. Totée, G. Silly, T. Lannuzel, et al., Influence of trans-1,3,3,3-tetrafluoropropene on the structure-properties relationship of VDF- and TrFE-based terpolymers, Macromolecules 50 (2017) 503–514. https://doi.org/10.1021/acs.macromol.6b02010 [85].*

150 MV/m), close to the E_c of the reference ferroelectric material (i.e., P(VDF_{65}-*co*-$TrFE_{35}$) copolymer) [85].

Thanks to the carboxylic acid side group of 2-trifluoromethacrylic acid (MAF), the P(VDF-*ter*-TrFE-*ter*-MAF) terpolymers could adhere well onto steel or gold electrodes. The coated electrodes were polarized at 200 MV/m up to five times and no delamination or loss of ferroelectric properties was noted [87].

1.4 Properties of PVDF and VDF copolymers

The properties of PVDF and VDF copolymers are dependent on the chemical compositions of the polymer, molar masses, and molecular architectures [88]. In fact, PVDF is considered as a polymer of choice for a wide range of scientific research due to its outstanding properties, including thermal stability, high mechanical strength, and chemical resistance [54,89] (Table 1.5).

TABLE 1.5 General engineering properties of PVDF.

Property	ASTM method	Value
Mechanical		
Tensile stress at yield	D638	30–60 MPa (25°C)
		20–40 MPa (100°C)
Tensile strength at break	D638	30–60 MPa (25°C)
		20–30 MPa (100°C)
Elongation at break	D638	25%–500% (25°C)
		300%–600% (100°C)
Tensile modulus	D638	1200–1600 MPa
Flexural strength	D790	55–75 MPa
Flexural modulus	D790	1200–1800 MPa
Compressive strength	D695	55–100 MPa
Compressive modulus	D695	1000–1400 MPa
Thermal		
Melting range		160°C–175°C
Heat distortion temperature	D648	150°C (66 psi)
		100°C (264 psi)
Thermogravimetric weight loss		0.5% (300°C, 1 h)
Specific heat		0.26–0.33 cal/g/°C
Dielectric		
Dielectric constant	D150	8.4–13.5 (60 Hz)
		7.4–13.2 (10^3 Hz)
		6.0–7.6 (10^6 Hz)
Dielectric strength	D149	100 kV/mm (0.10 mm thick)
		65 kV/mm (0.25 mm thick)
		20 kV/mm (1.00 mm thick)
Dissipation factor	D150	0.012–0.019 (60 Hz)
		0.013–0.019 (10^3 Hz)
		0.150–0.220 (10^6 Hz)

Reproduced with permission of A.J. Lovinger, Poly(vinylidene fluoride), in: Developments in Crystalline Polymers, Springer Netherlands, Dordrecht, 1982, pp. 195–273. https://doi.org/10.1007/978-94-009-7343-5_5 [38].

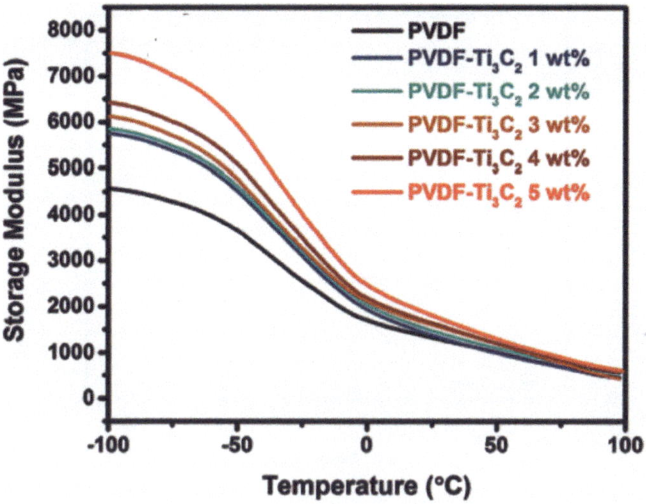

FIGURE 1.14 Storage modulus as a function of temperature for neat PVDF and PVDF composites with Ti_3C_2 of various amounts. Source: *Reproduced with permission of Y. Cao, Q. Deng, Z. Liu, D. Shen, T. Wang, Q. Huang, et al., Enhanced thermal properties of poly(vinylidene fluoride) composites with ultrathin nanosheets of MXene, RSC Adv. 7 (2017) 20494–20501. https://doi.org/10.1039/C7RA00184C [90].*

1.4.1 Mechanical and thermal properties

Among the common crystalline polymers (PE, polypropylene, PTFE, etc.), PVDF is considered to be an outstanding material due to its tensile, flexural, and compressive strength, stiffness, and hardness, and it presents the highest impact strength [38,54]. Nevertheless, major efforts have been conducted in order to further improve the mechanical properties of this polymer. For instance, Cao et al. [90] introduced Ti_3C_2 nanosheets into a PVDF matrix and prepared nanocomposites by varying the Ti_3C_2 content (0–5 wt.%). The storage modulus of neat PVDF at −80°C was 4362 MPa, while after the addition of only 5 wt.% of Ti_3C_2, the PVDF composite exhibited a storage modulus as high as 7501 MPa, corresponding to a 64% enhancement compared to that of neat PVDF (Fig. 1.14).

In addition to the remarkable and enhanced mechanical properties obtained for such PVDF composites, the authors also investigated the thermal conductivity and thermal diffusivity of these materials. The thermal properties are regarded as a key factor in determining the potential applications of PVDF and its copolymers, including thermal conductivity, which is the feature that determines the working temperature levels of a material [91]. Fig. 1.15 exhibits the thermal conductivity and thermal diffusivity of pure PVDF and PVDF/ Ti_3C_2 composites versus the Ti_3C_2 content [90]. Both studied properties were enhanced by

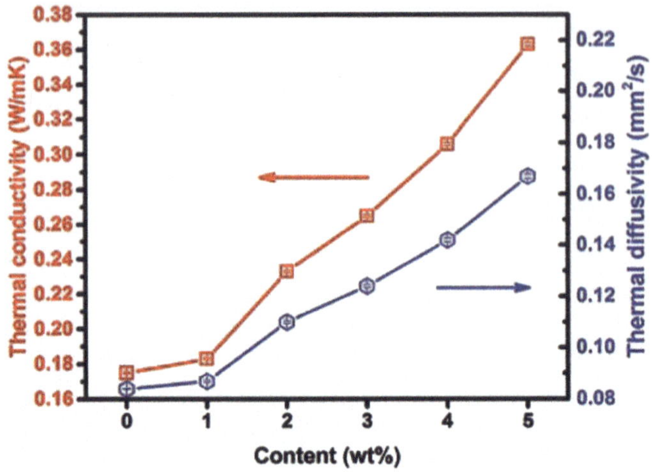

FIGURE 1.15 Thermal diffusivity and thermal conductivity of neat PVDF and PVDF composites with different Ti_3C_2 loadings. Source: *Reproduced with permission of Y. Cao, Q. Deng, Z. Liu, D. Shen, T. Wang, Q. Huang, et al., Enhanced thermal properties of poly(vinylidene fluoride) composites with ultrathin nanosheets of MXene, RSC Adv. 7 (2017) 20494–20501. https://doi.org/10.1039/C7RA00184C [90].*

increasing the Ti_3C_2 loading. The thermal diffusivity and thermal conductivity for pure PVDF were 0.084 mm^2/s and 0.175 W m/K, respectively, and they doubled to 0.167 mm^2/s and 0.363 W m/K after the addition of only 5 wt.% of Ti_3C_2 to the PVDF matrix, respectively. The enhancement in thermal conductivity of PVDF composites was assigned to the fact that Ti_3C_2 nanosheets can provide nucleation sites for the PVDF matrix, which is crucial for thermal conductivity improvement.

In another study, increases of both tensile strength and Young's modulus of PVDF were achieved by preparing PVDF/polyaniline (PANI) blend composites using chemical polymerization of aniline in a mixture of PVDF and DMF solvent [92]. Stress–strain curves for pure PVDF and PVDF/PANI blends containing 12 and 22 wt.% of dedoped PANI are shown in Fig. 1.16. Results revealed an increase of tensile strength from 40 MPa for pure PVDF to 50 and 55 MPa for blends with 12 and 22 wt.% of PANI, respectively. Thus the addition of PANI enabled the reinforcement of the PVDF matrix, by constraining the PVDF chain movements and increasing the tensile strength of the blend to a higher stiffness associated with the PANI backbone.

Sousa et al. [60] prepared P(VDF-*co*-CTFE) membranes by solvent casting using DMF as the solvent. The effects of polymer:solvent ratio and solvent evaporation temperature on the mechanical properties were investigated. Fig. 1.17 exhibits the strain–stress curves for P(VDF-*co*-CTFE) membranes at various initial polymer concentrations and solvent

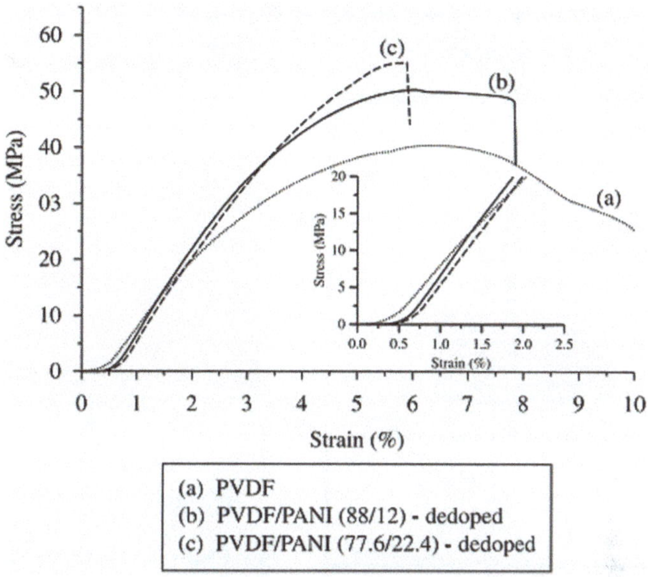

FIGURE 1.16 Stress-strain curves of: (A) PVDF; (B) PVDF/PANI (12 wt.% PANI); and (C) PVDF/PANI (22.4 wt.% PANI). The blends were dedoped. The inset is the expansion of stress versus the 0–2.5 strain range. Source: *Reproduced with permission of L.F. Malmonge, S. Do Carmo Langiano, J.M.M. Cordeiro, L.H.C. Mattoso, J.A. Malmonge. Thermal and mechanical properties of PVDF/PANI blends. Mater. Res. 13 (2010) 465–470. https://doi.org/10.1590/S1516-14392010000400007 [92].*

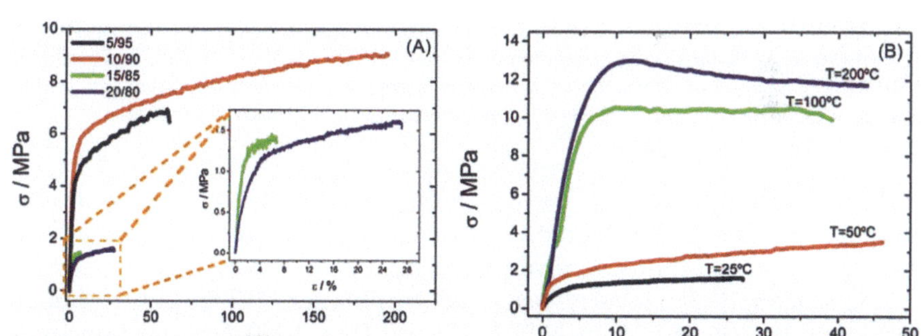

FIGURE 1.17 Stress–strain curves for (A) samples prepared after solvent evaporation at $T = 25°C$ for different initial polymer concentration in the P(VDF-*co*-CTFE)/DMF system, and (B) samples prepared from 20 wt.% P(VDF-*co*-CTFE) initial polymer concentration with solvent evaporation at different temperatures. Source: *Reproduced with permission of R.E. Sousa, J.C.C. Ferreira, C.M. Costa, A.V. Machado, M.M. Silva, S. Lanceros-Mendez, Tailoring poly(vinylidene fluoride-co-chlorotrifluoroethylene) microstructure and physicochemical properties by exploring its binary phase diagram with dimethylformamide, J. Polym. Sci. Part B Polym. Phys. 53 (2015) 761–773. https://doi.org/10.1002/polb.23692 [60].*

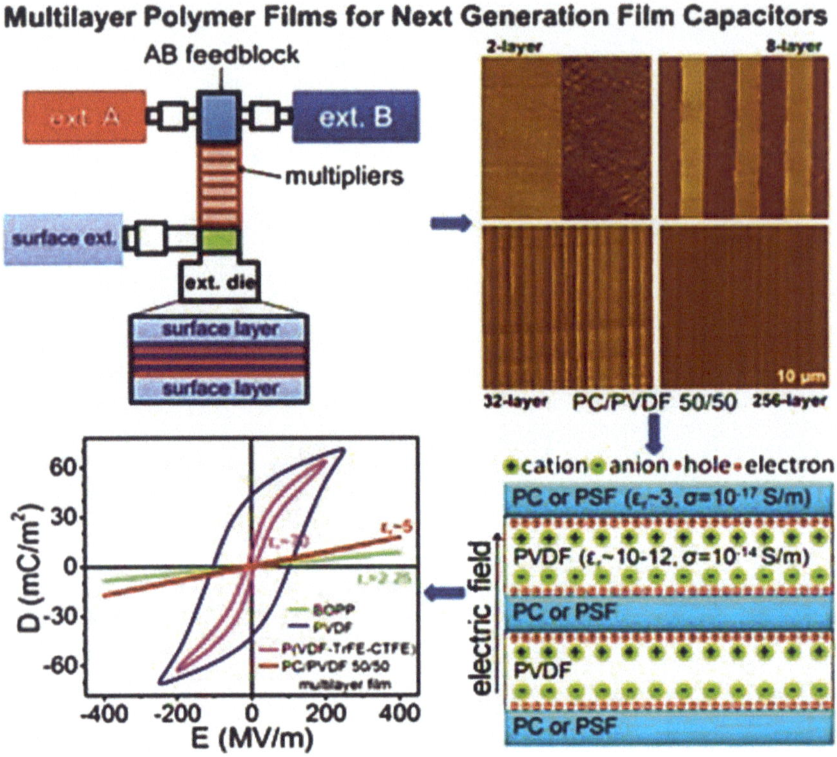

FIGURE 1.18 Multilayer PVDF/Polycarbonate coextruded blend. Scheme of the process (top left); AFM images of the multilayered films (from 2 to 256 layers, top right); D–E loops of PVDF and PVDF/PC blends (bottom left) and nanoconfinement picture (bottom right). Source: *Reproduced with permission of M. Mackey, D.E. Schuele, L. Zhu, L. Flandin, M. A. Wolak, J.S. Shirk, et al., Reduction of dielectric hysteresis in multilayered films via nanoconfinement, Macromolecules 45 (2012) 1954–1962. https://doi.org/10.1021/ma202267r [93].*

evaporation temperatures. These authors found that, as the solvent temperature increased, the degree of porosity in the membranes decreased, leading to increased mechanical properties.

Zhu's team [93] cleverly achieved multilayer PVDF/polycarbonate (PC) films by coextrusion of a 50/50 blend (Fig. 1.18). Several hundred/thousand nanolayers were produced in a 20–50 μm film. PC acts as a charge blocking layer and induces a nanoconfinement effect on ionic transport in PVDF nanolayers.

Hadji et al. [94] used alumina nanoparticles (Al_2O_3) dispersed in P(VDF-*co*-TrFE) copolymer matrix, and the resulting nanocomposite exhibits a high piezoelectric coefficient after polarization under high electric field without needing stretching during the polarization process. Piezoelectric measurements indicate that P(VDF-*co*-TrFE) copolymer

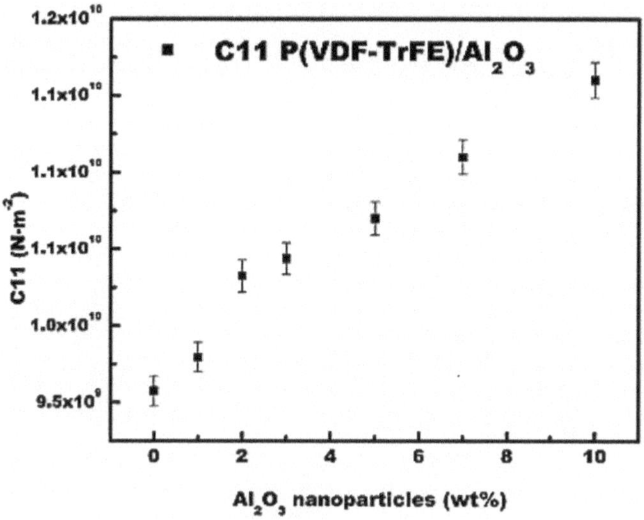

FIGURE 1.19 Variation of elastic constant (C_{11}) as a function of alumina concentration in poly(VDF-*co*-TrFE)/Al_2O_3 composites. Source: *Reproduced with permission of R. Hadji, V. S. Nguyen, B. Vincent, D. Rouxel, F. Bauer, Preparation and characterization of P(VDF-TrFE)/ Al_2O_3 nanocomposite, IEEE Trans. Ultrason. Ferroelectr. Freq. Control. 59 (2012) 163–167. https://doi.org/10.1109/TUFFC.2012.2168 [94].*

may be filled by up to 10 wt.% of alumina while retaining its high piezoelectric properties and increasing its elastic constant (C_{11}) by more than 20%, assessed by Brillouin spectroscopy (Fig. 1.19).

An enhancement in the mechanical properties of PVDF was also evidenced by Layek et al.'s work [95], who prepared graphene oxide functionalized with PMMA (MG), embedded within a PVDF matrix. Fig. 1.20 shows that for a fixed value of strain, an increase of tensile stress with the addition of MG content is noted: the Young's modulus increases from 1.44 GPa for pure PVDF to 6.07 GPa for the 5% addition of modified graphene (MG), indicating a stiffness increase of the film due to the addition of MG.

1.4.2 Electrical properties

PVDF presents a complex structure and can crystallize into five different crystalline phases, where the most studied ones are α, β, and γ phases [8–10]. In fact, the β phase in PVDF is considered to be that yielding the strongest piezoelectric and ferroelectric properties, making it the first known electroactive polymer [6,14,19,96,97].

Martins et al. [98] investigated the phase transformation from nonpolar α phase to polar β phase of PVDF. Fig. 1.21 exhibits the dielectric

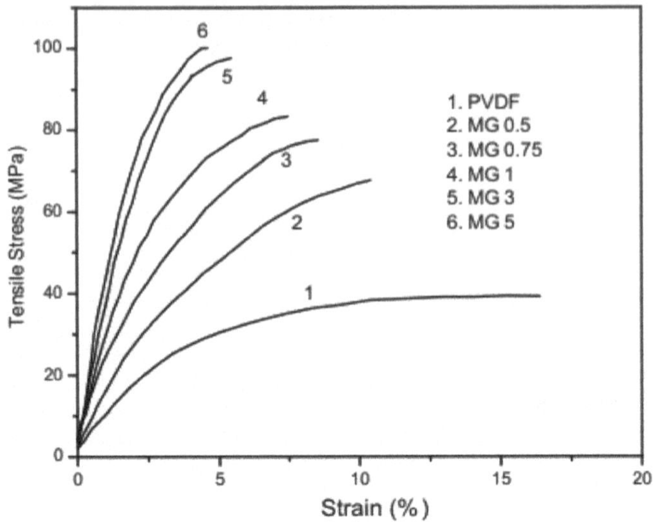

FIGURE 1.20 Stress-strain curves of the pure PVDF and PVDF-MG nanocomposite films with different contents of modified grapheme sheets. Source: *Reproduced with permission of R.K. Layek, S. Samanta, D.P. Chatterjee, A.K. Nandi. Physical and mechanical properties of poly(methyl methacrylate)-functionalized graphene/poly(vinylidine fluoride) nanocomposites: piezoelectric β polymorph formation, Polymer (Guildf.) 51 (2010) 5846–5856. https://doi.org/10.1016/j.polymer.2010.09.067 [95].*

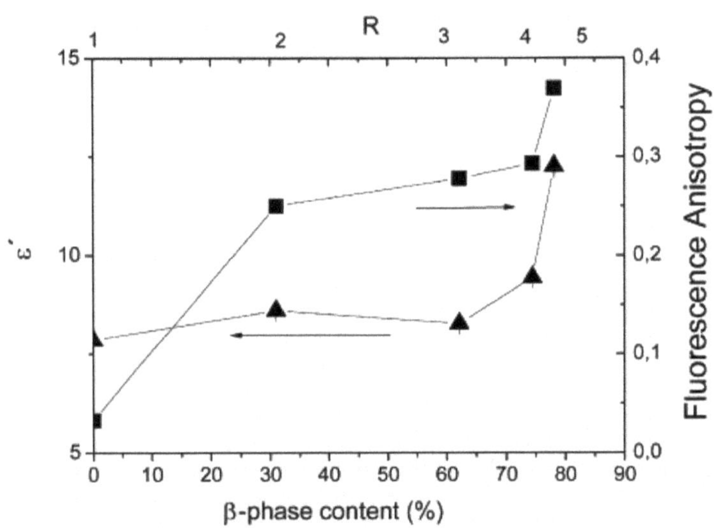

FIGURE 1.21 Variation in the fluorescence anisotropy and the dielectric constant (ε') as a function of β phase content. Source: *Reproduced with permission of P. Martins, J.S. Nunes, G. Hungerford, D. Miranda, A. Ferreira, V. Sencadas, et al., Local variation of the dielectric properties of poly(vinylidene fluoride) during the α- to β-phase transformation, Phys. Lett. A 373 (2009) 177–180. https://doi.org/10.1016/j.physleta.2008.11.026 [98].*

constant variation as a function of β phase content in PVDF. High dielectric constants (up to 12.5) could be obtained for higher β phase content, indicating the effect of the latter in the dielectric properties of PVDF.

The introduction section indicates that numerous methods are used to induce the β phase in PVDF [14]. Lanceros-Mendez's team [99] also studied the phase transformation from α to β phase in PVDF through the stretching process at different draw ratios ($R = L_{final}/L_{initial}$) and temperatures. Fig. 1.22 shows the infrared spectra of unstretched ($R = 1$) and stretched ($R = 2-5$) PVDF films at 80°C. The maximum β phase content was achieved at 80°C with a stretching ratio of 5, though such samples still showed 20% of the original α phase.

As the β modification could be also obtained directly from the melt-crystallization of PVDF, Gradys et al. [100] studied that process in nonisotherm mode at ultrahigh cooling rates ranging from 30 to 3000 K/s, as well as at constant temperature after quenching at 6000 K/s. The increase of the cooling rate above 150 K/s led to β phase formation, demonstrated by a low temperature shoulder of crystallization exotherm in addition to the α modification, while a cooling rate above 2000 K/s generated only a small exothermic peak, attributed to the crystallization of pure β phase.

In addition, polymer blends have been also reported to induce β crystal phase. Raquez et al. [13] used only 5 wt.% PMMA by melt blending and extrusion compression to get an electroactive PMMA/PVDF

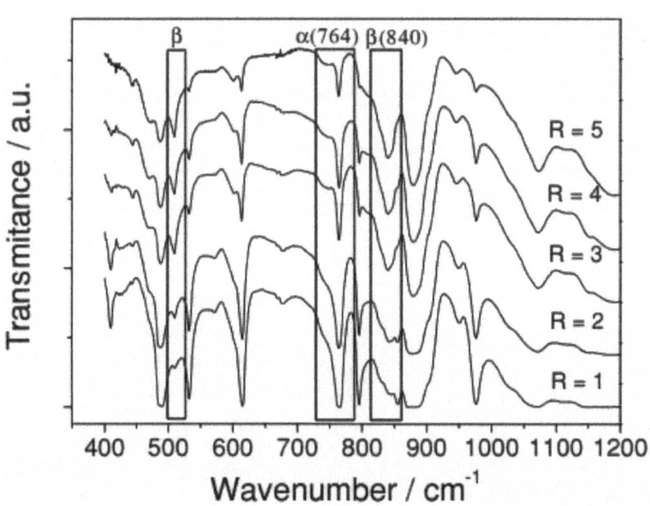

FIGURE 1.22 Infrared spectra of unstretched ($R = 1$), and stretched ($R = 2-5$) PVDF at 80°C. Source: *Reproduced with permission of V. Sencadas, R. Gregorio, S. Lanceros-Méndez, α to β Phase transformation and microestructural changes of PVDF films induced by uniaxial stretch, J. Macromol. Sci. Part B 48 (2009) 514–525. https://doi.org/10.1080/00222340902837527 [99].*

blend. PMMA plays a key role in the crystallization of PVDF polar crystals from the melt state.

As reported in Section 1.3.2, another way of promoting the formation of β phase is to consider VDF copolymers. For instance, P(VDF-*co*-TrFE) copolymer presents always the ferroelectric β phase [14,61,101]. In fact, the addition of the third fluoride (in the TrFE monomer unit) with a large steric hindrance favors the all-trans conformation and therefore induces the ferroelectric β phase independently of the processing method used, that is, either melt or solution casting, when the VDF content is between 50% and 80% [10].

An improvement in PVDF properties and its adaptation to the increasing technological demands could be achieved with VDF co/terpolymers (see Section 1.3.2), except that the final properties would depend on the monomer contents. A first example deals with P(VDF-*ter*-CTFE-*ter*-TrFE) terpolymer synthesized by Claude et al. [102] and Table 1.6 summarizes crystallinity rates, melting points, and dielectric constants of these prepared terpolymers. For the same VDF mol% (73.6), both crystallinity and melting temperature increased with the increase of the TrFE amount and the decrease of the CTFE content. Moreover, further incorporation of TrFE units tends to convert α and γ phases to β, and the lengthened sequences in the polar phase β increases the activation energy for dipole alignment and thus limits the polarization response, leading to a decreased dielectric constant [102]. Since then, Arkema company has been marketing poly(VDF-*ter*-TrFE-*ter*-M) terpolymers (where M is either CFE or CTFE of various contents).

TABLE 1.6 Characteristics of the prepared VDF-based ferroelectric co- and terpolymers.

Polymer	Composition (mol%)			χ_c (%)	T_m (°C)	ε_r
	VDF	TrFE	CTFE			
1	73.6	0	26.4	7.2		14.1
2	73.6	3.9	22.5	14.4	37.8	16.4
3	73.6	9.2	17.2	17.1	72.5	24.1
4	73.6	11.9	14.5	18.0	86.7	24.0
5	73.6	18.3	8.1	24.2	121	14.7
6	73.6	19.3	7.1	28.6	127	14.2
7	73.6	26.4	0	35.4	163	10.4

Reproduced with permission of J. Claude, Y. Lu, K. Li, Q. Wang. Electrical storage in poly(vinylidene fluoride) based ferroelectric polymers: correlating polymer structure to electrical breakdown strength, Chem. Mater. 20 (2008) 2078–2080. https://doi.org/10.1021/cm800160r [102].

1.5 Applications

Thanks to their outstanding properties, PVDF and VDF-based co-/terpolymers have attracted a growing interest in both industry and academia. Consequently, these polymers could be exploited in many technologies (including internet of things) and fields, such as sonars, actuators, sensors, energy storage devices, etc. [1–3,7,64,103,104].

1.5.1 Sonars

Sound navigation and ranging (SONAR) is a method for using sound to detect objects on or under the surface of water [105]. It consists of transmitting sound pulses into water by the transmitting element (such as ceramic material) and receiving echoes returned from underwater features such as fish and objects with the receiving element (PVDF polymer) [105].

Ahmad et al. [106] designed a receiver of piezoelectric acoustic transducer (PAT) in small scale using a printed circuit board as a conductive part and substrate, and the performances of the prepared PAT were studied using the pulse–echo method (Fig. 1.23). The designed PAT with 5-mm-diameter showed the highest sensitivity and a wide bandwidth, with a resonance frequency at 125 kHz and output voltage at 555 mV [106].

In the perspective to fabricate a complete ultrasonic system suitable for applications, both in the medical field and in robotics, Fiorillo et al. [107] prepared a low-frequency ultrasound sensor capable of emulating the biosonars of bats; the complete electronic system is depicted in Fig. 1.24. The produced system was light and suitable to be inserted into mobile robots. Moreover, the authors emphasized the ability of a PVDF curved transducer to operate in the constant frequency modality in order to measure the time-of-flight [107].

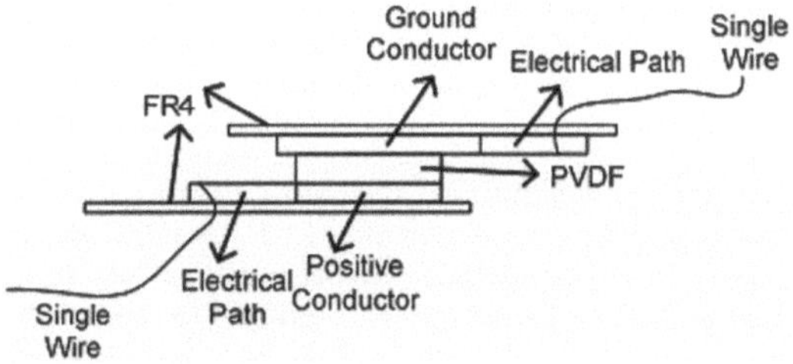

FIGURE 1.23 Schematic diagram of side piezoelectric ultrasonic transducer. Source: *Reproduced with permission of K.A. Ahmad, M.K. Osman, Z. Hussain, M.F. Abdullah, A.A. Manaf, N. Abdullah, Design and characterization piezoelectric acoustic transducer for sonar application, in: Proceedings—2018 8th IEEE International Conference on Control System, Computing and Engineering (ICCSCE), Institute of Electrical and Electronics Engineers Inc., 2019, pp. 233–237. https://doi.org/10.1109/ICCSCE.2018.8685014 [106].*

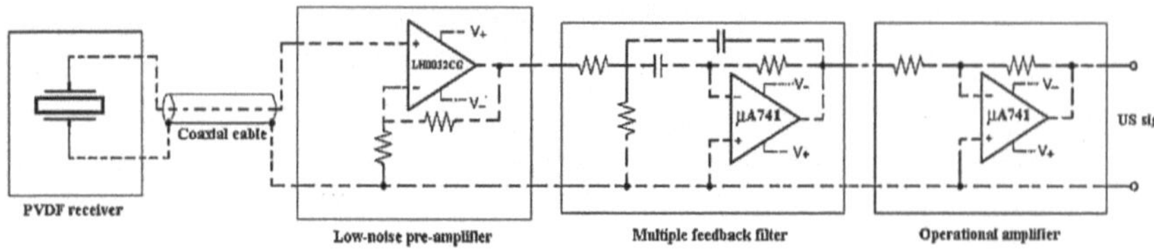

FIGURE 1.24 Block diagram of the complete receiving electronic unit including very low noise amplifier and active band pass filter. Source: *Reproduced with permission of A.S. Fiorillo, F. Lamonaca, S.A. Pullano. PVDF based sonar for a remote web system to control mobile robots, Sens. Transducers J. 8 (2010) 65–73 [107].*

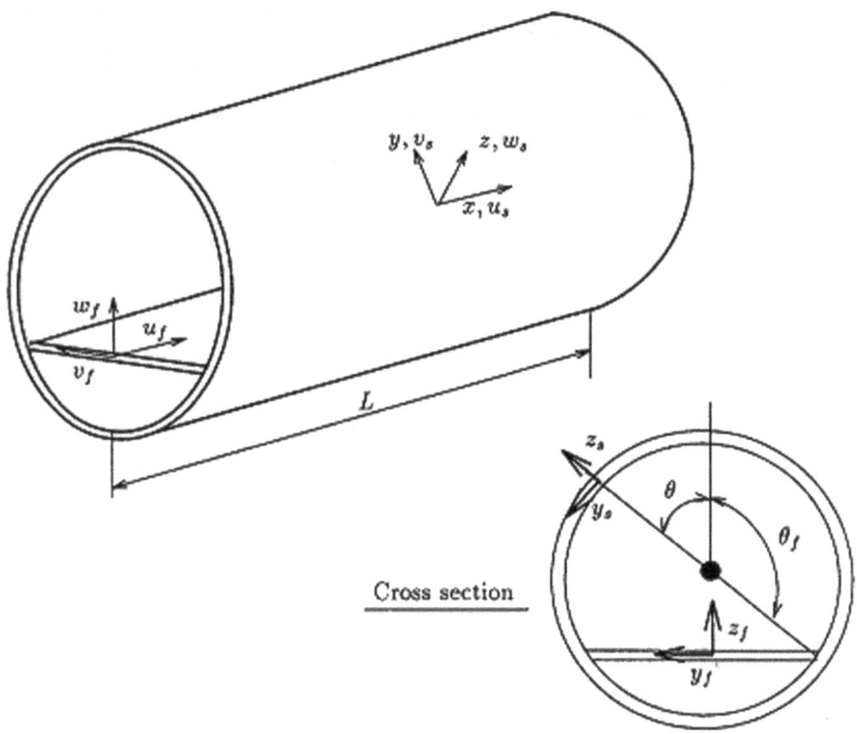

FIGURE 1.25 Schematic of a cylindrical shell with a floor partition and the coordinate system. Source: *Reproduced with permission of D.S. Li, L. Cheng, C.M. Gosselin, The design of structural acoustic sensors for active control of sound radiation into enclosures, Smart Mater. Struct. 13 (2004) 371. https://doi.org/10.1088/0964-1726/13/2/016 [108].*

Li et al. [108] proposed another method for the design of a PVDF structural acoustic sensor to control the sound radiation into enclosures. To optimize the number and locations of PVDF films, the authors applied the sensor to a cylindrical shell with floor partition (Fig. 1.25). The results revealed that the floor partition improved its low-frequency performance and the compared experiment demonstrated the effectiveness of the prepared sensors [108].

1.5.2 Actuators and sensors

An actuator (or artificial muscle) is a component that can convert a mechanical variable into a measurable electrical quantity [8,109]. Indeed, PVDF and VDF co-/terpolymers have been widely used in sensing [3] and actuators due to their inherent piezoelectric effect [67]. Moreover, due to their structural flexibility and processing simplicity, they can be easily processed into thin, light, and flexible thin-film systems [110,111].

Lee and Tarbutton [112] obtained PVDF actuators fabricated by the electric poling-assisted additive manufacturing (EPAM) process. The PVDF device was then tested for actuation applications based on the converse piezoelectric effect. The used manufacturing process kept PVDF dipoles well-aligned and uniform. High voltage was applied to the PVDF actuator in the positive and negative directions in order to perform the high-voltage polarity effect, and approximatively 3 and 10 μm displacements were measured in the positive and negative polarity conditions, respectively (Fig. 1.26) [112].

Yildirim et al. [113] designed a biocompatible piezoelectric membrane actuator (PMA) using a P(VDF-*co*-TrFE) copolymer. The fabrication process was performed on thick Silicium wafer coated with slow stress silicon nitride (SiNx) and silicone dioxide (SiO_2) films. The surface roughness of the prepared films was 7.9 nm while remanent polarization at 200 V was

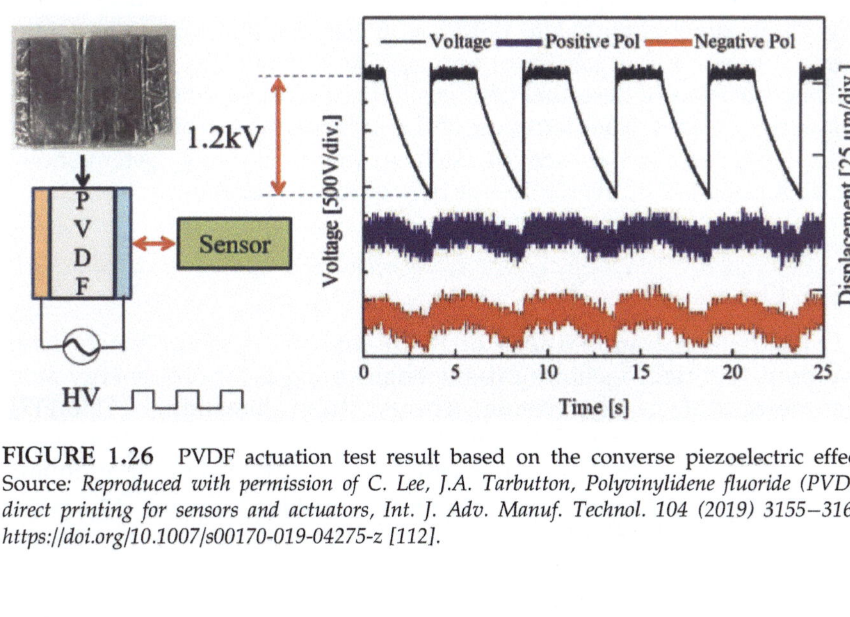

FIGURE 1.26 PVDF actuation test result based on the converse piezoelectric effect. Source: *Reproduced with permission of C. Lee, J.A. Tarbutton, Polyvinylidene fluoride (PVDF) direct printing for sensors and actuators, Int. J. Adv. Manuf. Technol. 104 (2019) 3155–3162. https://doi.org/10.1007/s00170-019-04275-z [112].*

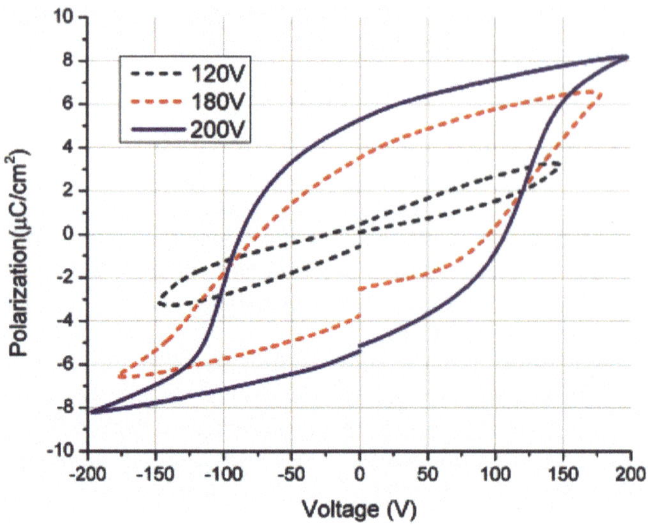

FIGURE 1.27 Polarization hysteresis loops obtained from the 1.03 mm^2 inner electrodes of a piezoelectric membrane actuator with 1750 μm diameter using bipolar triangular waveforms at 10 Hz at different voltages. Remnant polarization was measured as 5.38 $\mu C/cm^2$ at 200 V. Source: *Reproduced with permission of Y.A. Yildirim, A. Toprak, O. Tigli, Piezoelectric membrane actuators for micropump applications using PVDF-TrFE, J. Microelectromech. Syst. 27 (2018) 86–94. https://doi.org/10.1109/JMEMS.2017.2773090 [113].*

5.38 $\mu C/cm^2$ (Fig. 1.27). Moreover, high displacements were obtained without any fracture since all moving structures were fabricated from a polymer material [113].

Thevenot et al. [114] used P(VDF-*co*-TrFE) copolymer plasticized with 17 wt.% of diethyl phthalate (DEP) as a deformation sensor and integrated it in a living organism of Sprague Dawley rat (Fig. 1.28).

This consisted of detecting changes in the deformation of a real artery at low frequencies. The detection of the deformation of biological tissue and the precise measurement of the heart rate on a living system shows the high potential of the device for biomedical applications.

1.5.3 Others

Other potential applications of PVDF and its co-/terpolymers have also been reported, including tissue engineering [115] and energy storage devices (e.g., harvesting energy from breathing) [116,117]. Moreover, and because of their high flexibility and relevant processing, these copolymers can be easily inserted into a device configuration and thereby offer great promise for energy harvesting for flexible and wearable applications [118,119].

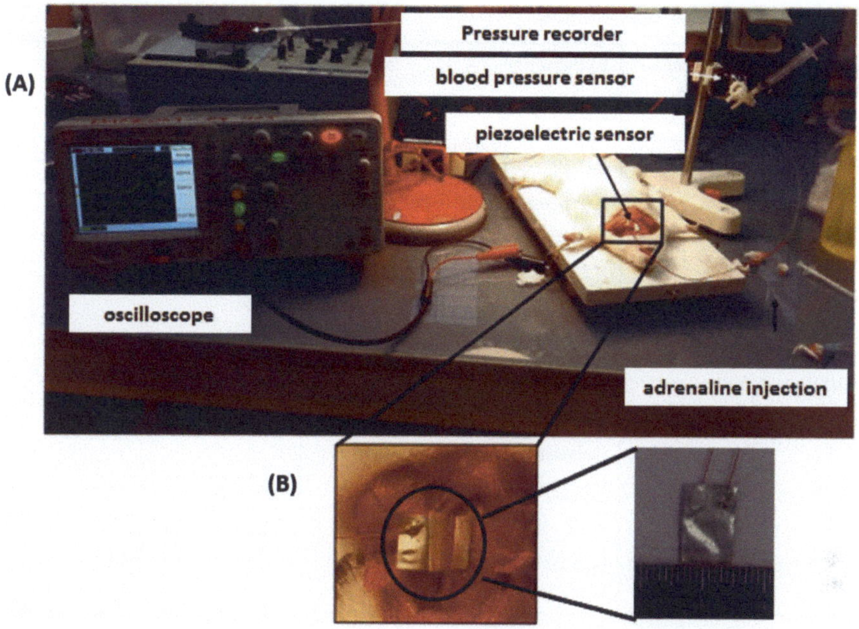

FIGURE 1.28 (A) Photograph of the in vivo experiment performed on a Sprague Dawley rat, and (B) P(VDF-*co*-TrFE) sensor. Source: *Reproduced with permission from Prof. Rouxel, C. Thevenot, Élaboration de membranes polymères piézoélectriques souples en vue d'applications biomédicales. Université de Lorraine, 2018 [114].*

Due to their higher piezoelectric responses, PVDF and VDF copolymers serve as ideal materials platforms for proving the concept of mechanoelectrical transductions for tissue engineering applications, mainly for bone, muscle, and nerve regeneration (Table 1.7) [115].

Nunes-Pereira et al. [120] tested PVDF and VDF copolymers as porous membranes for tissue engineering applications. Membranes based on PVDF, P(VDF-*co*-HFP), and P(VDF-*co*-TrFE) copolymers were prepared, then, two types of cells were seeded onto them. Cell viability was then tested to determine the suitability of the membranes for tissue engineering and biomedical applications. The authors showed that higher cell proliferation was found on the PVDF membranes with lower porosity and pore sizes, while lower proliferation was observed in P (VDF-*co*-HFP) copolymer (Fig. 1.29) [120].

Another promising application of PVDF and VDF copolymers deals with lithium ion batteries as a binder, separator [54], or even in electrolyte [55]. P(VDF-*co*-HFP) copolymer has been used as a binder to prepare cells based on $Li(Ni_{0.5}Mn_{0.3}Co_{0.2})O$ as the active material [51,54,55,121]. Compared to commercially available PVDF, this copolymer has a negative influence on cycling performances. However, it was found that it can

TABLE 1.7 Sensitivity of blood pressure sensors.

Sensitive element of the pressure sensor	Experience	Pressure range (mmHg)	Sensitivity (mV/mmHg)
P(VDF-*co*-TrFE) film	Air pressure chamber	0–300	0.099
P(VDF-*co*-TrFE) nanofibers	Air pressure chamber	0–300	0.280
P(VDF-*co*-TrFE)/ PEDOT: PSS nanofibers	Air pressure chamber	0–300	4
PVDF film	Simulation	0–225	10–200[a]
P(VDF-*co*-TrFE) film	Simulation	40–200	0.237–0.373[a]
P(VDF-*co*-TrFE) film	Sensor wrapped around a Polydimethylsiloxane tube	40–120	0.35
PVDF film	Sensor wrapped around a latex tube	100–170	173
PVDF film	Sensor wrapped around the aorta of a pig (in vivo)	160–220	14

[a]*According to sensors dimensions.*
Reproduced with permission of C. Thevenot, Élaboration de membranes polymères piézoélectriques souples en vue d'applications biomédicales. Université de Lorraine, 2018 [114].

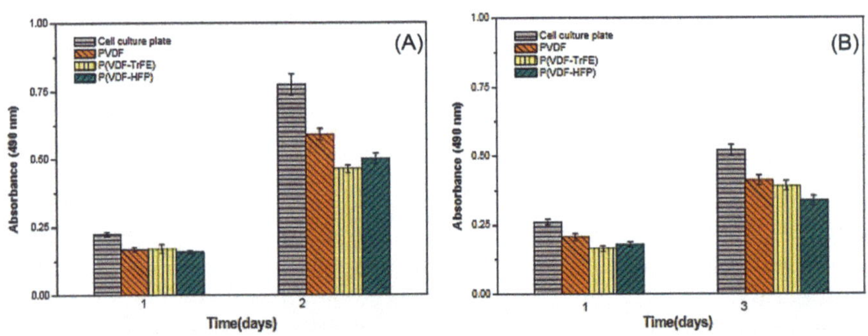

FIGURE 1.29 Cell viability of different cells growing on porous membranes made of PVDF and VDF copolymers as measured by MTS assay (it is a colorimetric method for sensitive quantification of viable cells in cell proliferation assay): (A) C2C12 (myoblast) cells up to 2 days of culture and (B) MC3T3-E1 (preosteoblast) cells up to 3 days of culture. Source: *Reproduced with permission of J. Nunes-Pereira, S. Ribeiro, C. Ribeiro, C.J. Gombek, F.M. Gama, A.C. Gomes, et al., Poly(vinylidene fluoride) and copolymers as porous membranes for tissue engineering applications, Polym. Test. 44 (2015) 234–241. https://doi.org/10.1016/j.polymertesting.2015.05.001 [120].*

efficiently reduce the peak temperature of preheated cells by up to 36% making it more suitable in applications where fire safety is highly prioritized [121].

1.6 Conclusion

This chapter gives a general overview on piezoelectric and ferroelectric polymers, mainly from PVDF and VDF co- and terpolymers. PVDF is a unique specialty polymer with exceptional thermal, chemical, and mechanical properties and five crystalline polymorphs. When processed accordingly, PVDF is considered to be one of the most frequently used ferroelectric polymers due to its outstanding electroactive properties, lightness, and softness. As a semicrystalline polymer, PVDF can exist in at least five crystalline phases, where the β one is considered to have the largest spontaneous polarization and is specifically targeted for achieving electroactive applications. Several methods have been developed to promote the formation of β conformation in PVDF such as stretching, pooling, annealing, and solvent casting. However, these methods ultimately influence the final properties. On the other hand, another promising method to endorse β phase formation is the introduction of a second polymer (such as PMMA or polycarbonate or P(VDF-*co*-TrFE), P(VDF-*ter*-TrFE-*ter*-CTFE) or P(VDF-*ter*-TrFE-*ter*-CFE) terpolymers) in blend. Actually, such fluorinated co/terpolymers are considered to be the most studied and used among all VDF copolymers for such electroactive applications. These relevant properties are triggered from β-crystal phase or ferroelectric phase, especially when the TrFE content is 20-50 mol%. On the other hand, the approach of polymer composite could also be considered. In fact, the incorporation of a ceramic filler with high dielectric constant into a polymer that exhibits good flexibility and high breakdown strength could lead to a material with improved features. In addition to the outstanding electroactive properties, PVDF and its co- or terpolymers display good mechanical and thermal properties as well as chemical inertness, making the resulting materials promising candidates in different fields such as smart magnetic sensors, energy storage devices, Internet-of-things, and mechanoelectrical transductions for tissue engineering applications, mainly for bones, muscles, and nerve regeneration.

Acknowledgments

The authors thank all coauthors cited in the following references, Arkema for financial support and supplying VDF and TrFE, and Prof. D Rouxel and Dr C. Thevenot (Université de Lorraine) for obtaining Fig. 1.28 and Table 1.7.

References

[1] H. Huang, J.F. Scott (Eds.), Ferroelectric Materials for Energy Applications, Wiley-VCH Verlag GmbH & Co. KGaA, Weinheim, Germany, 2018. Available from: https://doi.org/10.1002/9783527807505.

[2] N.R. Alluri, A. Chanderashkear, S.-J. Kim, Hybrid structures for piezoelectric nanogenerators: fabrication methods, energy generation, and self-powered applications, Energy Harvesting, InTech, 2018. Available from: https://doi.org/10.5772/intechopen.74770.

[3] B. Stadlober, M. Zirkl, M. Irimia-Vladu, Route towards sustainable smart sensors: ferroelectric polyvinylidene fluoride-based materials and their integration in flexible electronics, Chem. Soc. Rev. 48 (2019) 1787–1825. Available from: https://doi.org/10.1039/c8cs00928g.

[4] G. Ciofani, A. Menciassi, Piezoelectric Nanomaterials for Biomedical Applications, Springer, Berlin, Heidelberg, 2012.

[5] L.B. Kong, T. Li, H.H. Hng, F. Boey, T. Zhang, S. Li, Waste Energy Harvesting, 24, Springer Verlag, 2014. Available from: https://doi.org/10.1007/978-3-642-54634-1.

[6] N. Bizon, N.M. Tabatabaei, F. Blaabjerg, E. Kurt, Energy harvesting and energy efficiency: technology, Methods Appl. (2017). Available from: https://doi.org/10.1007/978-3-319-49875-1.

[7] X. Chen, X. Han, Q.D. Shen, PVDF-based ferroelectric polymers in modern flexible electronics, Adv. Electron. Mater. 3 (2017) 1600460. Available from: https://doi.org/10.1002/aelm.201600460.

[8] L. Ruan, X. Yao, Y. Chang, L. Zhou, G. Qin, X. Zhang, Properties and applications of the β phase poly(vinylidene fluoride), Polymers (Basel) 10 (2018) 228. Available from: https://doi.org/10.3390/polym10030228.

[9] X. Cai, T. Lei, D. Sun, L. Lin, A critical analysis of the α, β and γ phases in poly(vinylidene fluoride) using FTIR, RSC Adv. 7 (2017) 15382–15389. Available from: https://doi.org/10.1039/c7ra01267e.

[10] P. Martins, A.C. Lopes, S. Lanceros-Mendez, Electroactive phases of poly(vinylidene fluoride): determination, processing and applications, Prog. Polym. Sci. 39 (2014) 683–706. Available from: https://doi.org/10.1016/J.PROGPOLYMSCI.2013.07.006.

[11] M. Sharma, G. Madras, S. Bose, Process induced electroactive β-polymorph in PVDF: effect on dielectric and ferroelectric properties, Phys. Chem. Chem. Phys. 16 (2014) 14792–14799Royal Society of Chemistry. Available from: https://doi.org/10.1039/c4cp01004c.

[12] X. He, K. Yao, B.K. Gan, Phase transition and properties of a ferroelectric poly(vinylidene fluoride-hexafluoropropylene) copolymer, J. Appl. Phys. 97 (2005) 084101. Available from: https://doi.org/10.1063/1.1862323.

[13] A. De Neef, C. Samuel, G. Stoclet, M. Rguiti, C. Courtois, P. Dubois, et al., Processing of PVDF-based electroactive/ferroelectric films: importance of PMMA and cooling rate from the melt state on the crystallization of PVDF beta-crystals, Soft Matter 14 (2018) 4591–4602. Available from: https://doi.org/10.1039/c8sm00268a.

[14] F. Lederle, C. Härter, S. Beuermann, Inducing β phase crystallinity of PVDF homopolymer, blends and block copolymers by anti-solvent crystallization, J. Fluor. Chem. 234 (2020) 109522. Available from: https://doi.org/10.1016/j.jfluchem.2020.109522.

[15] C.R. Bowen, V.Y. Topolov, H.A. Kim, Modern Piezoelectric Energy-Harvesting Materials, 238, Springer Science Busssiness Media, 2016. Available from: https://doi.org/10.1007/978-3-319-29143-7.

[16] F.E. Bouharras, M. Raihane, B. Ameduri, Recent progress on core-shell structured BaTiO3@polymer/fluorinated polymers nanocomposites for high energy storage: synthesis, dielectric properties and applications, Prog. Mater. Sci. 113 (2020) 100670. Available from: https://doi.org/10.1016/j.pmatsci.2020.100670.

[17] A. Mayeen, M.S. Kala, M.S. Jayalakshmy, S. Thomas, D. Rouxel, J. Philip, et al., Dopamine functionalization of BaTiO3: an effective strategy for the enhancement of electrical, magnetoelectric and thermal properties of BaTiO3-PVDF-TrFE nanocomposites, Dalt. Trans. 47 (2018) 2039–2051. Available from: https://doi.org/10.1039/c7dt03389c.
[18] A. Mayeen, M.S. Kala, M.S. Jayalakshmy, S. Thomas, J. Philip, D. Rouxel, et al., Flexible and self-standing nickel ferrite-PVDF-TrFE cast films: promising candidates for high-end magnetoelectric applications, Dalt. Trans. 48 (2019) 16961–16973. Available from: https://doi.org/10.1039/c9dt02856k.
[19] A. Altomare, M. Bozorg, K. Loos, PVDF-based multiferroic, in: B. Ameduri, S. Fomin (Eds.), Fascinating Fluoropolymers and Their Applications: Progress in Fluorine Science, 2, Elsevier, Amsterdam, 2020, pp. 45–81. Available from: https://doi.org/10.1016/b978-0-12-821873-0.00002-3.
[20] E. Engineers, IEEE standard definitions of primary ferroelectric terms, East (1986) 986. Available from: https://doi.org/10.1109/IEEESTD.1986.81581.
[21] T. Furukawa, Ferroelectric properties of vinylidene fluoride copolymers, Phase Transit. 18 (1989) 143–211. Available from: https://doi.org/10.1080/01411598908206863.
[22] L. Zhu, Q. Wang, Novel ferroelectric polymers for high energy density and low loss dielectrics, Macromolecules 45 (2012) 2937–2954. Available from: https://doi.org/10.1021/ma2024057.
[23] L. Yang, X. Li, E. Allahyarov, P.L. Taylor, Q.M. Zhang, L. Zhu, Novel polymer ferroelectric behavior via crystal isomorphism and the nanoconfinement effect, Polymers (Guildf.) 54 (2013) 1709–1728. Available from: https://doi.org/10.1016/j.polymer.2013.01.035.
[24] L. Zhu, Exploring strategies for high dielectric constant and low loss polymer dielectrics, J. Phys. Chem. Lett. 5 (2014) 3677–3687. Available from: https://doi.org/10.1021/jz501831q.
[25] N. Izyumskaya, Y.I. Alivov, S.J. Cho, H. Morkoç, H. Lee, Y.S. Kang, Processing, structure, properties, and applications of PZT thin films, Crit. Rev. Solid. State Mater. Sci. 32 (2007) 111–202. Available from: https://doi.org/10.1080/10408430701707347.
[26] M. Ivanov, J. Macutkevic, R. Grigalaitis, J. Banys, General view of ferroelectrics: origin of ferroelectricity in metal oxide ferroelectrics and ferroelectric properties, Magnetic, Ferroelectric, and Multiferroic Metal Oxides, Elsevier, 2018, pp. 5–33. Available from: https://doi.org/10.1016/b978-0-12-811180-2.00001-3.
[27] Z. Lazarević, B.D. Stojanović, J.A. Varela, An approach to analyzing synthesis, structure and properties of bismuth titanate ceramics, Sci. Sinter. 37 (2005) 199–216. Available from: https://doi.org/10.2298/SOS0503199L.
[28] M.M. Vijatović, J.D. Bobić, B.D. Stojanović, History and challenges of barium titanate: Part I, Sci. Sinter. 40 (2008) 155–165. Available from: https://doi.org/10.2298/SOS0802155V.
[29] S.M. Said, M.F.M. Sabri, F. Salleh, Ferroelectrics and their applications, Reference Module in Materials Science and Materials Engineering, Elsevier, 2017. Available from: https://doi.org/10.1016/b978-0-12-803581-8.04143-6.
[30] R. Whatmore, Ferroelectric Materials, Springer Handbooks, Springer, 2017, p. 1. Available from: https://doi.org/10.1007/978-3-319-48933-9_26.
[31] Y. Wang, W. Chen, B. Wang, Y. Zheng, Ultrathin ferroelectric films: growth, characterization, physics and applications, Materials (Basel) 6 (2014) 6377–6485. Available from: https://doi.org/10.3390/ma7096377.
[32] F. Takeo, Structure and properties of ferroelectric polymersScientific.Net Key Eng. Mater. 92–93 (1994) 15–30.
[33] A.L. Moore, Fluoroelastomer monomers, Fluoroelastomers Handbook, Elsevier, 2006, pp. 25–36. Available from: https://doi.org/10.1016/b978-081551517-3.50006-x.
[34] B. Ameduri, From vinylidene fluoride (VDF) to the applications of VDF-containing polymers and copolymers: recent developments and future trends, Chem. Rev. 109 (2009) 6632–6686. Available from: https://doi.org/10.1021/cr800187m.

[35] G. Zapsas, Y. Patil, Y. Gnanou, B. Ameduri, N. Hadjichristidis, Poly(vinylidene fluoride)-based complex macromolecular architectures: from synthesis to properties and applications, Prog. Polym. Sci. 104 (2020) 101231. Available from: https://doi.org/10.1016/j.progpolymsci.2020.101231.
[36] H.S. Nalwa, Ferroelectric Polymers: Chemistry, Physics, and Applications, M. Dekker, Inc, New York, 1995.
[37] G.J. Puts, P. Crouse, B.M. Ameduri, Polytetrafluoroethylene: synthesis and characterization of the original extreme polymer, Chem. Rev. 119 (2019) 1763–1805. Available from: https://doi.org/10.1021/acs.chemrev.8b00458.
[38] A.J. Lovinger, Poly(vinylidene fluoride), Developments in Crystalline Polymers, Springer Netherlands, Dordrecht, 1982, pp. 195–273. Available from: https://doi.org/10.1007/978-94-009-7343-5_5.
[39] C. Boyer, D. Valade, L. Sauguet, B. Ameduri, B. Boutevin, Iodine transfer polymerization (ITP) of vinylidene fluoride (VDF). Influence of the defect of VDF chaining on the control of ITP, Macromolecules 38 (2005) 10353–10362. Available from: https://doi.org/10.1021/MA051349F.
[40] C. Boyer, B. Améduri, Iodine transfer copolymerization of vinylidene fluoride and α-trifluoromethacrylic acid in emulsion process without any surfactants, J. Polym. Sci. Part A Polym. Chem. 47 (2009) 4710–4722. Available from: https://doi.org/10.1002/pola.23525.
[41] C. Boyer, D. Valade, P. Lacroix-Desmazes, B. Ameduri, B. Boutevin, Kinetics of the iodine transfer polymerization of vinylidene fluoride, J. Polym. Sci. Part A Polym. Chem. 44 (2006) 5763–5777. Available from: https://doi.org/10.1002/pola.21654.
[42] A.D. Asandei, Photomediated controlled radical polymerization and block copolymerization of vinylidene fluoride, Chem. Rev. 116 (2016) 2244–2274. Available from: https://doi.org/10.1021/acs.chemrev.5b00539.
[43] S. Beuermann, M. Imran-ul-haq, Homogeneous phase polymerization of vinylidene fluoride in supercritical CO_2: surfactant free synthesis and kinetics, Macromol. Symp. 259 (2007) 210–217. Available from: https://doi.org/10.1002/masy.200751325.
[44] M. Imran-Ul-haq, N. Förster, R. Vukicevic, K. Herrmann, R. Siegmann, S. Beuermann, Iodine transfer radical polymerizations of vinylidene fluoride in supercritical carbon dioxide and polymer functionalization via click chemistry, ACS Symposium Series, 1024, American Chemical Society, 2009, pp. 233–243. Available from: https://doi.org/10.1021/bk-2009-1024.ch015.
[45] S. Banerjee, V. Ladmiral, A. Debuigne, C. Detrembleur, R. Poli, B. Améduri, Organometallic-mediated radical polymerization of vinylidene fluoride, Angew. Chem. Int. (Ed.) 57 (2018) 2934–2937. Available from: https://doi.org/10.1002/anie.201712347.
[46] P.G. Falireas, V. Ladmiral, A. Debuigne, C. Detrembleur, R. Poli, B. Ameduri, Straightforward synthesis of well-defined poly(vinylidene fluoride) and its block copolymers by cobalt-mediated radical polymerization, Macromolecules 52 (2019) 1266–1276. Available from: https://doi.org/10.1021/acs.macromol.8b02252.
[47] Y. Patil, B. Ameduri, First RAFT/MADIX radical copolymerization of tert-butyl 2-trifluoromethacrylate with vinylidene fluoride controlled by xanthate, Polym. Chem. 4 (2013) 2783–2799. Available from: https://doi.org/10.1039/c3py21139h.
[48] M. Guerre, B. Campagne, O. Gimello, K. Parra, B. Ameduri, V. Ladmiral, Deeper insight into the MADIX polymerization of vinylidene fluoride, Macromolecules 48 (2015) 7810–7822. Available from: https://doi.org/10.1021/acs.macromol.5b01528.
[49] M. Guerre, S.M.W. Rahaman, B. Améduri, R. Poli, V. Ladmiral, Limits of vinylidene fluoride RAFT polymerization, Macromolecules 49 (2016) 5386–5396. Available from: https://doi.org/10.1021/acs.macromol.6b01087.

[50] A. Manuel Stephan, D. Teeters, Characterization of PVdF-HFP polymer membranes prepared by phase inversion techniques I. Morphology and charge-discharge studies, Electrochim. Acta, 48, Pergamon, 2003, pp. 2143–2148. Available from: https://doi.org/10.1016/S0013-4686(03)00197-X.

[51] A.M. Stephan, K.S. Nahm, M. Anbu Kulandainathan, G. Ravi, J. Wilson, Poly(vinylidene fluoride-hexafluoropropylene) (PVdF-HFP) based composite electrolytes for lithium batteries, Eur. Polym. J. 42 (2006) 1728–1734. Available from: https://doi.org/10.1016/j.eurpolymj.2006.02.006.

[52] X. Tian, X. Jiang, Poly(vinylidene fluoride-co-hexafluoropropene) (PVDF-HFP) membranes for ethyl acetate removal from water, J. Hazard. Mater. 153 (2008) 128–135. Available from: https://doi.org/10.1016/j.jhazmat.2007.08.029.

[53] M.C. García-Payo, M. Essalhi, M. Khayet, Preparation and characterization of PVDF-HFP copolymer hollow fiber membranes for membrane distillation, Desalination 245 (2009) 469–473. Available from: https://doi.org/10.1016/j.desal.2009.02.010.

[54] J.T. Goldbach, R. Amin-Sanayei, W. He, J. Henry, W. Kosar, A. Lefebvre, et al., Chapter 6: Commercial synthesis and applications of poly(vinylidene fluoride), Fluorinated Polymers: Applications. RSC Polymer Chemistry Series, 2, Royal Society of Chemistry, 2017, pp. 127–157. Available from: https://doi.org/10.1039/9781782629368-00127.

[55] C.M. Costa, E. Lizundia, S. Lanceros-Méndez, Polymers for advanced lithium-ion batteries: state of the art and future needs on polymers for the different battery components, Prog. Energy Combust. Sci. 79 (2020) 100846. Available from: https://doi.org/10.1016/j.pecs.2020.100846.

[56] S. Fadhil, T. Marino, H.F. Makki, Q.F. Alsalhy, S. Blefari, F. Macedonio, et al., Novel PVDF-HFP flat sheet membranes prepared by triethyl phosphate (TEP) solvent for direct contact membrane distillation, Chem. Eng. Process. Process Intensif. 102 (2016) 16–26. Available from: https://doi.org/10.1016/j.cep.2016.01.007.

[57] B.S. Lalia, E. Guillen-Burrieza, H.A. Arafat, R. Hashaikeh, Fabrication and characterization of polyvinylidenefluoride-co-hexafluoropropylene (PVDF-HFP) electrospun membranes for direct contact membrane distillation, J. Memb. Sci. 428 (2013) 104–115. Available from: https://doi.org/10.1016/j.memsci.2012.10.061.

[58] P. Frübing, F. Wang, M. Wegener, Relaxation processes and structural transitions in stretched films of polyvinylidene fluoride and its copolymer with hexafluoropropylene, Appl. Phys. A 107 (2012) 603–611. Available from: https://doi.org/10.1007/s00339-012-6838-1.

[59] F. Boschet, B. Ameduri, Copolymers of chlorotrifluoroethylene: synthesis, properties, and applications, Chem. Rev. 114 (2014) 927–980. Available from: https://doi.org/10.1021/cr2002933.

[60] R.E. Sousa, J.C.C. Ferreira, C.M. Costa, A.V. Machado, M.M. Silva, S. Lanceros-Mendez, Tailoring poly(vinylidene fluoride-co-chlorotrifluoroethylene) microstructure and physicochemical properties by exploring its binary phase diagram with dimethylformamide, J. Polym. Sci. Part B Polym. Phys. 53 (2015) 761–773. Available from: https://doi.org/10.1002/polb.23692.

[61] Z. Li, Y. Wang, Z.Y. Cheng, Electromechanical properties of poly(vinylidene-fluoride-chlorotrifluoroethylene) copolymer, Appl. Phys. Lett. 88 (2006) 062904. Available from: https://doi.org/10.1063/1.2170425.

[62] Z. Zhang, T.C.M. Chung, Study of VDF/TrFE/CTFE terpolymers for high pulsed capacitor with high energy density and low energy loss, Macromolecules 40 (2007) 783–785. Available from: https://doi.org/10.1021/ma0627119.

[63] T. Yagi, M. Tatemoto, J.-I. Sako, Transition behavior and dielectric properties in trifluoroethylene and vinylidene fluoride copolymers, Polym. J. 12 (1980) 209–223.

[64] Prateek, V.K. Thakur, R.K. Gupta, Recent progress on ferroelectric polymer-based nanocomposites for high energy density capacitors: synthesis, dielectric properties,

and future aspects, Chem. Rev. 116 (2016) 4260–4317. Available from: https://doi.org/10.1021/acs.chemrev.5b00495.
[65] F. Bauer, E. Fousson, Q.M. Zhang, L.M. Lee, Ferroelectric copolymers and terpolymers for electrostrictors: synthesis and properties, IEEE Trans. Dielectr. Electr. Insul. 11 (2004) 293–298. Available from: https://doi.org/10.1109/TDEI.2004.1285900.
[66] T. Wongwirat, H. Manuspiya, L. Zhu, Electroactive fluoropolymers and polyamides, in: B. Ameduri, S. Fomin (Eds.), Fascinating Fluoropolymers and Their Applications: Progress in Fluorine Science, Elsevier, 2020, pp. 83–114. Available from: https://doi.org/10.1016/B978-0-12-821873-0.00003-5.
[67] T. Soulestin, V. Ladmiral, F. Domingues Dos Santos, B. Améduri, Vinylidene fluoride- and trifluoroethylene-containing fluorinated electroactive copolymers. How does chemistry impact properties? Prog. Polym. Sci. 72 (2017) 16–60. Available from: https://doi.org/10.1016/j.progpolymsci.2017.04.004.
[68] Y. Huang, G. Rui, Q. Li, E. Allahyarov, R. Li, M. Fukuto, et al., Enhanced piezoelectricity from highly polarizable oriented amorphous fractions in biaxially oriented poly(vinylidene fluoride) with pure β crystals, Nat. Commun. 12 (2021) 1–8. Available from: https://doi.org/10.1038/s41467-020-20662-7.
[69] V.S. Nguyen, D. Rouxel, M. Meier, B. Vincent, A. Dahoun, S. Thomas, et al., Effect of ultrasonication and other processing conditions on the morphology, thermomechanical, and piezoelectric properties of poly(vinylidene difluoride-trifluoroethylene) copolymer films, Polym. Eng. Sci. 54 (2014) 1280–1288. Available from: https://doi.org/10.1002/pen.23670.
[70] S. Wang, Q. Li, Design, synthesis and processing of PVDF-based dielectric polymers, IET Nanodielectrics 1 (2018) 80–91. Available from: https://doi.org/10.1049/iet-nde.2018.0003.
[71] W. Xia, Z. Zhang, PVDF-based dielectric polymers and their applications in electronic materials, IET Nanodielectrics 1 (2018) 17–31. Available from: https://doi.org/10.1049/iet-nde.2018.0001.
[72] B. Ameduri, A. Allaeddine. Controlled free-radical copolymerization of trifluoroethylene. US2017/9708419B2, 2013.
[73] S. Banerjee, S. Zaghloul, A. Alaaeddine, B. Ameduri, Kinetic and mechanistic aspects of the iodine transfer copolymerization of vinylidene fluoride with 2,3,3,3-tetrafluoro-1-propene and functionalization into ω-hydroxy fluorinated copolymers, Polym. Chem. 7 (2016) 6099–6109. Available from: https://doi.org/10.1039/c6py01152g.
[74] T. Soulestin, V. Ladmiral, T. Lannuzel, F. Domingues Dos Santos, B. Améduri, Differences in electroactive terpolymers based on VDF, TrFE and 2,3,3,3-tetrafluoropropene prepared by batch solution and semi-continuous aqueous suspension polymerizations, Polym. Chem. 8 (2017) 735–747. Available from: https://doi.org/10.1039/c6py01874b.
[75] A. Ghosh, L. Louis, A.D. Asandei, S. Nakhmanson, First-principles studies of spontaneous polarization in mixed poly(vinylidene fluoride)/2,3,3,3-tetrafluoropropene polymer crystals, Soft Matter 14 (2018) 2484–2491. Available from: https://doi.org/10.1039/c8sm00262b.
[76] T.C. Chung, A. Petchsuk, Ferroelectric polymers with large electrostriction; Based on semicrystalline VDF/TrFE/CTFE terpolymers, Ferroelectr. Lett. Sect. 28 (2001) 135–143. Available from: https://doi.org/10.1080/07315170108202957.
[77] Z. Zhang, Q. Meng, T.C.M. Chung, Energy storage study of ferroelectric poly(vinylidene fluoride-trifluoroethylene-chlorotrifluoroethylene) terpolymers, Polymer 50 (2009) 707–715. Available from: https://doi.org/10.1016/j.polymer.2008.11.005.
[78] Y. Lu, J. Claude, B. Neese, Q. Zhang, Q. Wang, A modular approach to ferroelectric polymers with chemically tunable Curie temperatures and dielectric constants, J. Am. Chem. Soc. 128 (2006) 8120–8121. Available from: https://doi.org/10.1021/ja062306x.

[79] Q. Li, G. Zhang, F. Liu, K. Han, M.R. Gadinski, C. Xiong, et al., Solution-processed ferroelectric terpolymer nanocomposites with high breakdown strength and energy density utilizing boron nitride nanosheets, Energy Environ. Sci. 8 (2015) 922–931. Available from: https://doi.org/10.1039/c4ee02962c.

[80] Z.-Y. Cheng, H. Li, F. Xia, H. Xu, D. Olson, C. Huang, et al., Electromechanical properties and molecular conformation in P(VDF-TrFE)-based terpolymer, in: Y. Bar-Cohen (Ed.), Smart Structures and Materials 2002: Electroactive Polymer Actuators and Devices (EAPAD), 4695, SPIE, 2002, pp. 167–175. Available from: https://doi.org/10.1117/12.475162.

[81] P. Mao, J. Wang, L. Zhang, Q. Sun, X. Liu, L. He, et al., Tunable dielectric polarization and breakdown behavior for high energy storage capability in P(VDF–TrFE–CFE)/PVDF polymer blended composite films, Phys. Chem. Chem. Phys. 22 (2020) 13143–13153. Available from: https://doi.org/10.1039/d0cp01071e.

[82] G. Casar, X. Li, B. Malič, Q.M. Zhang, V. Bobnar, Impact of structural changes on dielectric and thermal properties of vinylidene fluoride-trifluoroethylene-based terpolymer/copolymer blends, Phys. B Condens. Matter 461 (2015) 5–9. Available from: https://doi.org/10.1016/j.physb.2014.12.010.

[83] X. Zhang, Y. Shen, Z. Shen, J. Jiang, L. Chen, C.W. Nan, Achieving high energy density in PVDF-based polymer blends: suppression of early polarization saturation and enhancement of breakdown strength, ACS Appl. Mater. Interfaces 8 (2016) 27236–27242. Available from: https://doi.org/10.1021/acsami.6b10016.

[84] T. Soulestin, V. Ladmiral, T. Lannuzel, F. Domingues Dos Santos, B. Ameduri, Importance of microstructure control for designing new electroactive terpolymers based on vinylidene fluoride and trifluoroethylene, Macromolecules 48 (2015) 7861–7871. Available from: https://doi.org/10.1021/acs.macromol.5b01964.

[85] T. Soulestin, P. Marcelino Dos Santos Filho, V. Ladmiral, C. Totée, G. Silly, T. Lannuzel, et al., Influence of trans-1,3,3,3-tetrafluoropropene on the structure-properties relationship of VDF- and TrFE-based terpolymers, Macromolecules 50 (2017) 503–514. Available from: https://doi.org/10.1021/acs.macromol.6b02010.

[86] Y. Li, T. Soulestin, V. Ladmiral, B. Ameduri, T. Lannuzel, F. Domingues Dos Santos, et al., Stretching-induced relaxor ferroelectric behavior in a poly(vinylidene fluoride-co-trifluoroethylene-co-hexafluoropropylene) random terpolymer, Macromolecules 50 (2017) 7646–7656. Available from: https://doi.org/10.1021/acs.macromol.7b01205.

[87] T. Soulestin, P. Marcelino Dos Santos Filho, V. Ladmiral, T. Lannuzel, F. Domingues Dos Santos, B. Améduri, Ferroelectric fluorinated copolymers with improved adhesion properties, Polym. Chem. 8 (2017) 1017–1027. Available from: https://doi.org/10.1039/c6py02063a.

[88] B. Zaarour, L. Zhu, C. Huang, X. Jin, Fabrication of a polyvinylidene fluoride cactus-like nanofiber through one-step electrospinning, RSC Adv. 8 (2018) 42353–42360. Available from: https://doi.org/10.1039/c8ra09257e.

[89] Z. Yin, B. Tian, Q. Zhu, C. Duan, Characterization and application of PVDF and its copolymer films prepared by spin-coating and langmuir-blodgett method, Polymers (Basel) 11 (2019) 2033. Available from: https://doi.org/10.3390/polym11122033.

[90] Y. Cao, Q. Deng, Z. Liu, D. Shen, T. Wang, Q. Huang, et al., Enhanced thermal properties of poly(vinylidene fluoride) composites with ultrathin nanosheets of MXene, RSC Adv. 7 (2017) 20494–20501. Available from: https://doi.org/10.1039/C7RA00184C.

[91] W.N. dos Santos, C.Y. Iguchi, R. Gregorio, Thermal properties of poly(vinilidene fluoride) in the temperature range from 25 to 210°C, Polym. Test. 27 (2008) 204–208. Available from: https://doi.org/10.1016/j.polymertesting.2007.10.005.

[92] L.F. Malmonge, S. Do Carmo Langiano, J.M.M. Cordeiro, L.H.C. Mattoso, J.A. Malmonge, Thermal and mechanical properties of PVDF/PANI blends, Mater. Res. 13 (2010) 465–470. Available from: https://doi.org/10.1590/S1516-14392010000400007.

[93] M. Mackey, D.E. Schuele, L. Zhu, L. Flandin, M.A. Wolak, J.S. Shirk, et al., Reduction of dielectric hysteresis in multilayered films via nanoconfinement, Macromolecules 45 (2012) 1954–1962. Available from: https://doi.org/10.1021/ma202267r.

[94] R. Hadji, V.S. Nguyen, B. Vincent, D. Rouxel, F. Bauer, Preparation and characterization of P(VDF-TrFE)/Al2O3 nanocomposite, IEEE Trans. Ultrason. Ferroelectr. Freq. Control. 59 (2012) 163–167. Available from: https://doi.org/10.1109/TUFFC.2012.2168.

[95] R.K. Layek, S. Samanta, D.P. Chatterjee, A.K. Nandi, Physical and mechanical properties of poly(methyl methacrylate)-functionalized graphene/poly(vinylidine fluoride) nanocomposites: piezoelectric β polymorph formation, Polymer 51 (2010) 5846–5856. Available from: https://doi.org/10.1016/j.polymer.2010.09.067.

[96] A.C. Jayasuriya, A. Schirokauer, J.I. Scheinbeim, Crystal-structure dependence of electroactive properties in differently prepared poly(vinylidene fluoride/hexafluoropropylene) copolymer films, J. Polym. Sci. Part B Polym. Phys. 39 (2001) 2793–2799. Available from: https://doi.org/10.1002/polb.10035.

[97] J. Gomes, J.S. Nunes, V. Sencadas, S. Lanceros-Mendez, Influence of the β-phase content and degree of crystallinity on the piezo- and ferroelectric properties of poly (vinylidene fluoride), Smart Mater. Struct. 19 (2010) 065010. Available from: https://doi.org/10.1088/0964-1726/19/6/065010.

[98] P. Martins, J.S. Nunes, G. Hungerford, D. Miranda, A. Ferreira, V. Sencadas, et al., Local variation of the dielectric properties of poly(vinylidene fluoride) during the α- to β-phase transformation, Phys. Lett. A 373 (2009) 177–180. Available from: https://doi.org/10.1016/j.physleta.2008.11.026.

[99] V. Sencadas, R. Gregorio, S. Lanceros-Méndez, α to β Phase transformation and microestructural changes of PVDF films induced by uniaxial stretch, J. Macromol. Sci. Part B 48 (2009) 514–525. Available from: https://doi.org/10.1080/00222340902837527.

[100] A. Gradys, P. Sajkiewicz, S. Adamovsky, A. Minakov, C. Schick, Crystallization of poly(vinylidene fluoride) during ultra-fast cooling, Thermochim. Acta 461 (2007) 153–157. Available from: https://doi.org/10.1016/j.tca.2007.05.023.

[101] W. Künstler, M. Wegener, M. Seiß, R. Gerhard-Multhaupt, Preparation and assessment of piezo- and pyroelectric poly (vinylidene fluoride-hexafluoropropylene) copolymer films, Appl. Phys. A Mater. Sci. Process. 73 (2001) 641–645. Available from: https://doi.org/10.1007/s003390100991.

[102] J. Claude, Y. Lu, K. Li, Q. Wang, Electrical storage in poly(vinylidene fluoride) based ferroelectric polymers: correlating polymer structure to electrical breakdown strength, Chem. Mater. 20 (2008) 2078–2080. Available from: https://doi.org/10.1021/cm800160r.

[103] C. Wan, C.R. Bowen, Multiscale-structuring of polyvinylidene fluoride for energy harvesting: the impact of molecular-, micro- and macro-structure, J. Mater. Chem. A 5 (2017) 3091–3128. Available from: https://doi.org/10.1039/C6TA09590A.

[104] C.M. Costa, V.F. Cardoso, R. Brito-Pereira, P. Martins, D.M. Correia, V. Correia, et al., Electroactive poly(vinylidene fluoride)-based materials: recent progress, challenges, and opportunities, in: B. Ameduri, S. Fomin (Eds.), Fascinating Fluoropolymers and Their Applications: Progress in Fluorine Science, Elsevier, 2020, pp. 1–43. Available from: https://doi.org/10.1016/B978-0-12-821873-0.00001-1.

[105] A.F. Miller, J.B. Whiteside, C.L. Hicks, Pvdf sonar transducer system. US20140092709A1, 2013.

[106] K.A. Ahmad, M.K. Osman, Z. Hussain, M.F. Abdullah, A.A. Manaf, N. Abdullah, Design and characterization piezoelectric acoustic transducer for sonar application, Proceedings—2018 8th IEEE International Conference on Control System,

Computing and Engineering (ICCSCE), Institute of Electrical and Electronics Engineers Inc., 2019, pp. 233–237. Available from: https://doi.org/10.1109/ICCSCE.2018.8685014.

[107] A.S. Fiorillo, F. Lamonaca, S.A. Pullano, PVDF based sonar for a remote web system to control mobile robots, Sens. Transducers J. 8 (2010) 65–73.

[108] D.S. Li, L. Cheng, C.M. Gosselin, The design of structural acoustic sensors for active control of sound radiation into enclosures, Smart Mater. Struct. 13 (2004) 371. Available from: https://doi.org/10.1088/0964-1726/13/2/016.

[109] L. Ionov, Polymeric actuators, Langmuir 31 (2015) 5015–5024. Available from: https://doi.org/10.1021/la503407z.

[110] L. Xie, X. Huang, Y. Huang, K. Yang, P. Jiang, Core-shell structured hyperbranched aromatic polyamide/$BaTiO_3$ hybrid filler for poly(vinylidene fluoride-trifluoroethylene-chlorofluoroethylene) nanocomposites with the dielectric constant comparable to that of percolative composites, ACS Appl. Mater. Interfaces 5 (2013) 1747–1756. Available from: https://doi.org/10.1021/am302959n.

[111] B.J. Hansen, Y. Liu, R. Yang, Z.L. Wang, Hybrid nanogenerator for concurrently harvesting biomechanical and biochemical energy, ACS Nano 4 (2010) 3647–3652. Available from: https://doi.org/10.1021/nn100845b.

[112] C. Lee, J.A. Tarbutton, Polyvinylidene fluoride (PVDF) direct printing for sensors and actuators, Int. J. Adv. Manuf. Technol. 104 (2019) 3155–3162. Available from: https://doi.org/10.1007/s00170-019-04275-z.

[113] Y.A. Yildirim, A. Toprak, O. Tigli, Piezoelectric membrane actuators for micropump applications using PVDF-TrFE, J. Microelectromech. Syst. 27 (2018) 86–94. Available from: https://doi.org/10.1109/JMEMS.2017.2773090.

[114] C. Thevenot, Élaboration de Membranes Polymères Piézoélectriques Souples En Vue d'applications Biomédicales, Université de Lorraine, France, 2018.

[115] C. Ribeiro, V. Sencadas, D.M. Correia, S. Lanceros-Méndez, Piezoelectric polymers as biomaterials for tissue engineering applications, Colloids Surf. B Biointerfaces 136 (2015) 46–55. Available from: https://doi.org/10.1016/j.colsurfb.2015.08.043.

[116] J. Granstrom, J. Feenstra, H.A. Sodano, K. Farinholt, Energy harvesting from a backpack instrumented with piezoelectric shoulder straps, Smart Mater. Struct. 16 (2007) 1810. Available from: https://doi.org/10.1088/0964-1726/16/5/036.

[117] C. Sun, J. Shi, D.J. Bayerl, X. Wang, PVDF microbelts for harvesting energy from respiration, Energy Environ. Sci. 4 (2011) 4508–4512. Available from: https://doi.org/10.1039/c1ee02241e.

[118] J. Kim, J.H. Lee, H. Ryu, J.H. Lee, U. Khan, H. Kim, et al., High-performance piezoelectric, pyroelectric, and triboelectric nanogenerators based on P(VDF-TrFE) with controlled crystallinity and dipole alignment, Adv. Funct. Mater. (2017) 27. Available from: https://doi.org/10.1002/ADFM.201700702.

[119] S. Sukumaran, S. Chatbouri, D. Rouxel, E. Tisserand, F. Thiebaud, T. Ben Zineb, Recent advances in flexible PVDF based piezoelectric polymer devices for energy harvesting applications, J. Intell. Mater. Syst. Struct. 32 (2021) 746–780. Available from: https://doi.org/10.1177/1045389X20966058.

[120] J. Nunes-Pereira, S. Ribeiro, C. Ribeiro, C.J. Gombek, F.M. Gama, A.C. Gomes, et al., Poly(vinylidene fluoride) and copolymers as porous membranes for tissue engineering applications, Polym. Test. 44 (2015) 234–241. Available from: https://doi.org/10.1016/j.polymertesting.2015.05.001.

[121] A.V. Le, M. Wang, D.J. Noelle, Y. Shi, Y. Shirley Meng, D. Wu, et al., Using high-HFP-content cathode binder for mitigation of heat generation of lithium-ion battery, Int. J. Energy Res. 41 (2017) 2430–2438. Available from: https://doi.org/10.1002/er.3824.

CHAPTER

2

Characterization methods used for the identification of ferroelectric beta phase of fluoropolymers

K. Thiyagarajan[1], *T. Avanish Babu*[2], *W. Madhuri*[2] *and Arunkumar Chandrasekhar*[3]

[1]School of Electrical Engineering, Vellore Institute of Technology, Vellore, India, [2]Ceramic Composite Laboratory, Centre for Functional Materials, Vellore Institute of Technology, Vellore, India, [3]Nanosensors and Nanoenergy Lab, Department of Sensors and Biomedical Technology, School of Electronics Engineering, Vellore Institute of Technology, Vellore, India

2.1 Introduction

Faced with the issues of the energy crunch and conventional fossil energy usages, for example, greenhouse gas effects, people became aware of the value of creating sustainable fresh energy sources, such as tidal, wind, and solar power. Despite ample renewable energy reserves, there are several barriers in production and transition, which include expensive equipment, restricted topographical locations, and lack of efficient conversion techniques. Moreover, the broad utilization of smart and adaptable wearable gadgets in the clinical industry in combination of machines and artificial intelligence (AI) is the need of the hour. In the most recent decade, energy delivery and use remain the main restricting variables for the progress of wearable flexible electronics, regardless of significant advancements in the diminishing power utilization of these electric gadgets. Accordingly, it is unequivocally necessary to make a novel group of compact self-driven intelligent electronics that can give electron flow to micro/nanoelectronic gadgets

Engineered Polymer Nanocomposites for Energy Harvesting Applications
DOI: https://doi.org/10.1016/B978-0-12-824155-4.00002-8

with minimum power utilization, thus wiping out the utilization of hazardous chemical batteries. In 1880, Pierre and Jacques Curie, two French scientists, discovered the piezoelectric effect and the piezoelectric potential of piezoelectric materials, created by the absence of center of symmetry into a structure in the form of the crystal when a mechanical force is applied externally. At first, Professor Wang [1] proposed using ZnO nanowires to convert the MME (micromechanical energy) to electric energy and called it piezoelectric nanogenerator (PENG). The versatile properties, such as ferroelectric, dielectric, piezoelectric, and pyroelectric properties, in polyvinylidene fluoride (PVDF) and its copolymers are excellent choice for flexible electric gadgets. In contrast to rigid and brittle piezoelectric ceramics which are unfit for the processing and versatile electronics, PVDF and its copolymers [2] are the new focus of research because of their great flexibility and easy manufacturing processes. In addition, in most polymers, molecular weight also plays a major role. Some parameters, such as dielectric and energy storage, in addition to molecular weight, influence the performance of PVDF. Lower-molecular-weight samples contain high energy storage performance due to high resistivity and are suitable for capacitor appliccations compared to higher-molecular-weight PVDF. The piezoelectric properties of PVDF are due to the presence of polar crystalline beta phases, which show the strongest electrical dipole moment of all the crystalline phases. Thus an increase in β-phase contents in PVDF provides an efficient way to enhance the ability of PVDF for piezoelectric -based energy harvesters, actuators, and sensors. PVDF has high dielectric and piezoelectric activity due to its heterogeneity, which allows crystallization in four distinct phases, viz alpha (α), beta (β), gamma (γ), and delta (δ). The alpha phase is monoclinic with stability in thermodynamics and nonpolar with a TGTG chain conformation. The gamma stage is polar with the conformation of the TTTGTTTG chain. The beta phase has an orthorhombic unit cell with the conformation of the all-trans chains thereby crystallizing PVDF chains. They have a net dipole moment in parallel orientation: polar beta and gamma phases result in polarization. In antiparallel dipoles, the result is a nonpolar alpha phase, as the net dipole moment is zero. In general, the alpha or gamma-phases are not so polarized. For the β phase, there are two chains in a planar all-trans conformation with a zigzag. The resulting dipole moment generated by an applied electric field is important for the highest polarization. The polar β phase displays a trans-conformation composed of hydrogen and fluorine atoms on converse sides of the polymer base, leading to a nonnull dipole moment for the majority of applications. This dipole moment of PVDF gives strong dielectric, massive piezoelectric, pyroelectric, and ferroelectric responses. To improve the technical application of PVDF, the β phase should be attained; hence increasing the β phase was indeed of enormous importance [3]. Many researchers have published their works on beta phase PVDF due to the extensive applications. Researchers followed the different processes to get the high electroactive beta phase. Poling is one of the methods to get the ferroelectric beta phase in the presence of a static electric field when all the dipoles are oriented in the same direction. The tapped process is useful to convert the dipoles orientation process and the quenching process is a good method for getting the high electroactive beta phase. The selection of the preparation methods like quenching, poling, mechanical stretching, adding additives, addition of copolymers, and ceramics etc. depends on the application.

2.2 Processing of beta phase using different methods

The beta phase is one of the best phases in piezoelectric polymers that present high piezo-, pyro-, and ferroelectric properties. Most of the beta phase is formed by the mechanical stretching of PVDF films. Before stretching, the PVDF films bear alpha phase and after stretching the alpha phase is converted to the beta phase. The formation of the beta phase follows different methods [4,5], like mechanical stretching, pressing–folding–heating, and electrospinning [6]. Application of external electric fields, rapid cooling, additional additives, preparation of copolymer methods, and so on is shown in Fig. 2.1 [7]. Stepwise transformations to achieve the beta phase PVDF are classified into four

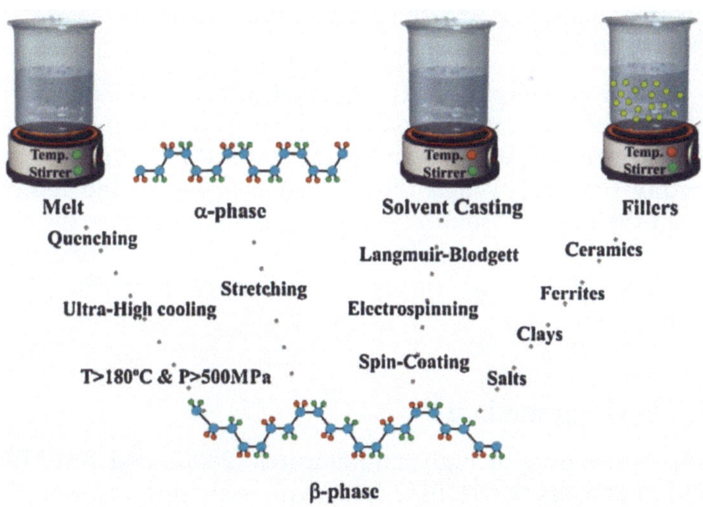

FIGURE 2.1 Different methods for getting beta phase of PVDF. Source: *Reproduced with permission P. Martins, A.C. Lopes, S. Lanceros-Mendez, Electroactive phases of poly(vinylidene fluoride): determination, processing and applications, Prog. Polym. Sci. 39 (4) (2014) 683–706. doi: 10.1016/j.progpolymsci.2013.07.006 [7]. Copyright Elsevier.*

types, based on the processing conditions: melt method, quenching method, pressing and folding (P&F) operation, and additives addition.

The melt process necessitates high pressure, high temperature, and other unique conditions [8,9]. In a very early analysis, Yang et al. discovered that quenching and then annealing PVDF films would cause the crystallization of phase directly from the melt [10]. The transformation of different phases into beta phase is shown in Fig. 2.1.

In this type of beta phase shape, the applied tension to the film allows polymer chains to align into crystals, resulting in an all-trans planar zigzag (TTT) conformation. This process causes the dipoles of polymer chains to arrange in a line naturally in the applied stress direction. The highest beta-phase content was stated to be obtained at 80°C with a value of 5 as the stretch ratio *R*, but only 20% of the original alpha-phase content in the sample will be obtained. The α to β transformation is improved by supplementary poling of the films. During the film extension, there is a disordering of crystallites and this has an impact on chain portability in scattered districts on beta stage crystallization.

The ferroelectric and piezoelectric switching behavior of PVDF film drawn at stretching ratios ranging from 1 to 5 for temperature ranging from 80°C to 140°C have been investigated and found that the two parameters influence the phase formation of PVDF [11]. High beta phase material means greater availability of oriented dipoles and high piezoelectric coefficients.

2.2.1 Melt method

Even though commonly used for applications, mechanically drawn PVDF film has inadequate piezoelectric activity and is thermally unstable. However, thermally stable and extra active PVDF films are required for everyday use [12]. When there is an increase in the temperature (lower than the melting point and the Curie temperature), the mechanically stretched PVDF depolarises. The mechanically stretched PVDF film is very difficult to incorporate directly into the device in microapplications.

2.2.2 Quenching method

Pressure quenching at high temperatures (280 C and 500 MPa) produced PVDF of a new crystalline form with a melting point of 185 C, 27 C higher than that of the -PVDF melt crystalized at ordinary pressures. The α-PVDF crystalline has a melting temperature of 27°C below compared to this at ordinary pressures [7]. On further research, the obtained crystalline structure was found to be a combination of alpha and beta forms and the β-PVDF alone can be obtained at a pressure above 500 MPa [13]. A meta-stable hexagonal crystallization process which consists of expanded chain lamellar crystals made from a mixture of beta and gamma PVDF forms was observed, which was later converted to β-PVDF at a high temperature of 120°C by the poling process.

The β phase content in the samples can be increased from 0% to 85% when the quenching pressure is increased from 200 to 700 MPa. Starting from an initial pressure of 100 MPa, it was found that the sample with 100% β-phase can be obtained at a pressure of 800 MPa [7]. Melt-crystallization of PVDF at extrahigh cooling speeds going from 30 to 3000 K/s was additionally examined utilizing nonisothermal crystallization and also at consistent temperatures after quenching at 6000 K/s [14]. A rise in cooling rates above 150 K/s contributes to the formation of the beta phase. At low temperatures with cooling speeds above 2000 K/s, the exothermic peak was observed because of the crystallization of the pure beta phase. The presence of vibration bands in the characteristic beta phase and the β phase of PVDF itself unequivocally relies upon the quenching temperature, as determined by an organized investigation of the impact of the quenching temperature on the heterogeneity in PVDF [15]. In order to perform the quenching process, initially the PVDF films are heated to a temperature at which there is no chance of the occurrence of polymerization and the polymer is cooled instantly. Distinct phases of PVDF are formed during the annealing process due to the occurrence of polymerization. The aggregation of PVDF molecules takes place at some local microscopic areas [10,11]. The nucleation of the β phase dominates at lower temperatures compared with the α phase nucleation.

2.2.3 Pressing and folding operation

The dimensional changes that occur during the P&F operation are of importance. At each cycle, the sample deforms plastically, but its volume remains unchanged. The maximal nominal stress (or strain) arises at the closing stages of each step in the process, as in Eq. (2.1).

$$P_a(c) = P_1 (2/a)^{n-1} \tag{2.1}$$

where c represents the number of cycles, P_1 denotes the pressure exerted at the first period, n represents the number of iterations, and a is a component describing how much the sample region expands during compression. The volume of the unit cell in the β phase is lower than that in the α phase. The β phase is less energetically stable. The increase in β phase (Δf_b) is proportional to the increase in plastic energy (Ω) adsorbed by the material. Based on these factors, and presuming that the portion of beta phase before beginning the P&F (f_{b0}) process is 8%, a differential approximation yields the following correlation between the volume of beta-phase and Ω [16]:

$$f_b(\Omega) = 1 - \left(1 - f_{b0}\right)e^{-a\,\Omega} \tag{2.2}$$

$$f_b(c) = 1 - \left(1 - f_{b0}\right)e^{-b[1-(2h)n]} \tag{2.3}$$

where a and b are proportionality constants.

Several factors can affect the relationship between f_b and Ω as well as the amount of Ω that can be produced in the P&F process. It is preferable to convert the maximum α phase as possible into β phase. For the least amount of Ω or minimum c, the molecular weight and annealing temperature of the PVDF decide the mobility of the chain and the phase of the PVDF. The β phase depends on several factors, such as molecular weight, folding and pressing, and annealing temperature.

It is assumed that polar structures of comparatively bigger size have higher thermal stability in P&F films with low molecular weight (M_w). The relevance of an external electric field is responsible for the reorientation of CF_2 dipoles and for the polymer chains to become more normal. Smaller polar structures with high M_w (534 and 670–700 kg/mol) are unstable in P&F films due to polarization randomization caused by thermal activation. Despite the effect of the applied field, the polar complexes remain surprisingly small in this situation.

Fig. 2.2 gives clear information about the beta phase [16]. The beta phase depends on the annealing temperature and the molecular weight of the sample. The low molecular weight and low annealing temperature contain high alpha phase and low beta phase and there are mixed phases like alpha, beta, and gamma at the five-degree temperature. The very high molecular weight (9534 kg/mol) contains the pure beta phase, mixed phase, and pure alpha phase. The beta phase of the high-molecular-weight samples possesses a good beta phase up to high annealing temperatures. The sample with high molecular weight (670–700 kg/mol) contains a better beta phase compared to the other samples, the beta phase will be stable at high annealing temperatures compared to other samples, and it does not have a mixed phase (alpha + beta + gamma).

2.2.4 Additives

The polar PVDF and nonpolar graphene form a homogenous composite when graphene nanosheets are used as fillers in PVDF. The homogenous dispersion in PVDF is a result of the surface functionalization of graphene. This is possible due to the advanced electrical,

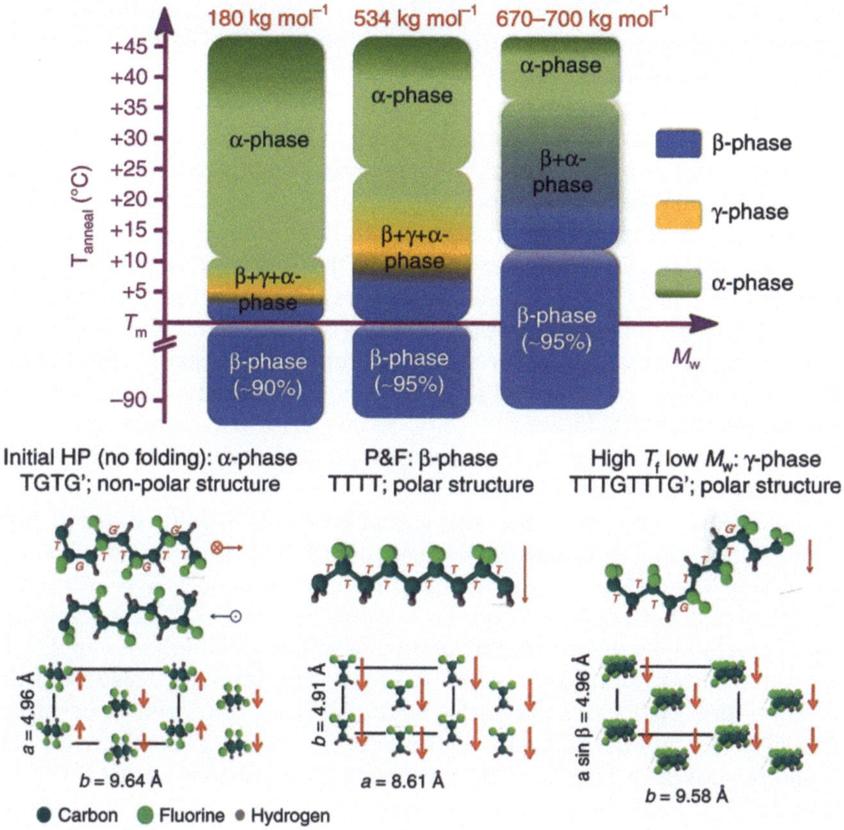

FIGURE 2.2 Different phases of PVDF at different molecular weights corresponding to annealing temperatures: initial hot press without folding alpha phase trans-gauche–trans-gauche chain structure; pressing and folding beta phase with trans chain structure and high temperature and low molecular weight gamma phase TTTG chain structure (In sequence from right to left). Source: *Reproduced with permission N. Meng, et al., Ultrahigh β-phase content poly(vinylidene fluoride) with relaxor-like ferroelectricity for high energy density capacitors, Nat. Commun. 10 (1) (2019) 1–9. doi: 10.1038/s41467-019-12391-3 [16]. Copyright Nature.*

mechanical, thermal, and ocular properties of the material [17]. The dielectric properties are also very important for piezoelectric materials. The dielectric losses play a major role in piezoelectric materials, especially when conducting graphene is integrated into PVDF and depending on which additive nanoparticles are chosen. By adding graphene into dielectric materials, the leakage currents are enhanced. This can degrade the properties of the compound for a lifetime due to the high conducting properties of graphene. To overcome this the oxide form of graphene is used, that is, GO. When adding GO to PVDF, the dispersibility of PVDF is very good but it degrades the mechanical and electrical properties because the oxygen bonds with carbon. Additive materials like graphene, GO, rGO, etc. are added in very small amounts because the properties of these materials are different. Carbon-related materials are conductive materials while PVDF is a dielectric material. Comparing the two materials, the properties of both materials are quite different. Hence the addition of large amounts of additives can degrade all the original properties of materials. Another solution to this problem is the use of reduced graphene oxide (rGO) which shows the properties of graphene. The rGO enhances the dispersion rate in the polymers and enhances the piezoelectric properties of PVDF [17]. The enhanced piezoelectric properties in the polar state are due to electrostatic interactions between positive metal ion charge centers and negative charge clouds in rGO, respectively. However, only certain critical weight% of rGO in PVDF is noticed to result in enhanced piezoelectric properties of the composite (1.0 rGO-Ag/PVDF).

Nucleation starts from the alpha phase by the locally ordered beta phase. When time progresses, the second cycle begins in which more beta phase crystals form the alpha phase crystals [17]. The attraction and repulsion between the dipoles present in PVDF + rGO-Ag are key to the mechanism of nucleation of piezoelectric polar phases in PVDF after the addition of additives like rGO-Ag. The final step results in the full transition of locally arranged beta phase to spherulite-type structures. The schematic diagram of the process of the piezoelectric polar phases is shown in Fig. 2.3. The positively charged ions of rGO/Ag are attracted by negatively charged ions in PVDF and the negatively charged ions in rGO/Ag are attracted to positively charged PVDF ions.

2.3 Characterization techniques

2.3.1 Differential scanning calorimetry

Differential scanning calorimetry (DSC) is a thermoanalytical procedure that is utilized in the identification of the crystalline form of PVDF. Various melting peaks occur in the DSC thermogram, depending on the crystalline level of the PVDF. Typically, materials are ferroelectric only below a certain phase transition temperature, called the Curie temperature (T_C), and are paraelectric above this temperature: the impulsive polarization vanishes, and the ferroelectric crystal structure converts into the paraelectric structure. Several ferroelectrics lose their piezoelectric properties above T_C entirely, because ferroelectrics have a noncentrosymmetric crystal structure below the transition temperature [18]. The characteristics of the DSC peaks depend on the crystalline phase as

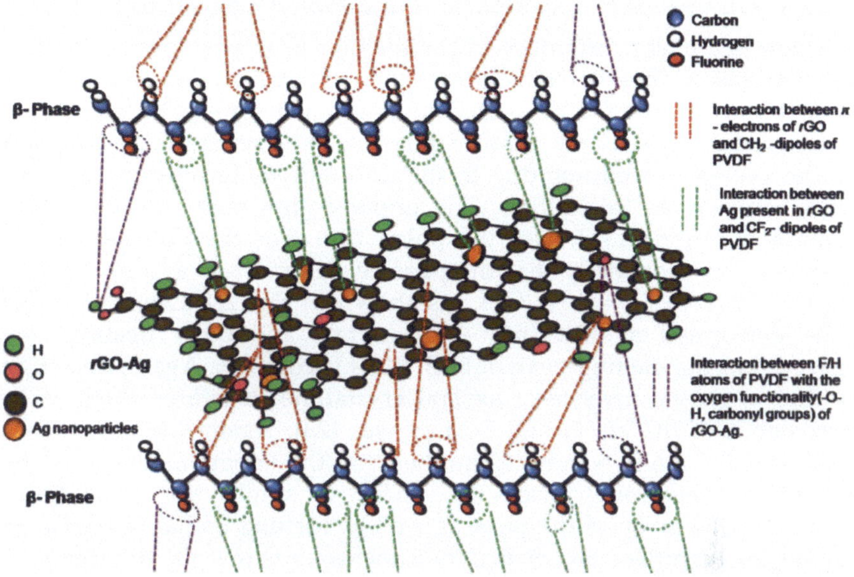

FIGURE 2.3 The schematic diagram of the process of the piezoelectric polar phases of the PVDF and rGO-Ag composite. Source: *Reproduced with permission M. Pusty, L. Sinha, P. M. Shirage, A flexible self-poled piezoelectric nanogenerator based on a rGO-Ag/PVDF nanocomposite, N. J. Chem. 43 (1) (2019) 284–294. doi: 10.1039/c8nj04751k [17]. Copyright Royal Society of Chemistry.*

well as the morphological properties that include the defects and crystalline dimensions. The predicted melting temperature for the alpha phase of PVDF by Prest and Luca [19] is 172°C, while Gregorio and Cestarini [20] reported at 167°C. Beta-crystallites have a melting temperature comparable with the alpha-PVDF [21]. DSC is not used to differentiate between α and β phases, but it quantifies the crystallinity of the film in terms of percentage. PVDF films crystallized in the gamma phase, depending on their origin, exhibit different endotherms than those found in the alpha and beta phases. For example, after isothermal crystallization at 166°C, quenching the sample results in the development of miniature alpha phase spherulites 8-μm in size where fusion happens at 160°C. Besides serving as an indication of the phase present in the polymer film, especially assessing the occurrence of beta phase PVDF, DSC should also be considered as a reference procedure to Fourier-transform infrared spectroscopy (FTIR) or X-ray diffraction (XRD). The features of DSC are influenced by the crystalline phase in addition to the crystalline defects exacerbated by the inclusion of fillers in polymer compounds [22].

2.3.1.1 *Difference between poled and unpoled DSC curves*

There is a major difference in the melting peak positions in the DSC pattern, based on several factors, as shown in Fig. 2.4, especially the onset temperatures for the poled and unpoled films. The mechanical features of the poled and unpoled films, such as strength, are enhanced by the poling procedures due to the different conformity in the interfaces between the crystallites. The unpoled film ought to have lower melting entropy (ΔS_m) than the poled film. For the poled film, it is observed experimentally that the phase equilibrium is at a lower melting temperature $T_m = \Delta H_m / \Delta S_m$ as shown in Fig. 2.4. Other investigations performed at 10°C/min heating rate and other heating rates proved that the melting peak of both films is complex. Again, this might be due to the occurrence of several crystalline structures, such as the coexistence of the alpha and beta phases [23]. Again, the closeness in peak area for the two samples implies that the variations in T_m are not related to enthalpic causes (i.e., ΔH_m is similar for both films). Therefore, the areas of the peak as well as melting enthalpy (ΔH_m) are similar for both the films, hence the conclusion is that the differences in T_m are not because of enthalpy. The poled peak demonstrates a melting peak at 167°C which is a high-temperature shoulder. A low-intensity endothermic process occurs in unpoled films at higher temperatures. The melting peak of both samples is confirmed to be complex from the experiments performed at heating rates of 10°C/min.

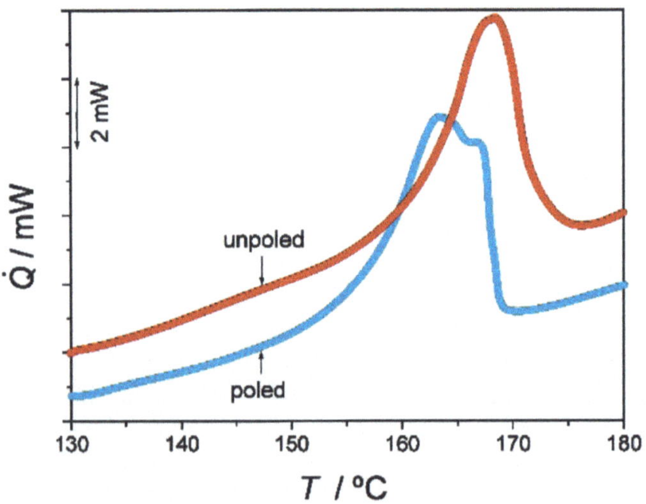

FIGURE 2.4 DSC patterns of poled and unpoled beta PVDF. Source: *Reproduced with permission V. Sencadas, S. Lanceros-Méndez, J.F. Mano, Characterization of poled and non-poled β-PVDF films using thermal analysis techniques, Thermochim. Acta 424 (1–2) (2004) 201–207. doi: 10.1016/j.tca.2004.06.006 [23]. Copyright Elsevier.*

2.3.2 Fourier-transform infrared spectroscopy

Many researchers have allocated the FTIR bands at about 840 and 510 cm^{-1} to the beta and gamma phases, respectively [11,24,25], even without having adequate data. The FTIR absorption peaks of the three primary PVDF polymorphs, alpha, beta, and gamma, can be categorized into three major groups: (1) common peaks found in α, β, and γ phases; (2) dual peaks that may be the product of two separate phases; and (3) unique peaks that appear only in one of the three phases. The FTIR patterns of PVDF at different phases are shown in Fig. 2.5. Peaks appear at the 876–885, 1067–1075, 1171–1182, and 1398–1404 cm^{-1} scale. On the other hand, they have identical properties in tests of alpha, beta, and gamma phases or other mixed structures [26–30]. The exceptional peaks for the alpha phase (alpha phase characteristic bands) are located at 410, 489, 532, 614, 763, 795, 854, 975, 1149, 1209, 1383, and 1423 cm^{-1}. The beta phase exclusive peaks are located at 445, 473, and 1275 cm^{-1}, whereas the gamma phase exclusive peaks are at 431, 482, 811, and 1234 cm^{-1}. The absorbance of the beta and gamma phases is considerably higher than that of the alpha phase. At the range of 1428–1432 cm^{-1}, beta and gamma phases have very similar peaks. Several findings show that the peaks at 1431 and 1429 cm^{-1} can be used to classify the beta and gamma stages, respectively [11,20,24,25,31–35]. Some authors have previously assigned the beta-phase at 600 cm^{-1} bands [17]. At high temperatures, the 776 cm^{-1} band, as well as

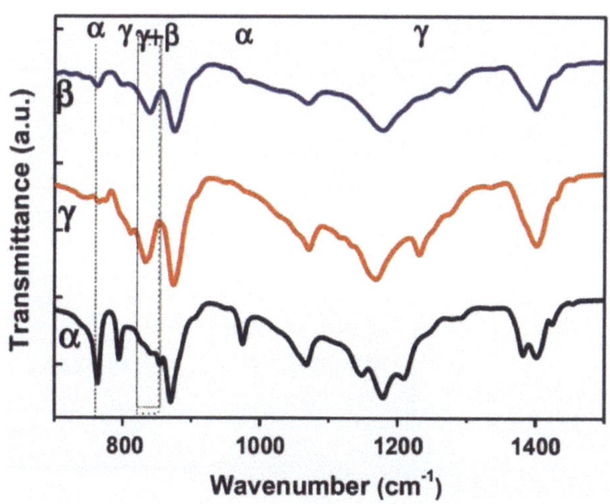

FIGURE 2.5 FTIR patterns of different phases in PVDF (alpha, beta, gamma). Source: *Reproduced with permission P. Martins, A.C. Lopes, S. Lanceros-Mendez, Electroactive phases of poly(vinylidene fluoride): determination, processing and applications, Prog. Polym. Sci. 39 (4) (2014) 683–706. doi: 10.1016/j.progpolymsci.2013.07.006 [7]. Copyright Elsevier.*

833 cm^{-1} band, will only be visible in the crystallization γ-phase. Based on the literature survey, 1275 cm^{-1} is the exclusive band for the beta phase and there is another exclusive band at 1234 cm^{-1} for the gamma phase. The 840 and 510 cm^{-1} bands are known for the beta phase when the 1275 cm^{-1} band occurs and the 1234 cm^{-1} band should be absent. On the other hand, the two bands (840 and 510 cm^{-1}) are regarded as the gamma phase. If both the 1275 and 1234 cm^{-1} bands exist at the same time, the two bands (840 and 510 cm^{-1}) are assumed to be both the beta and gamma phases, as shown in Fig. 2.6. In this situation, a higher resolution calculation may be used to separate the said two bands into two different bands for the beta and gamma phases, respectively.

2.3.2.1 *Calculation of individual beta and gamma phase*

As the 840 cm^{-1} bands can be allocated to the beta, gamma, or mutually to either phase depending on the other band details, the comparative fraction of the electroactive beta and gamma phases (FEA) as a function of crystalline structures in any sample such as one comprising

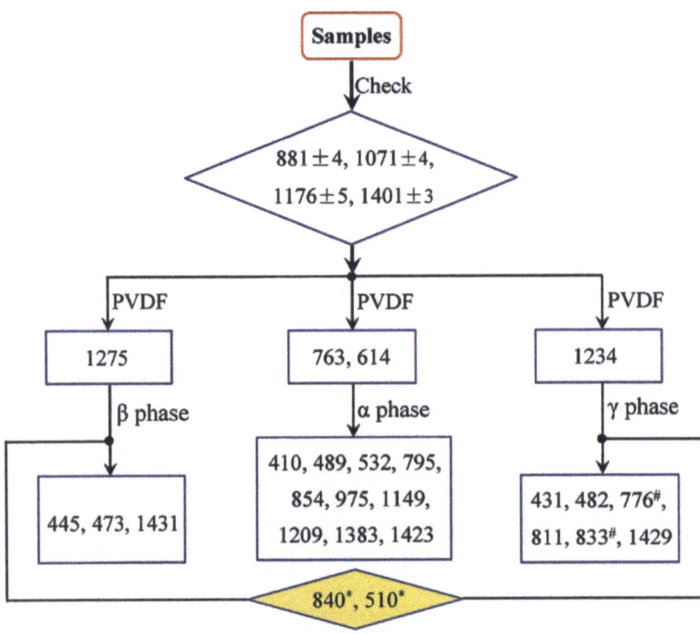

FIGURE 2.6 Identification of alpha beta and gamma phases. Source: *Reproduced with permission X. Cai, T. Lei, D. Sun, L. Lin, A critical analysis of the α, β and γ phases in poly (vinylidene fluoride) using FTIR, RSC Adv. 7 (25) (2017) 15382–15389. doi: 10.1039/c7ra01267e [26]. Copyright Royal Society of Chemistry.*

only two phases (alpha + beta, alpha + gamma, or beta + gamma), can be quantified as follows:

$$F_{\mathrm{EA}} = \frac{I_{\mathrm{EA}}}{\left(\frac{K_{840}}{K_{763}}\right) I_{763} + I_{\mathrm{EA}}} \times 100\% \tag{2.4}$$

where I_{EA} and I_{763} are the absorbances at 840 and 763 cm^{-1}, respectively; and K_{840} ($7.7 \times 10^4\ \mathrm{cm}^2/\mathrm{mol}$) and K_{763} ($6.1 \times 10^4\ \mathrm{cm}^2/\mathrm{mol}$) are the absorption coefficients at the corresponding wave numbers [26]. However, measuring the peak-to-valley height ratio between the peaks and their nearest valleys is a preferable approach. The equations are given below:

$$F_{(\mathrm{Beta})} = F_{\mathrm{EA}} \times \left(\frac{\Delta H_{\mathrm{B}}}{\Delta H_{\mathrm{B}} + \Delta H_{\mathrm{G}}}\right) \times 100\% \tag{2.5}$$

$$F_{(\mathrm{Gama})} = F_{\mathrm{EA}} \times \left(\frac{\Delta H_{\mathrm{r}}}{\Delta H_{\mathrm{B}} + \Delta H_{\mathrm{G}}}\right) \times 100\% \tag{2.6}$$

where ΔH_{B} and ΔH_{G} are the height variations (absorbance differences) between the nearby valley and the surrounding mountain. ΔH_{B} is the height difference between the two peaks to valley height and $\Delta \mathrm{H_G}$ is the difference between the near peaks to valley height. With the above equation we can find the value of the β-phase and γ-phase of PVDF.

2.3.3 X-ray diffraction

The XRD results show α, β, and γ phases of the PVDF. Various crystal structures contain analogous peaks even though some of them have a unique phase. Allowing uniqueness is very easy for the recognition of the phases in XRD [36]. Depending on the phase of PVDF, the position of the intensive peak always changes and hence, depending on the peak position, the phase can be easily found. For the alpha and gamma phase, the intensive peak is around 20 degrees and there is one more peak visible at 18 degrees. But for the beta phase there is no peak at 18 degrees and the main intensive peak is at 20.26 degrees which is a slight shift to the right side compared to other phases of PVDF, as shown in Fig. 2.7. The sum of diffraction planes of the beta phase is (200) and (110). The alpha phase contains many peaks compared to other phases with peak positions at 18.30, 17.66, and 26.56 degrees having the hkl planes of the peaks as (020), (100), and (021), respectively. The gamma phase has its intensive peak at 20.04 degrees with corresponding hkl planes as (110) and the other peaks are visible at 19.2 and 18.5 degrees with corresponding hkl planes as (002) and (020), respectively. The identification of phases in XRD is more comfortable than FTIR. All the results of FTIR

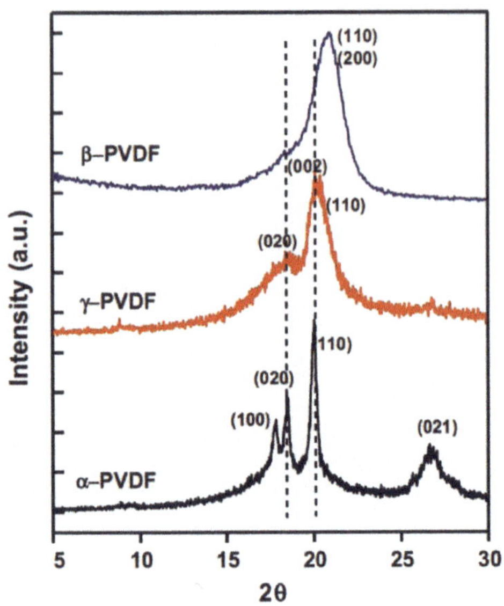

FIGURE 2.7 Alpha, beta, and gamma phase XRD patterns of PVDF. Source: *Reproduced with permission P. Martins, A.C. Lopes, S. Lanceros-Mendez, Electroactive phases of poly(vinylidene fluoride): determination, processing and applications, Prog. Polym. Sci. 39 (4) (2014) 683–706. doi: 10.1016/j.progpolymsci.2013.07.006 [7]. Copyright Elsevier.*

and XRD for phase formation of ferroelectric PVDF are similarly based on the survey.

The XRD pattern of pure PVDF and the addition of different ratios of rGO and Ag with PVDF are displayed in Fig. 2.8. The diffraction patterns of the pure PVDF at 17.7, 18.2, and 19.9 degrees correspond to hkl planes (100), (020), and (110) of the alpha crystalline PVDF, as shown in Fig. 2.8A. Fig. 2.8B–E represents the diffraction patterns of PVDF by adding additives at different rGO-Ag weight percentages. For the alpha phase of PVDF corresponding to 0.1% weight of rGO-Ag, as revealed in Fig. 2.8B, a peak appears at 17.7 and 19.9 degrees, whereas the peak at 18.2 degrees corresponding to the (020) hkl plane disappears. Fig. 2.8C gives the XRD pattern of PVDF when 0.5% weight of rGO-Ag is added. In the 0.5% weight rGO-Ag with PVDF diffraction patterns of alpha phase, 18.6 and 20.2 degrees peaks disappear and the phases are converted to polar beta and gamma phases. Fig. 2.8D shows the enhancement of intensity in polar phases and the diffraction broadening for the 1.0% weight rGO-Ag with PVDF XRD pattern. This composition gives a higher piezoelectric response.

Fig. 2.8E portrays the diffraction pattern of 2.0% weight rGO-Ag in PVDF. The higher accumulation of rGO and Ag reduces the intensity of β and γ phase. The rGO-Ag proportion in PVDF is the significant factor

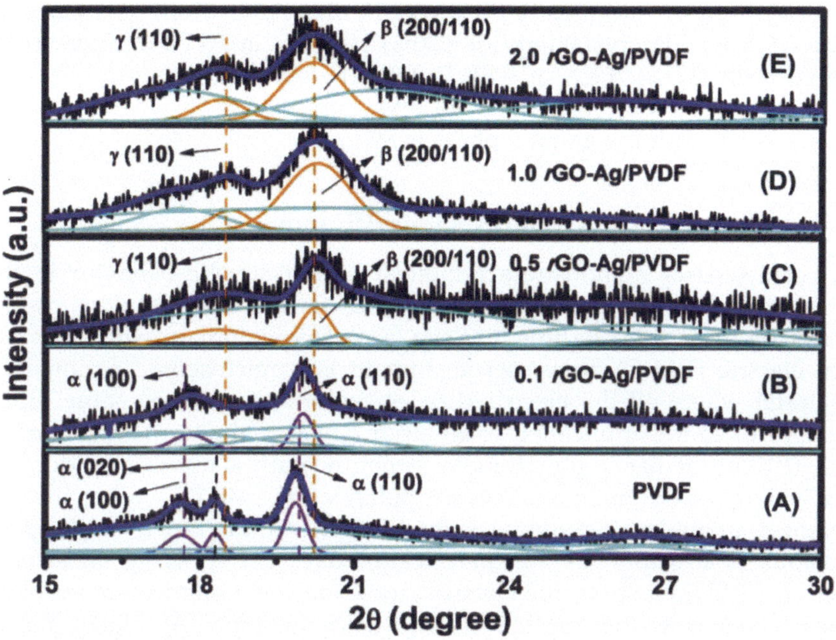

FIGURE 2.8 XRD pattern of pure (A) and different weight percentages of rGO 0.1 (B) 0.5 (C) 1.0 (D) and 2.0 (E). Source: *Reproduced with permission M. Pusty, L. Sinha, P.M. Shirage, A flexible self-poled piezoelectric nanogenerator based on a rGO-Ag/PVDF nanocomposite, N. J. Chem. 43 (1) (2019) 284–294. doi: 10.1039/c8nj04751k [17]. Copyright Royal Society of Chemistry.*

depending on the nucleation and stability of the polar beta and gamma piezoelectric polymers. The higher enhancement of beta and gamma phases is visible in 1.0% weight rGO-Ag with PVDF and is the reason for the good connection between polymer and rGO-Ag composite. The high weight percentage (2%) rGO-Ag in PVDF composite reduces the beta phase and rGO-Ag dominates the polymer chains. It will reduce the beta properties of the polymer chains. In high weight percentage samples (2% rGO-Ag sample), the conducting paths are increased, thereby reducing the nucleation of polar piezoelectric nature. The reason for the conversion of the alpha phase into beta and gamma phases is the addition of rGO-Ag at different proportions and not due to any applied pressure and temperatures. The following equation gives the crystallinity level of the rGO-Ag/PVDF nanocomposite:

$$\text{The degree of total crystallinity } X_{cr} = \frac{\varepsilon B_{cr}}{\varepsilon B_{cr} + \varepsilon B_{amr}} \times 100\% \qquad (2.7)$$

where B_{amr} and B_{cr} are areas under amorphous and crystalline regions, respectively, and the definite integral between B_{amr} and B_{cr} is considered.

The crystallinity increases in 0.1% weight rGO-Ag in PVDF compared to pure PVDF. The crystallinity of (2.0%) rGO-Ag in PVDF reduces compared with (1.0%) rGO-Ag with PVDF. The crystalline nature of (1.0%) rGO-Ag with PVDF increases the crystallinity to 46% where the beta phase is 38% and the gamma phase is 8%.

2.3.4 Ferroelectric hysteresis loop/PE loop

Ferroelectrics demonstrate nonlinear and hysteretic behavior in the existence of an applied electric field. This ferroelectric behavior results from the reversible alignment of the ferroelectric domain with an external electric field. A ferroelectric domain is a microscopic region in a material where all the electric dipoles are arranged in the same direction. Prior to the application of an electric field to the ferroelectric material (PVDF), it has a polydomain structure with a number of domain walls and each domain has its own polarization path [37]. Since the different domains are randomly aligned in general, the net/average spontaneous polarization of the dielectric is zero. On applying an electric field to PVDF material, the electric dipoles in the domains tend to align themselves with the applied electric field direction. At higher fields, polarization saturation occurs whose magnitude is determined by the point P_s (*B*), as shown in Fig. 2.9. The net polarization decreases as the field is inverted. On setting the applied field (*E*) to zero, a specific resultant polarisation is observed. The retained polarization is known as remnant polarization P_r. The applied electric field further is reduced

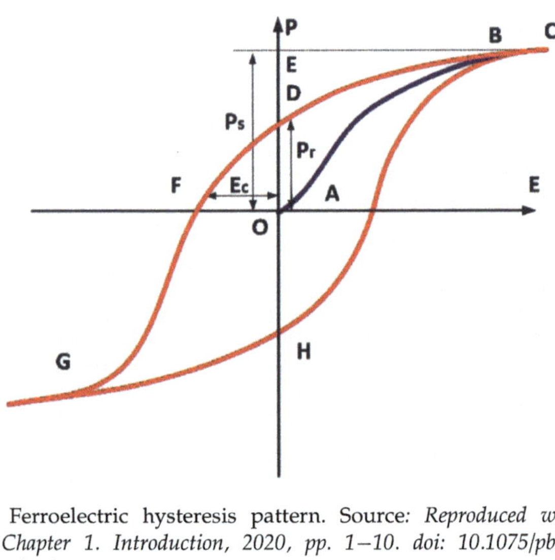

FIGURE 2.9 Ferroelectric hysteresis pattern. Source: *Reproduced with permission R. Laury, T. Ono, Chapter 1. Introduction, 2020, pp. 1–10. doi: 10.1075/pbns.315.01lau [37]. Copyright John Benjamins.*

more towards the negative values to decrease the net polarization to zero and is referred to as the coercive field, E_c [37]. The external negative field will induce polarization saturation at $-P_s$ (G), and the hysteresis loop will be closed upon field reversal, as shown in Fig. 2.9.

2.3.4.1 *Ferroelasticity*

Ferroelectric materials also exhibit ferroelasticity, which is a phenomenon that occurs when a material experiences spontaneous strain, i.e., strain that does not require any mechanical force. When the stress--strain behavior shows hysteresis, the domain composition of the crystal has a significant effect on the random strain of ferroelastic materials. Domains are the primary microstructures in this low-symmetry phase due to the conversion from the paraelastic phase to the ferroelastic phase. Ferroelastic domains are both attractive and popular in materials and minerals. When a structural phase transition occurs, nearly all the ferroelastic or coelastic materials exhibit domain patterns and these domain patterns affect their physical properties.

A comparison of dipolar displacement-electric field (D–E) loops of PVDF under an alternating current at 10 Hz and 150 MV/m electric fields is performed for the three different crystal phases: alpha, beta, and gamma, as shown in Fig. 2.10. The typical character of the β-PVDF in the D-E loop shows standard ferroelectric characteristics and has a strong remnant polarization, whereas the α-PVDF and γ-PVDF D-E loops are very narrow and the remnant polarization is smaller for a

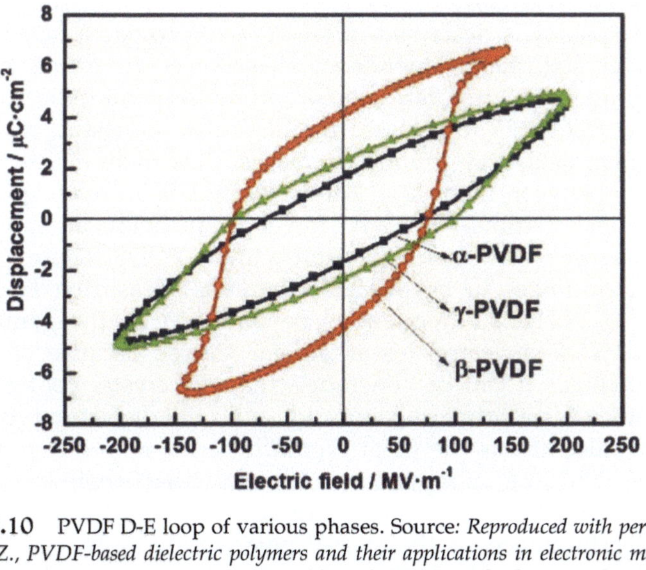

FIGURE 2.10 PVDF D-E loop of various phases. Source: *Reproduced with permission Xia, W., Zhang, Z., PVDF-based dielectric polymers and their applications in electronic materials, IET Nanodielectrics 1 (1) (2018) 17–31. doi: 10.1049/iet-nde.2018.0001 [38]. Copyright Wiley.*

200 MV/m electric field. Furthermore, for the γ-PVDF, the remnant polarization is significantly greater than that of alpha-PVDF because the T3G conformation of gamma-PVDF has a polarity higher than TGTG′ in alpha-PVDF. But for the same electric field, beta-PVDF has a higher overall displacement than alpha- and gamma-PVDF due to the high polarity of the TTTT (or GGGG) conformation.

2.3.5 Dielectric properties

PVDF has excellent thermal strength and is resistant to energy radiation, especially to ultraviolet and higher-order radiation. PVDF is also chemically resistant to most solvents and chemicals. PVDF has the water adsorption property with the ability to absorb 0.05% at room temperature, hence it is not hygroscopic. The properties of fluoropolymers depend on several factors, like chemical composition, molecular weight, and molecular structure. Most of the important properties of PVDF are determined by the crystalline material and the form of crystalline structures, all of which are influenced by manufacturing techniques and conditions. The most enticing property of PVDF and its polymers are their distinct dielectric properties in different electrical fields. The electrical properties of fluoro-PVDF were widely studied by researchers, who concentrated on two aspects: large-scale piezoelectric applications in energy transformation systems mainly during electrical to mechanical conversion and the use of high-pulse capacitors to produce large energy density.

The dielectric response of PVDF in alpha, beta, and gamma phases in terms of frequency is described in Fig. 2.11 to provide a better explanation of the crystal phase dependency based on their various properties, such as storage of energy, ferroelectric properties, and dielectric properties. If the frequency increases, the dielectric constant of the three phases: alpha, beta, and gamma decreases, thus there exists an inverse relationship between them [21]. Alpha-PVDF has a greater permittivity than beta- and gamma-PVDF. At lower frequencies, alpha-PVDF has higher dielectric losses than beta- and gamma-PVDF. There are significant dielectric losses in beta-PVDF when the measuring frequency is improved to 1 kHz. At frequencies beyond 1 MHz, the gamma-PVDF and alpha-PVDF dielectric losses appear to rise because of dielectric relaxation. Hence it can be concluded that the crystal configuration is the factor that has a primary connection with the dielectric properties of PVDF. This also affects the polar region in terms of polarity, polar area proportion, and the independence of molecular chains. The dipole movement beneath a high electric field is a crucial factor in deciding the dielectric performance, as the overall polarization system of PVDF will not be reflected under the lower electric field.

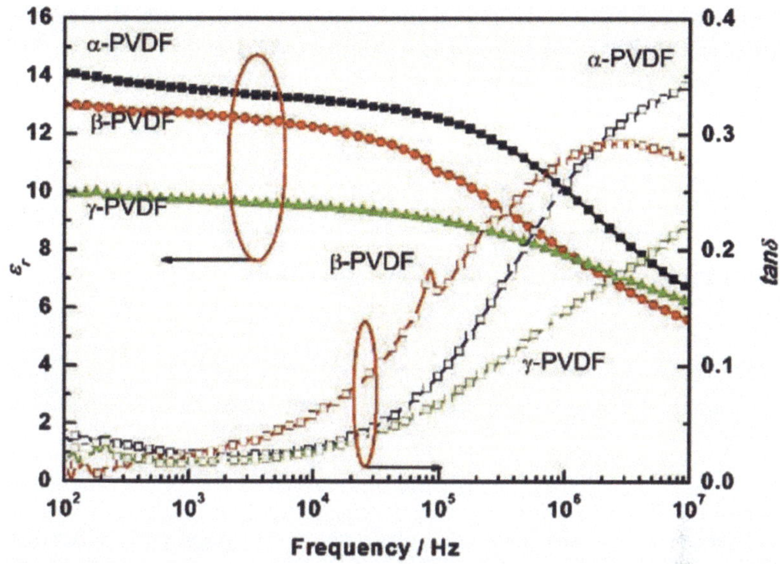

FIGURE 2.11 Dielectric loss and dielectric constant of PVDF in different phases. Source: *Reproduced with permission Xia, W., Zhang, Z., PVDF-based dielectric polymers and their applications in electronic materials, IET Nanodielectrics 1 (1) (2018) 17–31. doi: 10.1049/iet-nde.2018.0001 [38]. Copyright Wiley.*

2.4 Conclusion

Beta phase PVDF polymers are the most challenging and interesting polymers for advanced applications for future generations. They exhibit large piezoelectric coeffient, dielectric constant, and ferroelectric properties. PVDF has extensive applications in different areas like biomedical, energy harvesting, energy generation, storage, and sensors due to its electroactive phases in beta phase PVDF. The key properties of the electroactive phases of PVDF have been studied, and the experimental methods used to classify them were also discussed. Proper handling in addition to the various manufacturing techniques was used to obtain the desired material properties. The majority of recent studies are qualitative rather than quantitative. When used as a sensor, for example, the piezoelectric device can only satisfy estimated linear specifications within a certain range and is readily influenced by a number of external conditions which is not advantageous for true industrial growth in this direction. Furthermore, the lack of large-scale continuous manufacturing methods continues to severely restrict real-time engineering use. Meanwhile, the improvement of the electromechanical coupling coefficient and the energy transfer performance of materials is a significant area of research. It can be concluded that the beta phase PVDF, owing primarily to its outstanding electrical properties among the polymers, remains at the heart of a complex, fruitful, and exciting research area that will undoubtedly serve some of the most demanding applications in the near future.

Conflict of interest

The authors declare no conflict of interest.

References

[1] Z.L. Wang, J. Song, Piezoelectric nanogenerators based on zinc oxide nanowire arrays, Science (80-.) 312 (5771) (2006) 242–246. Available from: https://doi.org/10.1126/science.1124005.

[2] C. Miguel, V. Fernandes, Electroactive Poly (Vinylidene Fluoride)-Based Materials: Recent Progress, Challenges, and Opportunities, Elsevier Inc, 2020.

[3] P. Mao, et al., Tunable dielectric polarization and breakdown behavior for high energy storage capability in P(VDF-TrFE-CFE)/PVDF polymer blended composite films, Phys. Chem. Chem. Phys. 22 (23) (2020) 13143–13153. Available from: https://doi.org/10.1039/d0cp01071e.

[4] H. Kawai, The piezoelectricity of poly (vinylidene fluoride), Jpn. J. Appl. Phys. 8 (7) (1969) 975–976. Available from: https://doi.org/10.1143/jjap.8.975.

[5] K. Tashiro, H. Tadokoro, M. Kobayashi, Structure and piezoelectricity of poly(v1nylidene fluoride), Ferroelectrics 32 (1) (1981) 167–175. Available from: https://doi.org/10.1080/00150198108238688.

[6] J. Zheng, A. He, J. Li, C.C. Han, Polymorphism control of poly(vinylidene fluoride) through electrospinning, Macromol. Rapid Commun. 28 (22) (2007) 2159–2162. Available from: https://doi.org/10.1002/marc.200700544.

[7] P. Martins, A.C. Lopes, S. Lanceros-Mendez, Electroactive phases of poly(vinylidene fluoride): determination, processing and applications, Prog. Polym. Sci. 39 (4) (2014) 683–706. Available from: https://doi.org/10.1016/j.progpolymsci.2013.07.006.

[8] M.A. Bachmann, W.L. Gordon, J.L. Koenig, J.B. Lando, An infrared study of phase-III poly(vinylidene fluoride), J. Appl. Phys. 50 (10) (1979) 6106–6112. Available from: https://doi.org/10.1063/1.325780.

[9] W.M. Prest, D.J. Luca, The formation of the γ phase from the α and β polymorphs of polyvinylidene fluoride, J. Appl. Phys. 49 (10) (1978) 5042–5047. Available from: https://doi.org/10.1063/1.324439.

[10] T. Boccaccio, A. Bottino, G. Capannelli, P. Piaggio, Characterization of PVDF membranes by vibrational spectroscopy, J. Memb. Sci. 210 (2) (2002) 315–329. Available from: https://doi.org/10.1016/S0376-7388(02)00407-6.

[11] J. Gomes, J.S. Nunes, V. Sencadas, S. Lanceros-Mendez, Influence of the β-phase content and degree of crystallinity on the piezo-and ferroelectric properties of poly (vinylidene fluoride), Smart Mater. Struct. 19 (2010). Available from: https://doi.org/10.1088/0964-1726/19/6/065010.

[12] T. Hattori, M. Kanaoka, H. Ohigashi, Improved piezoelectricity in thick lamellar β-form crystals of poly(vinylidene fluoride) crystallized under high pressure, J. Appl. Phys. 79 (4) (1996) 2016–2022. Available from: https://doi.org/10.1063/1.361055.

[13] Y. Li, S. Tang, M.W. Pan, L. Zhu, G.J. Zhong, Z.M. Li, Polymorphic extended-chain and folded-chain crystals in poly(vinylidene fluoride) achieved by combination of

high pressure and ion-dipole interaction, Macromolecules 48 (23) (2015) 8565–8573. Available from: https://doi.org/10.1021/acs.macromol.5b01895.
[14] A. Gradys, P. Sajkiewicz, S. Adamovsky, A. Minakov, C. Schick, Crystallization of poly(vinylidene fluoride) during ultra-fast cooling, Thermochim. Acta 461 (1–2) (2007) 153–157. Available from: https://doi.org/10.1016/j.tca.2007.05.023.
[15] D. Yang, Y. Chen, β-phase formation of poly(vinylidene fluoride) from the melt induced by quenching, J. Mater. Sci. Lett. 6 (5) (1987) 599–603. Available from: https://doi.org/10.1007/BF01739296.
[16] N. Meng, et al., Ultrahigh β-phase content poly(vinylidene fluoride) with relaxor-like ferroelectricity for high energy density capacitors, Nat. Commun. 10 (1) (2019) 1–9. Available from: https://doi.org/10.1038/s41467-019-12391-3.
[17] M. Pusty, L. Sinha, P.M. Shirage, A flexible self-poled piezoelectric nanogenerator based on a rGO-Ag/PVDF nanocomposite, N. J. Chem. 43 (1) (2019) 284–294. Available from: https://doi.org/10.1039/c8nj04751k.
[18] U. De, K.R. Sahu, A. De, Ferroelectric materials for high temperature piezoelectric applications, Solid State Phenom. 232 (2015) 235–278. Available from: https://doi.org/10.4028/http://www.scientific.net/SSP.232.235. no. April 2015.
[19] M. Benz, W.B. Euler, Determination of the crystalline phases of poly(vinylidene fluoride) under different preparation conditions using differential scanning calorimetry and infrared spectroscopy, J. Appl. Polym. Sci. 89 (4) (2003) 1093–1100. Available from: https://doi.org/10.1002/app.12267.
[20] R. Gregorio, M. Cestari, Effect of crystallization temperature on the crystalline phase content and morphology of poly(vinylidene fluoride), J. Polym. Sci. Part B Polym. Phys. 32 (5) (1994) 859–870. Available from: https://doi.org/10.1002/polb.1994.090320509.
[21] Z. Yang, H. Peng, W. Wang, T. Liu, Crystallization behavior of poly(ε-caprolactone)/layered double hydroxide nanocomposites, J. Appl. Polym. Sci. 116 (5) (2010) 2658–2667. Available from: https://doi.org/10.1002/app.
[22] A. Rulka, Nanoscale advances retraction: a bio-based piezoelectric nanogenerator for mechanical energy harvesting, 2021, December 2020, p. 2021, doi: 10.1039/d0na90066d.
[23] V. Sencadas, S. Lanceros-Méndez, J.F. Mano, Characterization of poled and non-poled β-PVDF films using thermal analysis techniques, Thermochim. Acta 424 (1–2) (2004) 201–207. Available from: https://doi.org/10.1016/j.tca.2004.06.006.
[24] A. Salimi, A.A. Yousefi, FTIR studies of β-phase crystal formation in stretched PVDF films, Polym. Test. 22 (6) (2003) 699–704. Available from: https://doi.org/10.1016/S0142-9418(03)00003-5.
[25] Y. Bormashenko, R. Pogreb, O. Stanevsky, E. Bormashenko, Vibrational spectrum of PVDF and its interpretation, Polym. Test. 23 (7) (2004) 791–796. Available from: https://doi.org/10.1016/j.polymertesting.2004.04.001.
[26] X. Cai, T. Lei, D. Sun, L. Lin, A critical analysis of the α, β and γ phases in poly(vinylidene fluoride) using FTIR, RSC Adv. 7 (25) (2017) 15382–15389. Available from: https://doi.org/10.1039/c7ra01267e.
[27] D.M. Dhevi, A.A. Prabu, K.J. Kim, FTIR studies on polymorphic control of PVDF ultrathin films by heat-controlled spin coater, J. Mater. Sci. 51 (7) (2016) 3619–3627. Available from: https://doi.org/10.1007/s10853-015-9685-6.
[28] T. Lei, P. Zhu, X. Cai, L. Yang, F. Yang, Electrospinning of PVDF nanofibrous membranes with controllable crystalline phases, Appl. Phys. A Mater. Sci. Process. 120 (1) (2015) 5–10. Available from: https://doi.org/10.1007/s00339-015-9197-x.
[29] Y.L. Liu, Y. Li, J.T. Xu, Z.Q. Fan, Cooperative effect of electrospinning and nanoclay on formation of polar crystalline phases in poly(vinylidene fluoride), ACS Appl. Mater. Interfaces 2 (6) (2010) 1759–1768. Available from: https://doi.org/10.1021/am1002525.

[30] W.A. Yee, M. Kotaki, Y. Liu, X. Lu, Morphology, polymorphism behavior and molecular orientation of electrospun poly(vinylidene fluoride) fibers, Polymers (Guildf.) 48 (2) (2007) 512–521. Available from: https://doi.org/10.1016/j.polymer.2006.11.036.
[31] R. Gregorio, Determination of the α, β, and γ crystalline phases of poly(vinylidene fluoride) films prepared at different conditions, J. Appl. Polym. Sci. 100 (4) (2006) 3272–3279. Available from: https://doi.org/10.1002/app.23137.
[32] J.S. Andrew, D.R. Clarke, Effect of electrospinning on the ferroelectric phase content of polyvinylidene difluoride fibers, Langmuir 24 (3) (2008) 670–672. Available from: https://doi.org/10.1021/la7035407.
[33] R. Gregorio, D.S. Borges, Effect of crystallization rate on the formation of the polymorphs of solution cast poly(vinylidene fluoride), Polymers (Guildf). 49 (18) (2008) 4009–4016. Available from: https://doi.org/10.1016/j.polymer.2008.07.010.
[34] S. Yu, W. Zheng, W. Yu, Y. Zhang, Q. Jiang, Z. Zhao, Formation mechanism of β-phase in PVDF/CNT composite prepared by the sonication method, Macromolecules 42 (22) (2009) 8870–8874. Available from: https://doi.org/10.1021/ma901765j.
[35] P. Martins, C.M. Costa, M. Benelmekki, G. Botelho, S. Lanceros-Mendez, On the origin of the electroactive poly(vinylidene fluoride) β-phase nucleation by ferrite nanoparticles via surface electrostatic interactions, CrystEngComm 14 (8) (2012) 2807–2811. Available from: https://doi.org/10.1039/c2ce06654h.
[36] A.C. Lopes, C.M. Costa, C.J. Tavares, I.C. Neves, S. Lanceros-Mendez, Nucleation of the electroactive γ phase and enhancement of the optical transparency in low filler content poly(vinylidene)/clay nanocomposites, J. Phys. Chem. C. 115 (37) (2011) 18076–18082. Available from: https://doi.org/10.1021/jp204513w.
[37] R. Laury, T. Ono, Chapter 1. Introduction, 2020, pp. 1–10. doi: 10.1075/pbns.315.01lau.
[38] W. Xia, Z. Zhang, PVDF-based dielectric polymers and their applications in electronic materials, IET Nanodielectrics 1 (1) (2018) 17–31. Available from: https://doi.org/10.1049/iet-nde.2018.0001.

CHAPTER

3

Polymer/metal oxides nanocomposites-based piezoelectric energy-harvesters

Anshida Mayeen[1], *Sherin Joseph*[1] *and Honey John*[1,2]

[1]Inter University Centre for Nanomaterials and Devices, Cochin University of Science and Technology, Kochi, India, [2]Department of Polymer Science and Rubber Technology, Cochin University of Science and Technology, Kochi, India

3.1 Introduction

Piezoelectric materials are those materials which can produce electrical charge (q) due to the change in polarization (ΔP) when subjected to a mechanical deformation ($\Delta\sigma$). The word piezoelectricity is derived from the Greek word "piezo" and means "pressure electricity" [1,2]. Generally materials are restricted to organic crystals, ceramics, and polymers. Among the existing piezoelectric materials, the inorganic ceramics, such as lead zirconate titanate (PZT), lithium niobate, lead magnesium niobate lead titanate (PMN-PT), barium titanate ($BaTiO_3$), zinc oxide etc., display better piezoresponse, however they are brittle in nature and hence not suitable for flexible and wearable energy-harvesting technology [2–4].

On the other hand, piezoelectric polymers possess a high degree of flexibility, chemical stability, and resilient weathering characteristics. But the piezoresponse is comparatively lower than the ceramics. The incorporation of these piezoelectric ceramic fillers in the polymer matrix will result in a composite having excellent piezoresponse along with good mechanical strength. Polyvinyldene fluoride (PVDF) and its copolymers were widely used as polymer matrix, because of their

Engineered Polymer Nanocomposites for Energy Harvesting Applications
DOI: https://doi.org/10.1016/B978-0-12-824155-4.00009-0

outstanding piezoresponse when compared to other polymers, for the development of polymer/metal oxide-based piezoelectric nanogenerators (PENG)/energy-harvesting devices [4,5]. For the same purpose other polymers like odd nylons, cellulose, PLLA, and collagens were also used, but they exhibit comparatively lower piezoelectric output than PVDF. Generally, ferroelectric/piezoelectric oxides like PZT, lithium niobate, PMN-PT, $BaTiO_3$, zinc oxide, etc., were incorporated in these piezoelectric polymer matrix to enhance the piezoresponse of the system [5−9]. The resulting energy generator possesses excellent piezoelectric performance along with better flexibility. It could be a suitable choice for the development of smart flexible portable electronic devices [9,10].

In this context, most of the research works were focused on the development of PENG by embedding ceramic nanofillers of high piezoelectric coefficients in various flexible polymeric matrix of good processability. The simple structure, comparatively smaller size, and long-standing stability of the PENGs permits them to harvest energy by exploiting abundant energy sources from our living surroundings, even from sound waves, flowing air, mechanical impacts, etc. [9].

3.2 Polymer-based nanogenerators

Among piezoelectric materials, piezoelectric polymers have gained considerable attention, because they possess excellent electroactive behavior along with a high degree of flexibility and hence they find application in flexible electronic devices. They possess spontaneous and thermodynamically stable polarization states, which can be switched from one state to the other by applying a strong external electric field [11]. They are widely used in nonvolatiles for the development of memory devices, piezoelectric and pyroelectric sensors, high energy density capacitors, organic solar cells, etc. [11−14].

Piezoelectricity in polymers was first discovered in PVDF during the 1970s. After this several new ferroelectric polymers were also discovered that possess ferroelectric, piezoelectric, and pyroelectric properties. After the discovery of ferroelectricity in PVDF, ferroelectric behavior in its copolymers was also reported [i.e., polyvinlydene fluoride- hexafluoro propylene (PVDF-HFP) and polyvinyldene fluoride-trifluoroethylene (PVDF-TRFE)]. Later piezoelectric properties in other polymers like odd-numbered nylons, polyureas, poly-l-lactic acid, and polyvinylidene cyanide were also reported [12,13].

Ferroelectric polymers hold a wide range of functionalities, such as ease of processing, light weight, and greater flexibility compared with ferroelectric ceramics. Hence ferroelectric polymers offer a wide range of opportunities in various fields including random acess memory devices, piezoelectric and pyroelectric devices, capacitors, artificial skins, electronic textiles, etc. [14–17]. Among the ferroelectric polymers, PVDF and its copolymers were considered as having the best electroactive behavior [16]. Hence energy-harvesting devices using PVDF-based systems were more exploited than the others. A comprehensive discussion about polymer-based energy-harvesting devices is given in the following sections.

3.2.1 Polyvinyldene fluoride-based piezoelectric nanogenerators

PVDF is a well-known piezoelectric polymer formed by the repeated arrangement of $-CH_2$-CF_2- units. These monomers possess high dipole moments ranging from $5\times\ 10^{-30}$ Cm to $8\ \times\ 10^{-30}$ Cm. These monomer units try to polymerize mainly in "head to tail" arrangement, that is $-CH_2$-CF_2-CH_2-CF_2-, and as a result a large net dipole moment is exhibited. Each chain of the PVDF possesses a definite dipole moment, which is perpendicular to the entire polymer chain [18]. PVDF and its copolymers are widely studied as piezoelectric polymers because of their high spontaneous polarization, low temperature processing, nontoxicity, and chemical stability [18,19].

PVDF is a semicrystalline polymer with five crystalline phases: α, β, γ, δ, and ε phases. The β phase corresponds to the all-trans planar zigzag arrangement; α phases correspond to the trans-gauche-trans-gauche′ arrangement; and γ and ε phases correspond to the T_3GT_3G' arrangement. Among the different phases, the β phase is the one with highest dipole moment of around $8\ \times\ 10^{-30}$ Cm. Fig. 3.1 shows the schematic representation of the spatial arrangement of molecular chains of α, β, and γ phases. Out of the five phases, theα phase is a nonferroelectric phase since it is nonpolar, and β is the strongest ferroelectric phase [19]. The γ, δ, and ε phases possess half the ferroelectric performance of the β phase. Since the

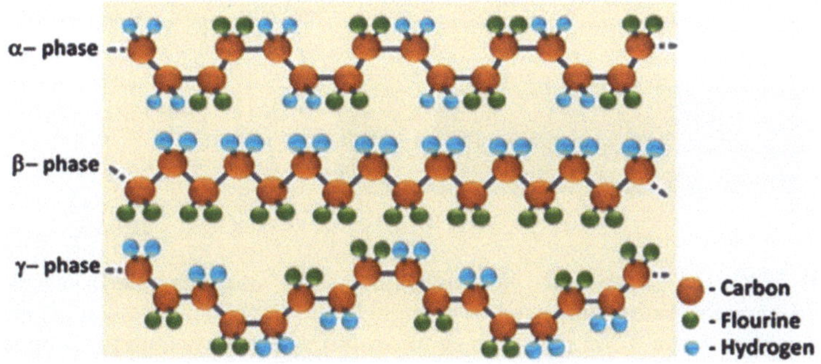

FIGURE 3.1 Schematic representation of the chain conformation for α, β, and γ phases of PVDF. Source: *Reproduced with permission from T. Tansel, High beta-phase processing of polyvinylidenefluoride for pyroelectric applications. J. Polym. Res. 27 (4) (2020) 1–5.*

β phase is more electroactive, there has been a great interest in β phase PVDF in the area of sensors, actuators, energy-harvesting devices, batteries, filters, ferroelectric random access memory devices, etc. [18–20]. The β phase of PVDF is obtained by mechanical drawing and poling of the α phase PVDF [17].

In PVDF the degree of crystallinity is in the range of 50%–60% and the polarization value continuously varies with temperature because of the disordered structure. To increase the crystallinity of the PVDF, the host polymer needs to be copolymerized with trifluoroethylene (TrFE), resulting in the production of PVDF-TrFE copolymer, which has been studied by many authors [17–20].

There are plentiful research works related to the fabrication of PENGs based on PVDF in the literature because of its excellent piezoresponse and high degree of flexibility and biocompatibility. Chieh Chang et al. fabricated a PVDF nanofiber-based PENG by means of near-field electrospinning method [21]. Fig. 3.2A shows the electrospinning setup followed by Chieh Chang et al. for the preparation of PVDF nanofibers. Theses electrospun nanofibers, with diameters varying from 500 nm to

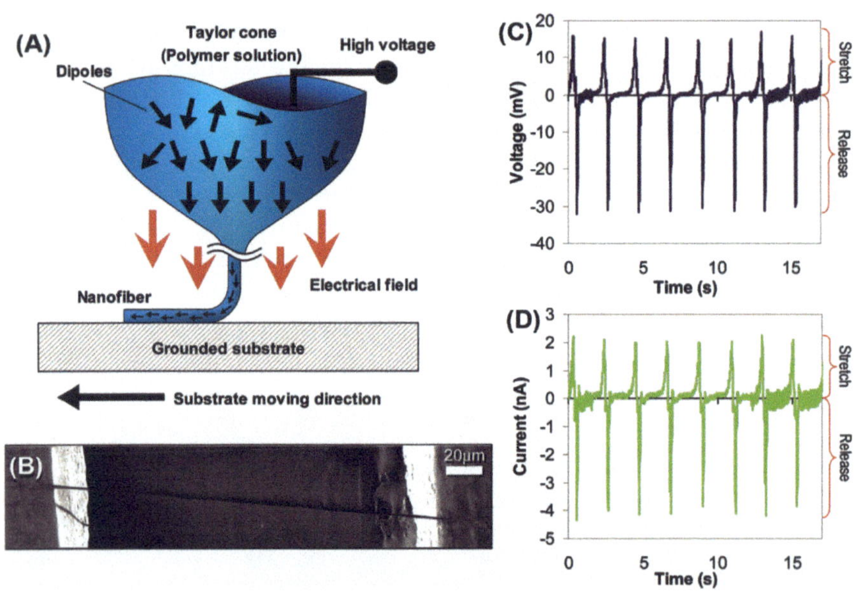

FIGURE 3.2 (A) Near-field electrospinning; (B) scanning electron microscope image of a nanogenerator comprised of a single PVDF nanofiber, two contact electrodes, and a plastic substrate; (C) output voltage measured versus time under an applied strain at 2 Hz; and (D) output current measured versus time under applied strain at 2 Hz. Source: *Reproduced with permission from C. Chang, V.H. Tran, J. Wang, Y.K. Fuh and L. Lin, Direct-write piezoelectric polymeric nanogenerator with high energy conversion efficiency. Nano Lett. 10 (2) (2010) 726–731.*

6.5 μm, were collected over a substrate. In this work, the plastic deposition substrate was bent by applying an axial stress and hence generated a piezoelectric potential. To measure the piezoelectric output of the developed PVDF nanofiber-based nanogenerator, the substrate was stretched and released repeatedly [21]. The voltage and current produced by the nanogenerator were recorded. Fig. 3.2C shows the variation of voltage versus time for each stretch and release. The output voltages generated were found to vary from 5 to 30 mV. Fig. 3.2D shows the variation of current versus time for every stretch and release [21].

Fig. 3.3A shows the electrical output generated from the PVDF nanofiber-based PENG. Here the green line corresponds to the output current produced by the nanogenerator under different strain rates, and the yellow line depicts the total charges produced by the nanogenerator under different strain rates [21].

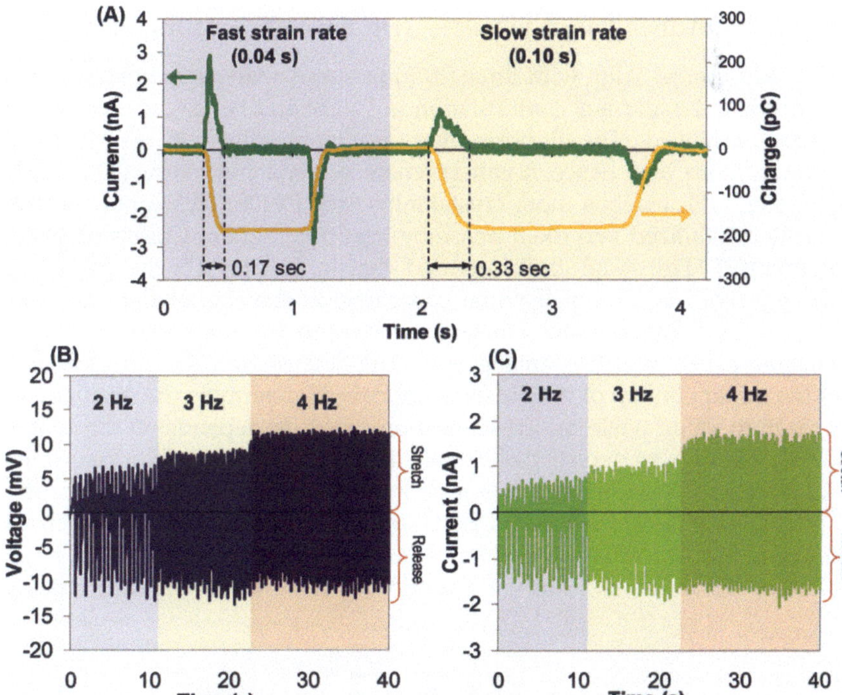

FIGURE 3.3 Electric output of a piezoelectric PVDF nanogenerator. (A) Output current (green line) of a PVDF nanogenerator subject to different strain rates under the same applied total strain. (B) Output voltage and (C) output current from PVDF nanogenerator subject to different stretch–release cycling frequencies of 2–4 Hz. Source: *Reproduced with permission from C. Chang, V.H. Tran, J. Wang, Y.K. Fuh and L. Lin, Direct-write piezoelectric polymeric nanogenerator with high energy conversion efficiency. Nano Lett. 10 (2) (2010) 726–731.*

Fig. 3.3B and C shows the piezoresponse of the nanogenerator under different cycling frequencies varying from 2–4 Hz under same applied strain. Here Chen et al. found that as the frequency increases, the output voltage and output current over a stretch cycle increases [21]. On the other hand during the release cycle, both current and voltage remain almost constant. In the above experimental setup, the stretching profile was controlled by loading machines and hence it can increase both strain rate and cycling frequency for a given applied strain [21]. Here the release process is due to the release of the cantilever beam from the strained location. During the release process, the nanogenerator returns to its original position and here the substrate in which PVDF nanofibers were placed will control the speed in which entire nanogenerator reaches its original location [21].

3.2.2 Polyvinyldene fluoride–trifluoroethylene/multiwalled carbon nanotubes-based piezoelectric nanogenerators

The addition of TrFE, with three fluoride atoms with each monomer unit, will promote the all-trans conformation in PVDF and hence the β phase can be easily achieved. The fluorine atom is comparatively bigger than the hydrogen atom and hence it can promote a steric hindrance [22,23]. The copolymer TrFE exhibits more crystallinity than PVDF. This copolymer can be easily crystallized into the β phase by heat treatment at Curie's transition temperatures (Tm). Yagi et al. reported that in PVDF-TrFE the size of the hydrogen atom is too small when compared to the size of fluorine atom; VDF and TrFE can randomly orient resulting in the formation of a single crystalline phase, which is analogous to the beta phase [24]. In this copolymer the molar content of vinylidene fluoride will determine the spontaneous polarization value while the crystallization structure depends on the content of TrFE [24]. Fig. 3.4 depicts the crystal structure of PVDF-TrFE repeat units.

Chaoxian Zhao et al. developed multiwalled carbon nanotubes (MWCNT)-PVDF-TrFE-based PENG and they evaluated the electrical output of the prepared nanogenerator. For the development of a nanogenerator Zhao et al. initially prepared MWCNT-PVDF-TrFE coaxial

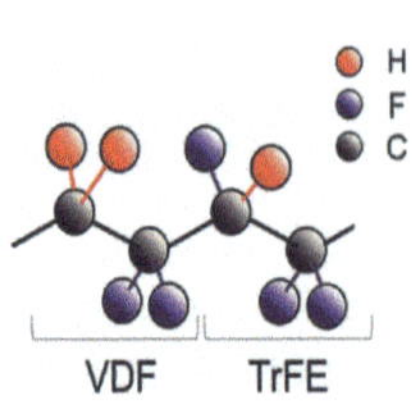

FIGURE 3.4 Chemical structure of PVDF-TrFE. Source: *Reproduced with permission from T. Sekine, R. Sugano, T. Tashiro, J. Sato, Y. Takeda, H. Matsui, et al. Fully printed wearable vital sensor for human pulse rate monitoring using ferroelectric polymer. Sci. Rep. 8 (1) (2018) 1–10.*

composite nanofibers via electrospinning. Here Zhao et al. incorporated MWCNT in the PVDF-TrFE matrix, in order to enhance the crystallinity and mechanical strength of the polymer [25].

For improving the bulk density and to reduce porosity MWCNTs/P (VDF-TrFE) fibrous membranes were cold-pressed. As a result of cold-pressing nanofibers were converted into thin films. These films were then covered with copper electrodes and a PENG was fabricated using the entire arrangement [25], as shown in Fig. 3.5A. The photograph of the developed PENG is shown in Fig. 3.5B. Fig. 3.5C1 and C2 show the mechanical and electrical conversion performances of the nanogenerator under periodic external force induced by a stepping motor where the device was unfolded and clamped by fixtures and then it was bent and released. The average peak values of the open circuit voltage (*V*oc) and short circuit current (Isc) produced by the electrospun nanofibers were shown in Fig. 3.5D1 and D2, respectively. From the above figures it was clear that both voltage and current exhibit a nonmonotonic evolution with an increase in CNT content and the average peak values were found to be increasing for neat polymer and 3wt.% MWCNTs/PVDF-TrFE composite system, and further addition of CNT in PVDF-TrFE matrix reduces both voltage and current [25]. The average peak values of output voltage and current for neat PVDF-TrFE were found to be

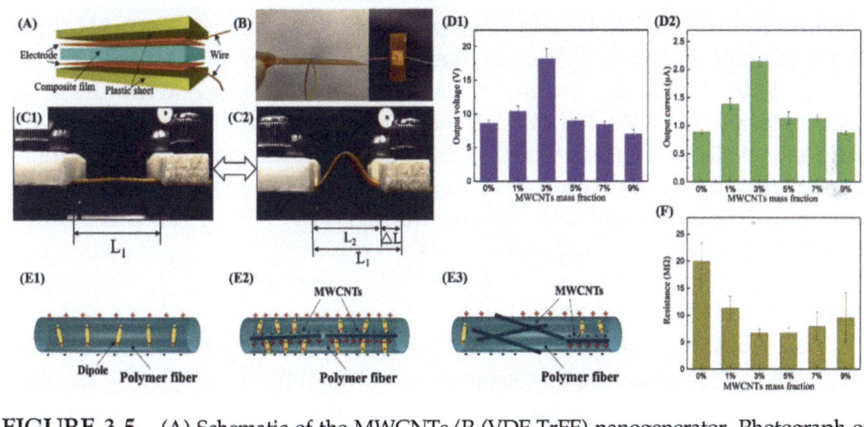

FIGURE 3.5 (A) Schematic of the MWCNTs/P (VDF-TrFE) nanogenerator. Photograph of the fabricated nanogenerator (B), photograph of nanogenerator under releasing mode (C1), photograph of nanogenerator under bending mode (C2) in the mechanical-to-electrical conversion measurement. The average peak values of output voltage (D1) and output current (D2) as a function of MWCNTs content. Schematic of dipoles and inductive charges distribution inside the nanofibers of (E1) neat P(VDF-TrFE), (E2) low loading samples and (E3) high loading samples. (F) The electric resistance of the MWCNTs/P(VDF-TrFE) fibrous membranes. Source: *Reproduced with permission from C. Zhao, J. Niu, Y. Zhang, C. Li, and P. Hu, Coaxially aligned MWCNTs improve performance of electrospun P (VDF-TrFE)-based fibrous membrane applied in wearable piezoelectric nanogenerator. Compos. B: Eng. 178 (2019) 107447.*

8.68 V and 0.89 μA, respectively, and the average peak values of output voltage and current for 3wt.% MWCNTs/P(VDF-TrFE) fibrous membrane were 18.23 V and 2.14 μA. For 9wt.% MWCNTs/P(VDF-TrFE) nanofibrous membrane the output voltage and current were found to be 7.09 V and 0.87 μA, which was lower than the previous ones [25]. Here the mechanical-to-electrical conversion capacity of the nanofibers was related to the amount of inductive charges on the top surface of the nanofibers under stress. This inductive charge formation depends upon the piezoelectric property of the polymer and the property of the filler incorporated [25]. Fig. 3.4E discloses the mechanism of the nonmonotonic evolution on the output of MWCNTs-loaded PVDF-TrFE PENGs.

Fig. 3.5E1–E3 shows the relation connecting the amount of inductive charges at the surface and the intrinsic content piezoelectric phase within the neat PVDF-TrFE nanofibers, low loading samples, and high loading samples, respectively. Here filler (MWCNT) does not possess any piezoelectric property but it exhibits a capability to distribute the inductive charges within the polymer matrix. At a low filler loading (1 and 3wt.%) CNTs were coaxially well distributed in the polymer matrix; this can be achieved during the electrospinning process [25]. During the elecrospinning process, under external electric field the Coulomb force induces high crystalline polar β formation in PVDF-TrFE and also MWCNTs within the solution can trap free charges and can gather all free charges at the interfaces. High filler loading above 3wt.% of CNT in polymer matrix weakens the entanglement of carbon nanotubes (CNTs) and hence the crystallization in the polymer PVDF-TrFE will be reduced [25].

Fig. 3.5F shows the variation of electrical resistance with the filler concentration in MWCNT/PVDF-TrFE composite nanofibers. Since the conductive fillers were embedded inside individual polymer nanofibers, instead of long-range conductive network, here the percolation behavior was found to be absent [25]. Among the MWCNT/PVDF-TrFE nanocomposites, neat polymer nanofibers exhibit relatively high resistance. This behavior in PVDF-TrFE can be attributed to the difficulty in transferring inside inductive charges to the surface due to the absence of CNT [25].

The variation of electric current and instantaneous power generated by the PENG for different load resistance was shown in Fig. 3.6A. The relationship of the instantaneous output power on the resistance was calculated using the equation $P = I^2R$, where I is the current and R is the load resistance. Fig. 3.6B shows the voltage evolution of 3wt.% MWCNTs/PVDF-TrFE-based nanogenerator when connected to a bridge rectifier circuit with a 1 μF capacitor under a strain rate of 20% and a frequency of 1 Hz, and the capacitor can be charged above 16 V within 15 s. The peak electrical output generated by the nanogenerator under bending and unbending the finger, was found to be 6 V and the peak voltage generated during stressing–releasing wrist and clenching–relaxing were found to be 1 and 0.6 V, respectively

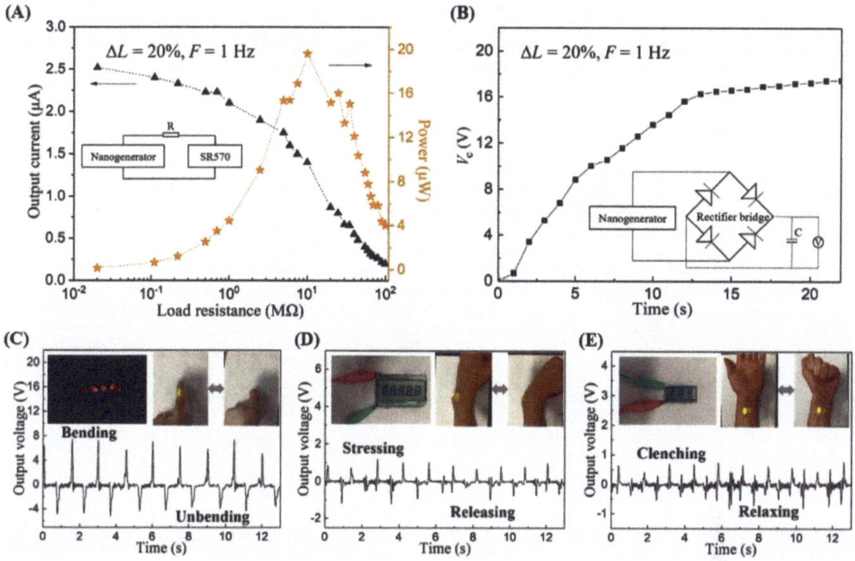

FIGURE 3.6 (A) The variation of current and instantaneous power generated by the nanogenerator for different load resistance, (B) output voltage generated by 3wt.% MWCNTs/PVDF-TrFE-based nanogenerator, under 20% strain and 1 Hz external stress, when operated by 1 μF capacitor. The electrical output signals generated by the piezoelectric nanogenerator: (C) bending-unbending finger, (D) stressing-releasing wrist, and (E) clenching-relaxing fist by pasting it on finger, wrist, and pulse point, respectively. Source: *Reproduced with permission from C. Zhao, J. Niu, Y. Zhang, C. Li, and P. Hu, Coaxially aligned MWCNTs improve performance of electrospun P (VDF-TrFE)-based fibrous membrane applied in wearable piezoelectric nanogenerator. Compos. B: Eng. 178 (2019) 107447.*

[25]. The maximum voltage was generated by the nanogenerator during the bending and unbending of fingers (Fig. 3.6C–E) and even tiny body movements, such as finger bending–unbending, can power LED bulbs or liquid crystal display (LCD) viewing screens with high sensitivity [25]. From the above results it was clear that MWCNTs/PVDF-TrFE nanofibers-based nanogenerators were very promising for self-powered wearable applications.

3.2.3 Polyvinylidene fluoride–hexafluoropropylene/multiwalled carbon nanotubes-based piezoelectric nanogenerator

PVDF-HFP copolymer consists of the amorphous phase of hexafluoropropylene along with the semicrystalline phase of PVDF homopolymer. PVDF-HFP is chemically inert and possesses poor crystallinity when compared to PVDF, which is attributed to the presence of bulky CF_3 groups. PVDF-HFP are widely used in the area of polymer electrolytes in rechargeable lithium batteries, for the fabrication of organophilic

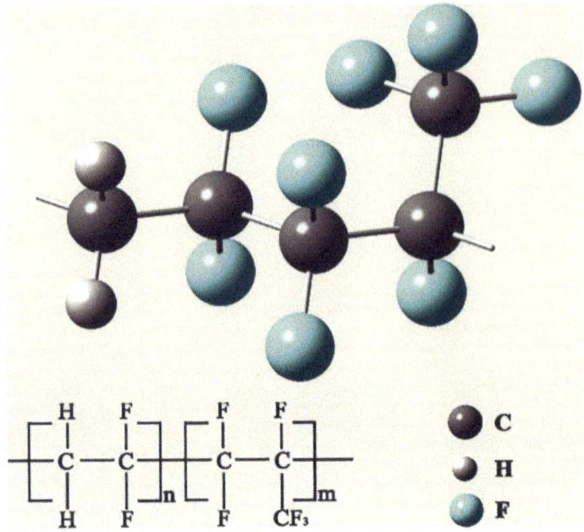

FIGURE 3.7 Schematic representation of the P(VDF-HFP) repeat units. Source: *Reproduced with permission from T. Liu, Z. Chang, Y. Yin, K. Chen, Y. Zhang, X. Zhang, The PVDF-HFP gel polymer electrolyte for Li-O2 battery. Solid State Ionics 318 (2018) 88–94.*

pervaporation [20,26]. Generally piezoelectric behavior can be imparted on PVDF-HFP based on the preparation methods. For example, with the aid of the electrospinning method a more ferroelectric β-crystalline phase can be achieved. If PVDF-HFP films are prepared by means of slow-cooling method, then the obtained PVDF-HFP films do not possess ferroelectric behavior. On the other hand, if PVDF-HFP films are prepared by means of solvent-casting/quenching, they will exhibit a good ferroelectric signature [20,26,27]. Fig. 3.7 shows the schematic of PVDF-HFP repeated units [27].

The highest remnant polarization (P_R) values reported for PVDF-HFP solvent-cast film are found to be 80/mCm^2 for 5% of HFP content in PVDF and this value increases with an increase in HFP content. Direction X, Y, or Z is represented by the subscript 1, 2, or 3, respectively. For d_{31}, the induced polarization in direction 3 (parallel to direction in which ceramic element is polarized) per unit stress applied in direction 1 (perpendicular to direction in which ceramic element is polarized) or the induced strain in direction 1 per unit electric field applied in direction 3 [28].

Haoxuan Li et al. fabricated a flexible, breathable, and washable piezoelectric sensor based on PVDF-HFP-MWCNT nanocomposites [29]. They adopted electrospinning method for the development of nanocomposites in fiber form. Here a strong interaction between the filler and polymer was achieved by a thermal-induced welding method. The thermal-induced welding method will wrap the MWCNT with the polymer matrix and also it welds filler within the nanofibers at their cross-point. Hence the resistance between the adjacent nanofibers can be reduced [29]. According to Haoxuan Li et al. MWCNT/PVDF-HFP nanofibers possess multifunctional properties, that is, they can simultaneously act as a flexible electrode to power light-emitting diodes (LEDs) and can generate piezoelectric signals for analyzing tactile stimulants and to observe human movements with a rapid response. Haoxuan Li et al. states that PVDF-HFP-MWCNT nanofiber-based nanocomposites possess a better intrinsic conductivity when compared to the traditional nanofiber-based piezoelectric materials because of their capability to transfer the electrons directly without being shielded by the electrode layer [29].

In order to realize the potential application of the MWCNT/PVDF-HFP-based nanofibrous membrane as a wearable smart device, these nanofibers were designed as an electric skin, so that they can sense tactile stimulants and hence can monitor human motions. The sensing device fabricated using MWCNT/PVDF-HFP nanofibers is shown in Fig. 3.8A. Here MWCNT/PVDF-HFP composite nanofibrous membranes were pressed continuously with the human finger and the corresponding

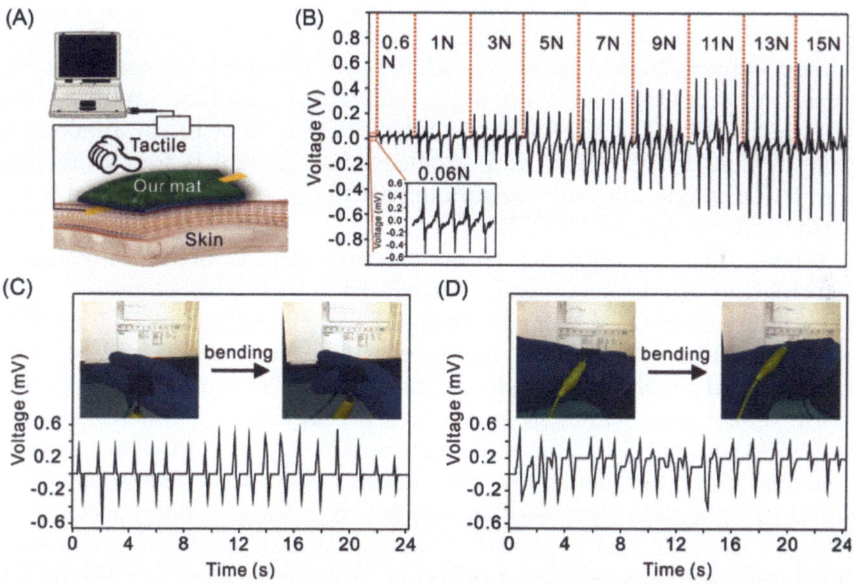

FIGURE 3.8 (A) Conductive MWCNT/PVDF-HFP nanofibrous mat with good breathability and flexibility could be attached to human skin to detect tactile stimulants directly. (B) Output voltages of the MWCNT/PVDF-HFP nanofibrous mat under different pressures. (C) finger joint bending and (D) wrist bending. The insets show the detailed photographs of the wearable and breathable motion sensor. Source: *Reproduced with permission from H. Li, W. Zhang, Q. Ding, X. Jin, Q. Ke, Z. Li, et al., Facile strategy for fabrication of flexible, breathable, and washable piezoelectric sensors via welding of nanofibers with multiwalled carbon nanotubes (MWCNTs). ACS Appl. Mater. Interfaces 11 (41) (2019) 38023–38030.*

variations of the electric signals with the applied pressure were evaluated, as shown in Fig. 3.8B. From the experimental results it was evident that the output voltage was found to be increased from 0.57 to 0.62 V when the pressure applied varied from 0.06 to 13 N, and also it could be noted that, the pressure beyond 13 N did not alter the maximum voltage generated [29]. The above results reveal that microstructural deformation of the MWCNT/PVDF-HFP nanofibrous membrane gets saturated at 13 N and further increases in pressure cannot further influence the output electrical voltage. These MWCNT/PVDF-HFP nanofibrous mat can sense tiny human motions [29]. Fig. 3.8C and D show the output voltage variation produced by the sensor against body movement. This voltage variation according to finger and wrist bending reveals that MWCNT/PVDF-HFP nanofibrous sensors can be used for human health monitoring processes [29].

3.2.4 Poly-l-lactic acid nanofiber-based piezoelectric nanogenerator

Polylactic acid (PLA) is an all-purpose biodegradable polymer with a comparable strength and formability to petroleum resins. It is also an ecofriendly polymer with high light transmittance when compared with other polymers. In addition to the above properties, PLLA possess a piezoelectric property [30] with a piezoelectric coefficient ranging from 7–12 pC/N [31]. The piezoelectric output constants of PLLA are comparable to that of PVDF but its piezoelectric coefficient is four times less than PVDF. Generally, the crystalline nature of PLLA will determine the piezoelectric response [31–33].

When PLLA is subjected to an external electric field it undergoes deformation due to an inverse piezoelectric effect. This phenomenon is due to the existence of shear piezoelectricity within PLLA [33]. Lee et al. prepared PLLA electrospun nanofibers and fabricated a piezoelectric sensor based on the prepared nanofibers [34]. The piezoelectric output of the sensor was evaluated and it was found that I_{max} (maximum current) = 30 nA and $V_{p\text{-}p}$ (peak to peak voltage) = 7.2 V in series connection. On the other hand, the piezoelectric output of the sensor was found to be slightly decreased in parallel connection in comparison, that is, $I_{max} = 3.25$ nA and $V_{p\text{-}p} = 5.5$ V. Lee et al. studied the possibility of developing a PLLA sensor-based power source by attaching the sensor along with a parallel electrode [34]. For this, a circuit for LED operation was designed, and the entire arrangement for LED powering is shown in Fig. 3.9A. According to their study, during charging the voltage increases and reaches a value equivalent to the driving voltage of LEDs; the LEDs were turned on [34]. These PLLA-based electrospun nanofibers possess excellent piezoresponse and hence can be used for the fabrication of smart self-powered electronic devices.

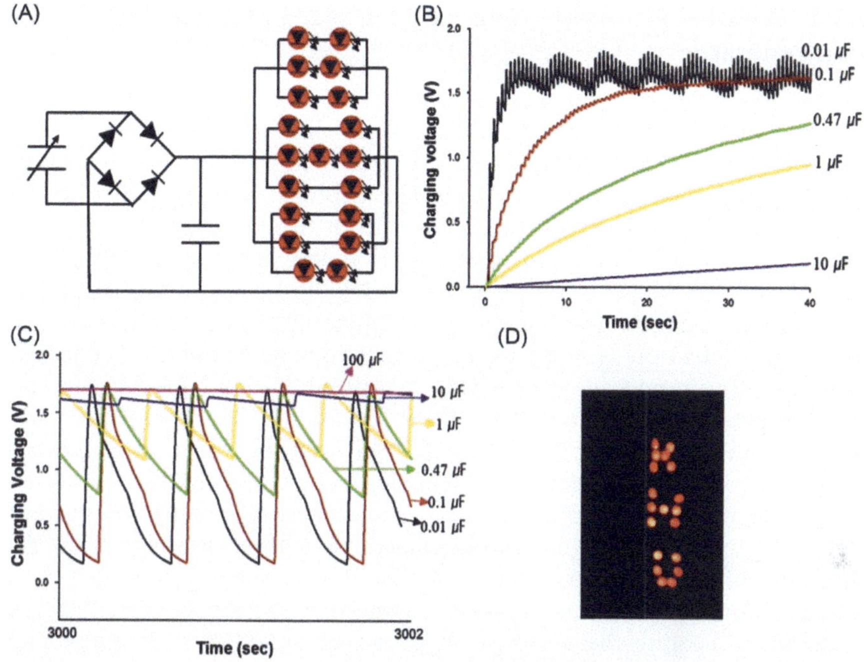

FIGURE 3.9 (A) Equivalent circuit for powering LEDs; (B) charging voltage versus time graph when the LEDS were not connected; (C) charging voltage versus time when LEDs are connected for a time period more than 3000 s using different capacitors; (D) photograph of LED powering using 100 μF capacitor. Source: *Reproduced with permission from S. J. Lee, A.P. Arun, K.J. Kim, Piezoelectric properties of electrospun poly (l-lactic acid) nanofiber web. Mater. Lett. 148 (2015) 58–62.*

In order to evaluate the practical applications of the developed piezoelectric sensor a bridge rectifier circuit was connected to the PLA sensor and capacitors were charged as a function of time. Fig. 3.9B shows the charging of capacitors (having different capacitance) using a PLA-based sensor as a function of time. From Fig. 3.9B, it is clear that the capacitors having small capacitance charges discharge quickly and get saturated within a small time period [34]. But when the capacitor size was increased the charging becomes comparable and discharge is less. Hence to operate LEDs, capacitors with higher capacitance values were preferred. Fig. 3.9C shows the variation of charging voltage versus time for different capacitors, when the LEDs are connected for a time period more than 3000 s [34]. Fig. 3.9D shows the photograph of LED operation using a 100 μF capacitor. From the above results it is clear that PLLA-based PENGs possess an excellent degree of piezoelectric response [34].

3.2.5 Nylon 11/cellulose nanocrystal-based piezoelectric nanogenerator

The piezoelectric property in odd-numbered nylon was initially reported by Scheinbeim and Newman. According to Scheinbeim the piezoelectric property in odd nylon is due to a high degree of hydrogen bonding and the orientation of electric dipoles, resulting due to the arrangement of polyamide molecules within nearby chains upon crystallization [35,36]. Compared with PVDF, nylon possesses a high Curie temperature or melting temperature. PVDF possesses a comparatively low thermal stability for higher temperature application. For example, early-fault detection systems of heavy machinery need a wide temperature operating range. Nylon 11 has great potential for high-temperature application. In this regard odd nylons like nylon 11 and nylon 9 have been gaining considerable interest [37,38]. Among the odd-numbered nylons, nylon-11 (polyamide-11 = $[C_{11}H_{21}ON]_n$) shows higher piezoelectric and ferroelectric responses, which are comparable to PVDF. Usually Nylon 11 exhibits five crystalline phases: α phase, attained from the annealing of the quenched polymer or solution casting from m-cresol; α' phase, attained from melt crystallization; the pseudohexagonal phase, obtained by heating the α' phase; δ' phase, obtained by melt quenching; and γ phase, obtained by solution casting from Trifluoro acetic acid [38].

Farza et al. fabricated a PENG-based on nylon 11/cellulose nanocrystal-based film with nylon 11 as the polymer matrix and cellulose nanocrystal as the nanofiller [39]. The photograph of nylon 11/cellulose nanocrystal composite films was shown in Fig. 3.10. The addition of cellulose nanocrystals in the polymer matrix improved the piezoelectric property of the

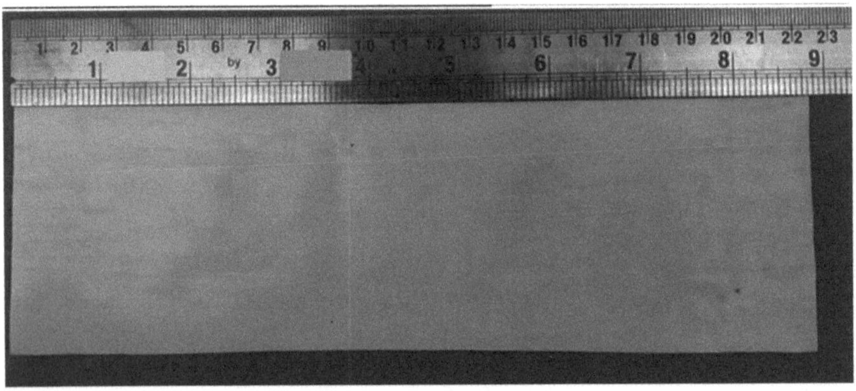

FIGURE 3.10 Photographic image showing nylon 11/Cellulose nanocrystal composite film. Source: *Reproduced with permission from F. Ram, S. Radhakrishnan, T. Ambone, K. Shanmuganathan, Highly flexible mechanical energy harvester based on nylon 11 ferroelectric nanocomposites. ACS Appl. Polym. Mater. 1 (8) (2019) 1998–2005.*

composite system. PENG devices were prepared by attaching adhesive copper tapes on both the surfaces of nylon 11/cellulose nanocrystals-based nanocomposite films and thin copper wires were tightly connected to the copper electrodes. After that the entire device were covered in an insulating matrix to reduce damage and to improve electrostatic contribution [39]. The neat nylon 11-based nanogenerator and 5wt.% of cellulose nanocrystals-loaded nylon 11 composite films generated an *V*oc of 2.68 and 6.95 V, respectively, at ~23 N force [39].

For illustrating the practical application of the developed nanogenerator, Farza et al. charged a polarized capacitor by means of the developed nanogenerator and a bridge rectifier [39]. They found that the current produced by the nanogenerator was ac current, and hence in order to convert ac into dc, a bridge rectifier circuit was used. The rectified output current was then used for charging capacitor using a 10 μF polarized capacitor. Here the dc output was stored in the capacitor and the voltage generated from the nanogenerator was measured by using a multimeter [39].

The *V*oc generated by neat nylon 11-based nanogenerator and 5wt.% cellulose nanocrystal/nylon 11-based nanogenerator are shown in Fig. 3.11A and B, respectively. From Fig. 3.11A and B, it is clear that 5wt.% cellulose nanocrystals/nylon11-based systems possess better charging rate and output voltage compared to the neat nylon 11-based nanogenerators. Piezoelectric voltage generated by the neat nylon11-based nanogenerator and 5wt.% cellulose nanocrystal/nylon11-based nanogenerator were found to be 2.65 and 6.87 V, respectively [39]. Nylon11/cellulose nanocrystal-based low-cost, flexible, and durable nanogenerators have potential applications in self-powering smart electronic devices [39].

3.2.6 Cellulose-based energy generator

Cellulose is a piezoelectric polymer that is infinitely available in nature itself, and it displays an interesting natural structure [40]. It is an ecofriendly and biocompatible product [41]. Cellulose can be easily obtained from various plants as microfibrils, consisting of glucose–glucose linkages with linear chains [42]. Cellulose comprises crystallites and scattered amorphous regions of low degree of order. It exhibits both direct and inverse piezoelectric effects. Native cellulose is also known as cellulose 1, which is crystalline cellulose. On the other hand regenerated cellulose is known as cellulose 2. In order to extract cellulose 2, chemical modifications are essential [41–43].

Bairagi et al. developed a self-poled, sustainable, biowaste derived, biocompatible cellulose-based PENG [44]. These celluloses-based PENGs were made from pomelo fruit membrane. For the fabrication of PENG based on pomelo membrane, Bairagi et al. initially dried the

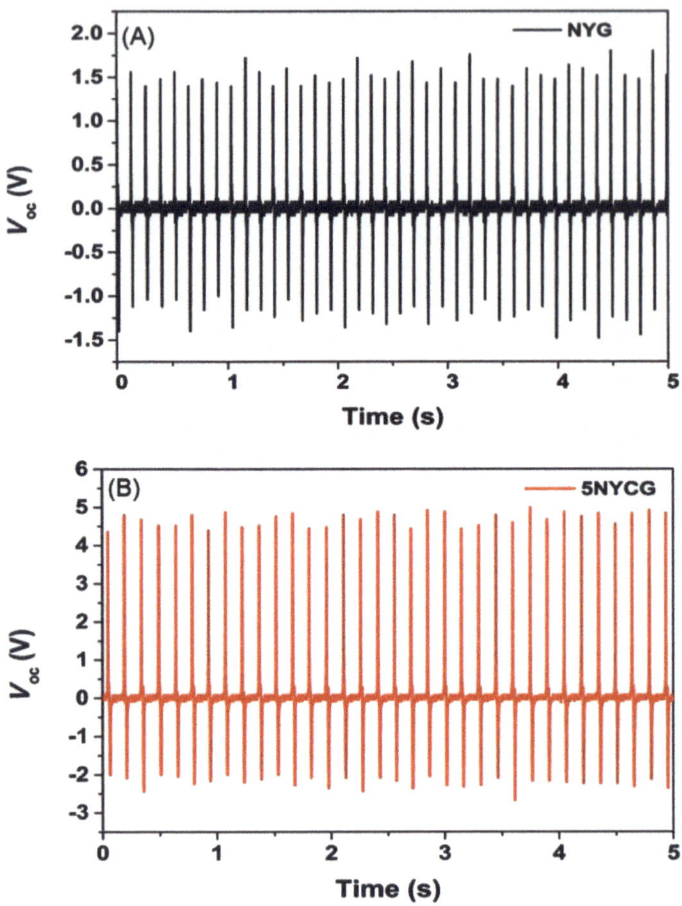

FIGURE 3.11 Output voltage generated by the nanogenerator made of (A) Nylon 11, and (B) 5wt.% of cellulose nanocrystal/nylon 11 composite films. Source: *Reproduced with permission from F. Ram, S. Radhakrishnan, T. Ambone, K. Shanmuganathan, Highly flexible mechanical energy harvester based on nylon 11 ferroelectric nanocomposites. ACS Appl. Polym. Mater. 1 (8) (2019) 1998–2005.*

pomelo fruit membrane and they cut it into small pieces with dimensions of $2 \times 2 \times 0.045\ cm^3$ [41–43]. A pomelo fruit membrane-based PENG was fabricated without any chemical treatment. Conducting aluminum tape was pasted on both sides of the membrane and it acted as the electrode. For measuring the piezoelectric output generation, copper wires were fixed over both sides of the electrodes and the entire arrangement were placed in a polypropylene tape to protect the device from environmental issues [44]. The schematic diagram in Fig. 3.12 shows the fabrication of the PENG.

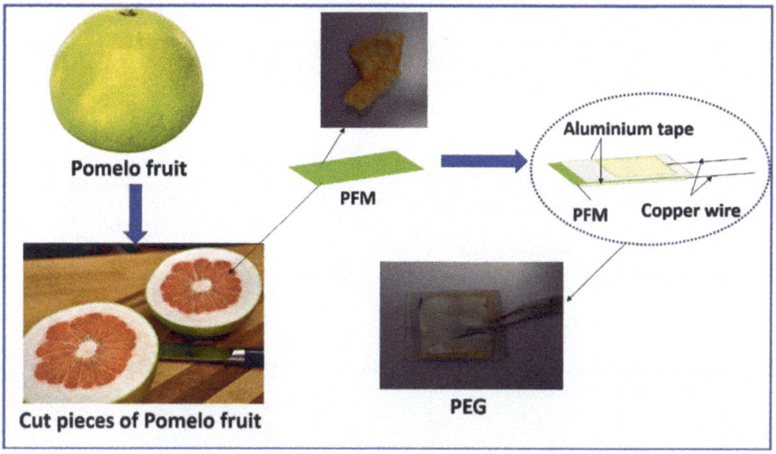

FIGURE 3.12 Schematic illustration of the fabrication of piezoelectric nanogenerator developed from pomelo fruit membrane. Source: *Reproduced with permission from S. Bairagi, S. Ghosh, S.W. Ali, A fully sustainable, self-poled, bio-waste based piezoelectric nanogenerator: electricity generation from pomelo fruit membrane. Sci. Rep. 10(1) (2020) 1–13.*

Fig. 3.13A and B represent the output voltage generated by pomelo fruit membrane-based PENGs by simple finger tapping. The magnitude of the applied pressure is around 1.1 kPa. Piezo voltage generated by the nanogenerator was found to be 6.4 V. The positive and negative peaks in Fig. 3.13A are due to the applied pressure on the device and the release of pressure from the device, respectively [44]. The output voltage generated by the nanogenerator is due to the displacement and reorientation of the dipoles present in the crystal lattice of the pomelo fruit membrane under the application of external pressure [44]. In addition, the microfibrils of cellulose pomelo fruit membrane possess hemicellulose and pectin components, but here both they act as sources of dipoles. These dipoles are mainly formed due to intramolecular and intermolecular hydrogen bonding between the cellulose, hemicellulose, and pectin components [44]. Since a large number of hydroxyl and carbonyl groups were present in pomelo fruit membrane when compared to other reported biomaterials, the total number of electric dipoles is higher here, and as a result the piezoelectric response of this membrane will be higher when compared to the other reported biopiezoelectric materials [44].

Fig. 3.13C shows the output current produced by the piezoelectric energy generator when 220 kΩ resistances were connected parallel to it. Here the output current was calculated by the equation $V = IR$, where V is the output voltage and R is the resistance [44]. Fig. 3.13D shows the output voltage and current generated by the pomelo fruit membrane-based energy

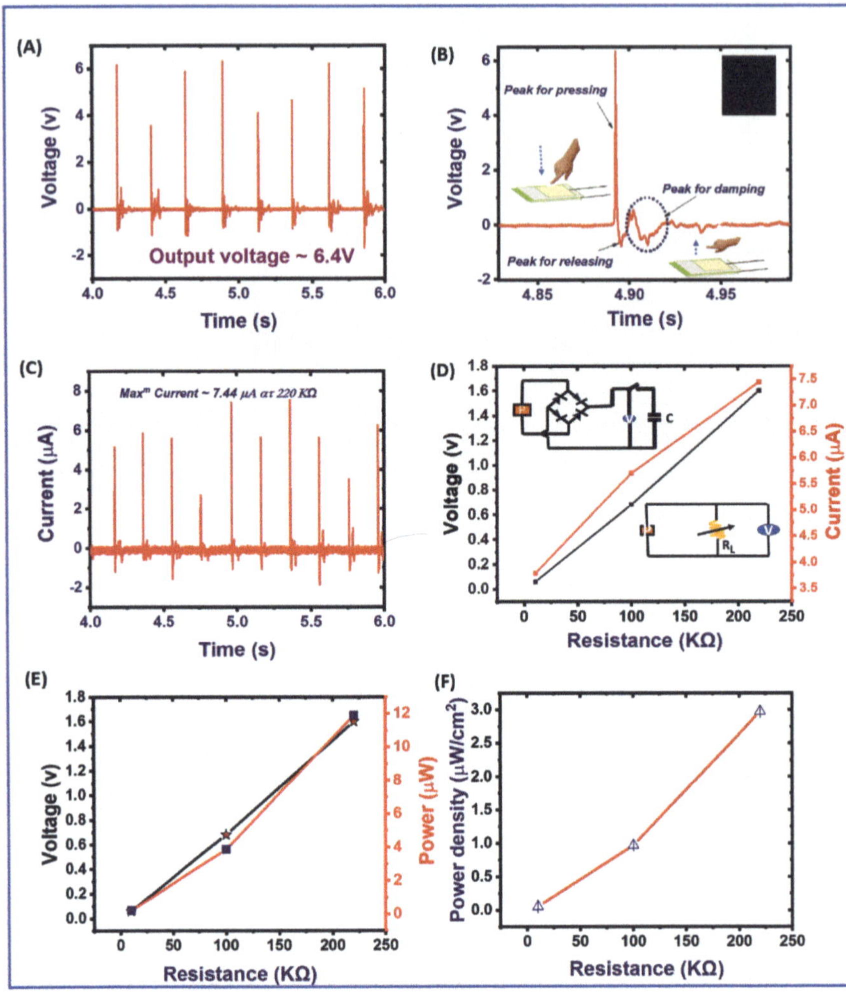

FIGURE 3.13 (A) piezoelectric voltage generated by the generator; (B) enlarged view of the selected part in (A); (C) output current produced by the generator at 220 kΩ resistance; (D) output voltage and current with different resistances; (E) voltage and power with different resistances; (F) power density of the generator with respect to different resistances. Source: *Reproduced with permission from S. Bairagi, S. Ghosh, S.W. Ali, A fully sustainable, self-poled, bio-waste based piezoelectric nanogenerator: electricity generation from pomelo fruit membrane. Sci. Rep. 10(1) (2020) 1–13.*

generator with different capacity of resistors. From the figure it is clear that when the resistance capacity is increased, both output voltage and current were found to be enhanced [44].

Fig. 3.13D shows the variation in output voltage and current produced by the nanogenerator for varying resistance values. From the figure it is

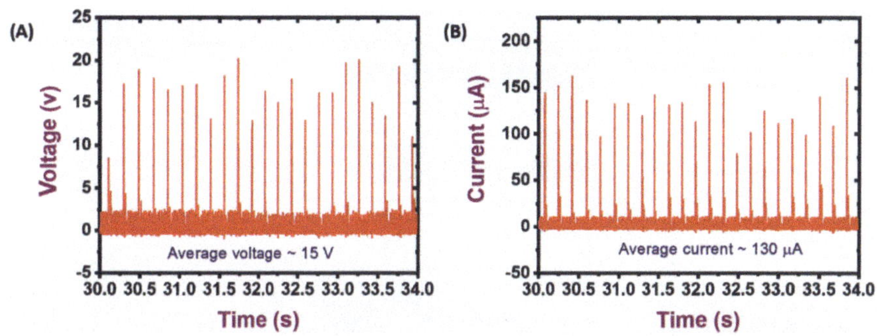

FIGURE 3.14 (A) Output voltage, (B) output current of the pomelo fruit membrane-based piezoelectric generator when connected to a full wave rectifier to convert AC voltage to DC voltage. Source: *Reproduced with permission from S. Bairagi, S. Ghosh, S.W. Ali, A fully sustainable, self-poled, bio-waste based piezoelectric nanogenerator: electricity generation from pomelo fruit membrane. Sci. Rep. 10(1) (2020) 1–13.*

clear that as the resistance increases the output signals also increases. The circuit diagram for measuring the variation in current with respect to resistance is shown in the inset of Fig. 3.13D. Fig. 3.13E and F shows the calculated power and power density for different resistance levels (10, 100, and 220 kΩ at 50 Hz excitation frequency) [44]. Here both power and power density increase with the increase in load resistances. Here the output voltage, current, power, and power density of the developed nanogenerator was found by connecting it to a bridge rectifier circuit [44] (Fig. 3.14).

Fig. 3.15(A) and (B) shows the output voltage and output current produced by the energy generator when a bridge rectifier circuit is connected to it. Average output voltage and current produced by the energy generator were found to be 15 V and 130 μA [44].

3.2.7 Gelatin nanofiber-based piezoelectric pressure sensor

Gelatin is a kind of protein produced by the partial hydrolysis of native collagen. Based on the hydrolysis procedure gelatin can be classified into two types, namely type A, synthesized by acid hydrolysis, and type B, synthesized by alkaline hydrolysis [45]. Gelatin is biocompatible and can be available at low cost [46]. Gelatin can be made from the by-products of meat and leather industries. Fish gelatin contains more amino acid residues with electron donar groups than mammalian gelatin, which is highly preferred for piezoelectric applications [47,48].

Sujoy et al. fabricated piezoelectric self-powered bio-e-skin for healthcare monitoring using electrospun gelatin nanofibers [49]. For the fabrication of the piezoelectric bio-e-skin sensor, the copper (Cu)–nickel (Ni)-plated polyester fabric was placed over the top and bottom of the

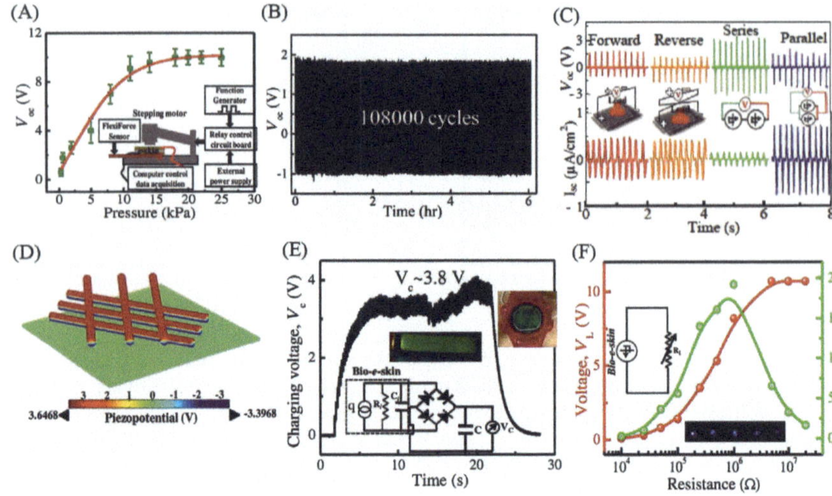

FIGURE 3.15 The pressure-sensing performance of the bio-e-skin sensor. (A) Open circuit output voltage (*V*oc) with the variation of pressure (σa) rendered by the stepping motor probe revealed by the schematic illustration in the inset. (B) Durability test over 108,000 cycles by applying the pressure of 0.6 kPa at a frequency of 5 Hz. (C) The *V*oc and *I*sc (short circuit current) from single bio-e-skin when it is forward and reverse biased and from two bio-e-skins when connected in series and parallel configurations with the schematic circuit diagrams in the inset. (D) Simulation result of the piezopotential distribution in the network structure of GNFs. (E) Transient output response of 1 μF capacitor charging with associate circuit diagram and driving. Source: *Reproduced with permission from S.K. Ghosh, P. Adhikary, S. Jana, A. Biswas, V. Sencadas, S.D. Gupta, et al., Electrospun gelatin nanofiber based self-powered bio-e-skin for health care monitoring. Nano Energy 36 (2017) 166–175.*

gelatin nanofibers as electrodes. Finally, the entire arrangement was encapsulated with PDMS (Sylgard, 184 Silicone Elastomer), for mechanical robustness and for moisture protection. For the development of a sensor, a relay circuit board-based custom-made striking system was designed to apply definite pressure and the force pressure sensor was placed underneath the piezoelectric bio-e-skin to record the applied pressure [49]. The material was subjected to pressures at various amplitudes and frequencies and the response was recorded for different time periods. [49].

The output voltage and current produced by the bio-e-skin under different dynamic pressures were documented using National Instruments (NI) DAQ board (USB 6000) through an online interface with a PC with a sampling rate of 1000 samples/s and a picometer (Keithley 6485) respectively. The real-time wrist pulses were documented using a digital oscilloscope (Agilent DSO3102A) with 2-mV measurement resolution [49]. Fig. 3.15 shows the pressure-sensing performance of the gelatin nanofibers. The controlled sensing performance of the nanofibers was estimated under the dynamic stimulus ranging (pressure σa) between 0.3 and 25 kPa. A relay circuit board-based custom-built system handling the stepping motor probe (area $\sim 25\ mm^2$) was used for applying periodic pressures (inset of Fig. 3.15A) [49].

*V*oc produced by gelatin-based fibers were found to be increasing linearly with respect to σa in a low pressure regime (< 10 kPa). Further increase of σa leads to nonlinear behavior in a medium pressure regime (> 10 kPa) with a saturation of *V*oc of ~ 10 V at 20 kPa [49].

In order to check the reliability of the piezoelectric sensing response under low pressure (such as $\sigma a \sim 0.6$ kPa), Sujoy et al. undertook cyclic stability testing over an extended period of time (~ 6 h) at the frequency of 5 Hz (Fig. 3.15B). The bio-e-skin unveiled outstanding operational stability of *V*oc of ~ 1.8 V over 108,000 cycles. Fig. 3.15C displays the *I*sc and *V*oc measurements under forward connection and reverse biasing. The bio-e-skin showed a well–behaved periodic alternation of positive and negative peaks of output voltage (*V*oc) and short-circuit current density (*I*sc) (Fig. 3.15C) [49].

A theoretical study was conducted using the Finite Element Method in order to find a quantitative explanation of the piezopotential distribution within the deformed gelatin nanofibers of the bio-e-skin under uniaxial $\sigma a \sim 0.6$ kPa. It is presented by six nanofibers making a network structure on metal electrode (Fig. 3.15D). The experimentally evaluated piezopotential (~ 1.8 V) was somewhat lower than the simulated result of 3.6 V. This deviation is probably because of the voltage drop due to internal leakage paths and charge loss in the metal–insulator–metal structure [49].

According to Sujoy et al., during applied stress, maximum deformations ($\sim 0.01\ \mu m$) of gelatin nanofibers occur in the contact region between the lower and upper layer of the fibers. In the noninteractive regions, the deformations gradually weaken [49]. This indicates that the network structure of gelatin nanofibers improves the resulting polarization in comparison to single fiber owing to the cooperative electromechanical mutual interaction between the nearby fibers during applied pressure. Also, Sujoy et al. carried out the switching polarity test, by reversing the electrode connections [49]. The results obtained reveal the identical amplitude of *V*oc with reversed polarization, as shown in Fig. 3.15C. It proves the authenticity of piezoelectric output signals from gelatin nanofibers themselves and not due to electrostatic interactions. In other words, this is the indirect proof of dipole alignment [49].

Fig. 3.15E shows the transient response (having time constant, $\tau \sim 3$ s) of the externally connected capacitor. From Fig. 3.15E, it was clear that the steady-state charging voltage (*Vc*) across the capacitor reaches 3.8 V within 20 s. The stored energy ($\sim 7.2 \times 10-6$ J) of the capacitor can be used to drive commercial electronic devices like a digital wrist watch and LCD during discharging of the capacitor (inset of Fig. 3.15E), signifying the feasibility of bio-e-skin as a power generator. Also the variation of power (P) with load resistance R_L is shown in Fig. 3.15F. From the figure it is clear that the high-output power density of 19.2 $\mu W/cm^2$ was obtained for a load resistance R_L of 1 MΩ [49].

Besides these, Sujoy et al. found that a gelatin nanofiber-based pressure sensor possesses higher instantaneous piezoelectric energy conversion efficiency (η *piezo*) even up to 52.5%. Hence this gelatin nanofiber-based e-skin with high power and η *piezo* can be used for powering five blue LEDs by repeated finger tapping [49].

3.3 Polymer–metal oxide nanocomposites-based piezoelectric energy harvesters

Among the wide variety of piezoelectric materials available for energy-harvesting applications, complex metal oxides or piezoceramics (such as PZT, barium titanate) have dominated the industry over the years due to their excellent piezoelectric properties compared to their counterparts [50]. The salient features include high dielectric constants, greater piezoelectric charge constant, high coupling coefficients, and high crystallinity. In spite of their wide acceptance, piezoceramics also have certain inherent limitations, such as a rigid and brittle nature, high density, and requirement of high temperature for processing, toxicity of lead-based components, etc., which can restrict their design flexibility and affect the performance of the fabricated device. Piezoelectric polymers such as PVDF have come up as an attractive material for the development of energy-harvesting devices or nanogenerators due to their flexible, lightweight, and robust nature [51]. The piezoelectric polymers also have higher piezoelectric stress/voltage coefficients, lower acoustic and mechanical impedance, and higher operating field strength than ceramics. Moreover, they are cheap, easy to fabricate, and readily available. However, polymers have inferior energy conversion efficiency, lower dielectric constants, and limited crystallinity when compared to other inorganic piezoelectric materials. Despite the shortcomings, polymers have emerged as a major competitor of piezoceramics for various key applications. Property comparison between a piezoceramic (PZT) and piezopolymer (PVDF) is given in Table 3.1.

To strike a balance between the properties and achieve a synergistic effect, piezoelectric polymer–ceramic/metal oxide composites or nanocomposites have been developed. The aim is to combine the excellent piezoelectric properties of the ceramics/metal oxides with the mechanical flexibility of the polymers [31]. It has been found that the composites fabricated have superior properties and show enhanced energy-harvesting capability when compared to the individual components.

TABLE 3.1 Comparison between a piezoceramic (PZT) and piezopolymer (PVDF) [50].

Piezoelectric material	Piezoelectric strain constant d_{ij} (pC/N)	Piezoelectric stress constant g_{ij} (pC/N)	Electromechanical coupling factor k_{ij} (% at 1 kHz)	Salient features
Polyvinyldene fluoride	$d_{31} = 23$	$g_{31} = 216$	$k_{31} = 12$	Flexible, light weight, low acoustic, and mechanical impedance
	$d_{33} = -33$	$g_{33} = -330$		
Lead zirconate titanate	$d_{31} = 110$	$g_{31} = 10$	$k_{31} = 30$	Brittle, heavy, and toxic
	$d_{33} = 225–590$	$g_{33} = 26$		

Researchers around the world have been trying to develop piezoelectric nanocomposites by different combinations of polymers and ceramics/metal oxides to derive improved energy-harvesting performance.

3.3.1 Lead zirconate titanate–polymer nanocomposites

PZT is one most significant and highly explored piezoceramic materials due to its outstanding piezoelectric performance in varied applications. Being a lead-based material and a brittle ceramic, PZT has many drawbacks but until now no material has been able to practically replace PZT in various commercial applications due to its high piezoelectric constants and lower cost. However, PZT–polymer composites have been developed to overcome the hard and brittle nature of PZT and facilitate the fabrication of flexible devices [52].

Hilczer, Kułek et al. in 2002 studied the dielectric properties of PZT–PVDF nanocomposites over a wide range of temperature and frequencies [53]. Nanosize PZT powder was synthesized using a sol–gel method, and then mixed with PVDF and carefully hot-pressed to obtain nanocomposite disks. Dielectric measurements have been performed at frequencies ranging from 100 Hz to 1 MHz and the temperatures varied from 100K to 450K [53]. Experimental and theoretical dielectric studies of PZT–PVDF nanocomposites were reported by Zak et al. in 2011 where PVDF solutions containing fine PZT nanoparticles were spin-coated to obtain the nanocomposite films [54]. The effective dielectric constants of the nanocomposites determined are 40% higher than pristine PVDF. Also, the experimental results have been found to match with theoretical model data at relatively high frequencies. Moreover, Revathi et al. in 2018 also prepared PZT−PVDF nanocomposites as suitable candidates for piezoelectric energy-harvesting [55]. The work gives detailed characterization of the nanocomposites including microstructure analysis, phase determination, thermal studies, and dielectric measurements. Furthermore, Pradhan et al. in 2017 fabricated PZT−PVDF nanocomposites with an emphasis on ferroelectric properties [56]. It has been observed that the polar β-phase formation in the composite enhances the ferroelectric properties, such as polarizability, remnant polarization, coercive field, etc.

Wankhade et al. in 2020 described the preparation of a nanohybrid of PVDF and PZT by a solvent-casting method. PZT has been synthesized by a solid state method and its concentration was varied to optimize the properties of the nanohybrid. A nanogenerator device has been fabricated using the PZT−PVDF nanohybrid which showed very high mechanical energy conversion efficiency with an output voltage of 55 V and power density of 36 $\mu W/cm^2$ [57].

Zhang et al. in 2014 have described the advantages of using self-charging power cells (SCPCs) over the conventional energy-harvesting systems [58]. The proposed SCPCs have increased conversion and storage efficiency owing to lower energy wastage on the rectifying circuit. They fabricated the SCPCs using a PZT−PVDF hybridized nanocomposite film as a piezoseparator in a Li battery where mechanical deformations can generate piezoelectric potential in the film in order to drive the diffusion of Li ions. Therefore the mechanical energy gets converted to electrochemical energy and is stored in the SCPCs [58].

Electrospinning is another important method which can be effectively used for the fabrication of piezoelectric nanofiber composites. In this regard, Yun et al. in 2016 fabricated piezoelectric PZT−PVDF nanofiber composites by electrospinning and studied their mechanical and piezoelectric properties for application in energy-harvesting nanogenerator applications [59]. Furthermore, the development of a flexible pressure sensor for mechanical energy-harvesting was discussed by Chamankar et al. in 2020 [60]. They made use of PVDF nanofibers doped with different PZT volume fractions using an electrospinning technique. The piezoelectric properties and the sensitivity of the nanocomposites were improved with the increase in PZT volume fraction, as it acted as a dielectric nanofiller, which in turn improved the crystallinity and phase of the PVDF nanofibers. The fabricated pressure sensor generated an output voltage of 184 mV for an applied force of 2.125 N with a piezoelectric sensitivity of 173.507 mV/Nμm [60].

In addition, Koç et al. in 2020 prepared and characterized a PZT−PVDF-based flexible PENG for vibrational energy applications [61]. The polyethylene naphthalate (PEN) has been fabricated by placing the electrospun PZT−PVDF nanofiber mat in-between two aluminum

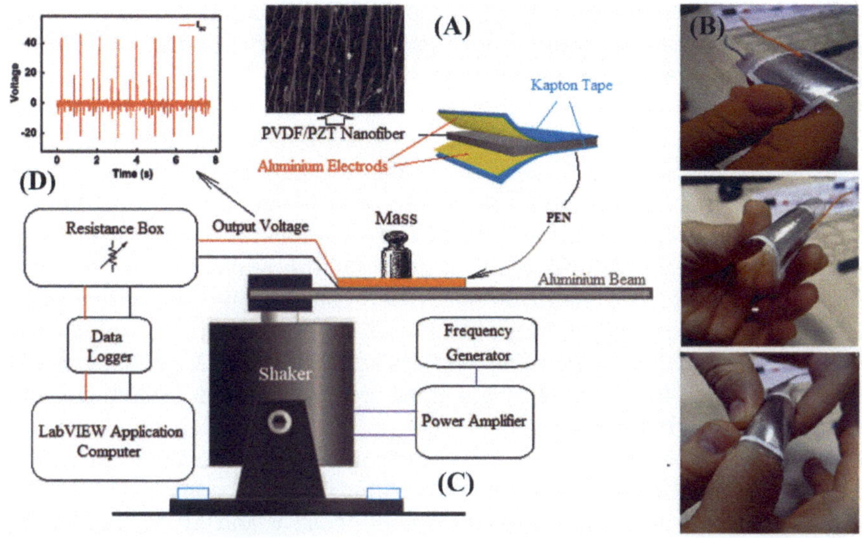

FIGURE 3.16 Schematic presentation of (A) piezoelectric nanogenerator (PEN). (B) Actual view of flexible PEN. (C) Vibrational harvesting system. (D) Symbolic output voltage changes on PEN. Source: *Reproduced with permission from M. Koç, L. Paralı, O. Şan Fabrication and vibrational energy harvesting characterization of flexible piezoelectric nanogenerator (PEN) based on PVDF/PZT. Polym. Test. 90 (2020) 106695.*

electrodes and the device fully covered with a thin Kaptan tape (Fig. 3.16). Moreover, an experimental setup was also devised to evaluate the energy-harvesting performance of the fabricated PEN. Under a resistive load of 1 MΩ, the fabricated nanocomposite-based PEN generated a maximum electric power of 6.35 μW at a vibration frequency of 15 Hz.

Apart from PVDF, other polymers have also been used to prepare piezocomposites along with PZT, such as PZT–polyamide-11 composites [62], PZT–polyurethane composites [63], and PZT–PMMA composites [64].

3.3.2 Barium titanate–polymer nanocomposites

$BaTiO_3$ is a piezoceramic metal oxide with high piezoelectric properties and, being a lead-free material, it can be considered to be an environment-friendly alternative to PZT. Preparing polymeric composites with $BaTiO_3$ can result in flexible, scalable, and environment-friendly mechanical harvesting solutions. In view of this Zhao et al. in 2015 proposed a simple, low-cost, and highly efficient PENG from $BaTiO_3$–PVDF nanocomposite films prepared by the solvent-casting method. The nanogenerator gave a stable performance under continuous mechanical stress and a remarkably high output voltage of 150 V and 1500 nA *Isc* for an applied stress of 10 MPa [65].

In 2015, Nunes-Pereira et al. prepared spin-coated $BaTiO_3$–PVDF-TrFE nanocomposites and investigated the dependence of energy-harvesting performance on ceramic filler content. It has been observed that for large (100 and 500 nm) and small (10 nm) average filler sizes, the optimum output response was recorded at 5% and 20% filler concentrations, respectively [66]. This behavior can be attributed to the increase in mechanical stiffness of the composite by the addition of ceramic fillers. More studies on $BaTiO_3$–PVDF-TrFE nanocomposites were reported by Siddiqui et al. in 2015 for the realization of lead-free flexible PENGs for mechanical energy-harvesting from body movements [67]. The nanocomposites were prepared by spin-coating a solution of $BaTiO_3$ nanoparticles (BT NPs) and PVDF-TrFE with high loadings of BT NPs up to 40%. The PENGs have been designed by placing the nanocomposite between the top and bottom electrodes of aluminum and nickel, respectively, on a polyimide substrate. The performance of the fabricated PENGs has found to be on a par with conventional PZT-based PENGs with a high output voltage of 9.8 V and an output power density of 13.5 $\mu W/cm^2$. The enhanced performance can be due to the superior piezoelectric properties and high crystallinity of the PVDF-TrFE copolymer and increased loading of the BT NPs [67].

The major properties which can enhance the energy-harvesting performance of piezoelectric materials are dielectric constant (k) and piezoelectric charge constant (d_{ij}). Although the ceramic component improves the k and d_{ij} of a ceramic–polymer composite, they can be further enhanced by the addition of conductive fillers such as graphene, CNT, Ag, Al, Ni, etc. Yaqoob and Chung in 2017 have studied the effect of reduced graphene oxide (rGO) on $BaTiO_3$–PVDF-TrFE nanocomposite properties [68]. rGO improves the piezoelectric properties by improving the dipole alignment in PVDF-TrFe and forming a microcapacitor model in the composite. Also, the carbonyl and carboxyl groups in the rGO basal plane help in dipole alignment by attracting all the fluorine groups to one side. Furthermore, $BaTiO_3$/PVDF-TrFE/rGO nanocomposite-based PENGs have been fabricated and a maximum output power of 4.5 μW was obtained at 0.5% rGO content [68]. Similarly, Dudem et al. in 2018 reported the use of Ag nanowires as conducting filler to fabricate $BaTiO_3$/PVDF/Ag nanowires composite for harvesting mechanical energy from automotive vehicles and human body movements [69]. Also, other nanofillers, such as hexagonal boron nitride (h-BN) (also known as white graphene), have been used to prepare hybrid nanocomposites with $BaTiO_3$ and poly (vinylidene fluoride–hexafluoro propylene) (P(VDF-HFP)) [70]. The studies indicate a high electroactive phase and α to β phase transition in P(VDF-HFP) resulting in a high dielectric constant value up to 45 and an output voltage of 2.4 V for compositions of 3% $BaTiO_3$ and 1% h-BN [70].

3.3.3 Zinc oxide−polymer nanocomposite

Zinc oxide is an important material of choice for the fabrication of highly efficient lead-free, nontoxic, biocompatible, cost-effective piezoelectric nanocomposites for self-powered nanogenerator applications. ZnO is characterized by moderate piezoelectric properties due to its wurtzite crystal structure and nonferroelectric nature. Many studies were reported on ZnO−PVDF nanocomposites, such as those by Cherumannil Karumuthil et al. in 2020, who utilized ZnO−PVDF-TrFE nanocomposites for atmospheric air pressure sensor fabrication [71]. The paper discussed in detail the film preparation by spin coating, device patterning by photolithography and etching process, and detailed studies on sensor performance. Among the various reported polymer−ZnO composites, those with ZnO nanorods are the most significant as they facilitate the separation of static ionic charges when a stress is applied, resulting in large piezoelectric potential and also offer low electric resistance. In this regard, Anand and Bhatnagar in 2019 prepared PVDF−ZnO nanorod-based composites and studied the role of vertically aligned and randomly placed ZnO nanorods on energy-harvesting performance [72]. In addition, Sun et al. in 2019 synthesized ZnO−PVDF composite fiber membranes with a hierarchical structure to develop acoustoelectric nanogenerators for noise energy-harvesting applications [73]. The PVDF fiber membrane was prepared by electrospinning and ZnO seeds applied on it by spin-coating followed by a hydrothermal process resulted in the growth of hexagonal shaped ZnO nanorods on the PVDF nanofiber axis. The ZnO nanorods in the hierarchical structure deflect and slide against each other which strengthens the deformation of ZnO nanorods and PVDF nanofibers, which leads to superior piezoelectric performance [73].

Use of doped ZnO can further increase the electrical conductivity and piezoelectric coefficients and also decrease the dissipation factor of the piezoelectric composites. Aleksandrova et al. in 2019, reported the preparation of novel composites between Gallium (GZO)-doped ZnO and PVDF on a flexible PEN and the piezoelectric response was studied for different metal electrodes [74]. Similarly, Ca-doped ZnO−PVDF-TrFE and Fe-doped ZnO−PVDF-TrFE nanocomposites were prepared for piezoelectric energy-harvesting [75]. It has been noted that doping of ZnO improves the dielectric properties by increasing the number of dipoles which further enhances the dipole polarization of the composite. The addition of graphene also improves dipole orientation and therefore Cherumannil Karumuthil et al. in 2020 reported the preparation of hybrid nanocomposites of ZnO−graphene oxide−PVDF-TrFE by an electrospinning method for PENG applications [71].

3.3.4 Lead magnesium niobate–lead titanate–polymer nanocomposites

PMN–PT are a type of piezoelectric ceramics which belong to the special class of relaxor ferroelectrics characterized by their exceptionally high piezoelectric coefficients (d_{33} ~2500 pC/N). To obtain flexible energy-harvesting devices to get high voltage output, piezoelectric polymers are used along with relaxor ferroelectrics with high piezoelectric constants. In this regard, Li et al. in 2016 reported a nanocomposite-based on PMN–PT nanorods and PVDF for flexible self-powered sensing applications. The PMN–PT nanorods were synthesized by a hydrothermal method, mixed with PVDF and tape-cast on to an electrode polyimide substrate. Similarly tape-cast and spin-coated PMN–PT–PVDF nanocomposites were prepared by Dietze and Es-Souni in 2019 for piezoelectric and pyroelectric energy-harvesting [76]. As in the case of other piezocomposites, the addition of nanofillers can improve the dielectric and piezoelectric properties, acting as a dispersant and strengthening agent. Hence, studies were reported on the addition of CNT fillers in PMN–PT–PVDF composites [77]. Also, rGO is used as a conducting filler for the development of rGO–PMN–PT–PVDF–TrFE composites for flexible energy-harvesting [78]. The dielectric properties and dielectric dispersion was improved with the addition of rGO and PMN–PT in the PVDF–TrFE polymer. Moreover, rGO can attract electrons in the nanocomposite which enhances the energy-harvesting performance.

3.3.5 Other metal oxide–polymer nanocomposites

Sodium potassium niobates (NKN) have been widely used as an alternative to lead-based piezoelectric materials as the dielectric and piezoelectric properties are comparable to PZT. NKN–PVDF nanocomposites can be prepared by embedding NKN nanoparticles in PVDF nanofibers for flexible nanogenerator applications [79]. Other prominent lead-free ferroelectric metal oxides for energy-harvesting include bismuth sodium titanate (BNT), barium titanate zirconate (BTZ), bismuth ferrite (BFO), etc. Mahdi et al. [80] in 2015 studied the ferroelectric and pyroelectric properties of PNT–PVDF-TrFE nanocomposites for sensor applications. BNT was chosen because of its large remnant polarization, high Curie temperature and lead-free nature. Saba et al. in 2020 synthesized BFO–PVDF nanocrystalline composites by a sol–gel auto combustion method followed by solution casting. BFO has an edge over other ceramics as it is a multiferroic material above room temperature where the coexistence of ferroelectricity and antiferromagnetism is observed [81].

3.4 Conclusion

Piezoelectric energy-harvesting devices have gained huge attention among the scientific community because of their capability to harvest electrical energy even from tiny body movements. The main concerns regarding the development piezoelectric energy-harvesting devices are piezoelectric performance, flexibility, and biocompatibility of the materials. In this scenario polymer/metal oxide-based energy-harvesting systems with better piezoresponse along with better flexibility were widely preferred over bare polymer-based and metal oxide-based energy-harvesting devices. The output performance of polymer-based nanogenerators or energy-harvesting devices is very weak, when compared to the metal oxide-based systems. On the other hand, metal oxide-based energy-harvesting devices possess better output performance but it is brittle, so that it is difficult to apply a large force on that material. For the fabrication of energy-harvesting devices polymer systems like PVDF, PVDF-TRFE, PVDF-HFP, Odd Nylons, PLLA, and cellulose were preferred. Ceramics like $BaTiO_3$, PZT, ZnO, and $LiNbO_3$ were also added to the polymer matrix in order to tune the piezoelectric output. These piezoelectric energy-harvesting devices can be used for powering smart electronic devices and for the development of actuators and sensors.

References

[1] S.R. Anton, H.A. Sodano, A review of power harvesting using piezoelectric materials (2003–2006), Smart Mater. Struct. 16 (3) (2007). p.R1.

[2] H.A. Sodano, D.J. Inman, G. Park, A review of power harvesting from vibration using piezoelectric materials, Shock Vibrat. Digest 36 (3) (2004) 197–206.

[3] A. Mayeen, N. Kalarikkal, Development of ceramic-controlled piezoelectric devices for biomedical, in: Fundamental Biomaterials: Ceramics, 47, 2018.

[4] A. Mayeen, N. Kalarikkal, Piezoelectric polymer nanocomposites for energy-scavenging applications, in: Polymeric and Nanostructured Materials: Synthesis, Properties, and Advanced Applications, pp: 273, 2018.

[5] J. Rödel, W. Jo, K.T. Seifert, E.M. Anton, T. Granzow, D. Damjanovic, Perspective on the development of lead-free piezoceramics, J. Am. Ceram. Soc. 92 (6) (2009) 1153–1177.

[6] Z. Suo, C.M. Kuo, D.M. Barnett, J.R. Willis, Fracture mechanics for piezoelectric ceramics, J. Mech. Phys. Solids 40 (4) (1992) 739–765.

[7] S.K. Ghosh, T.K. Sinha, B. Mahanty, D. Mandal, Self-poled efficient flexible "Ferroelectretic" nanogenerator: a new class of piezoelectric energy harvester, Energy Technol. 3 (12) (2015) 1190–1197.

[8] A. Mayeen, M.S. Kala, M.S. Jayalakshmy, S. Thomas, D. Rouxel, J. Philip, et al., Dopamine functionalization of BaTiO 3: an effective strategy for the enhancement of electrical, magnetoelectric and thermal properties of $BaTiO_3$-PVDF-TrFE nanocomposites, Dalton Trans. 47 (6) (2018) 2039–2051.

[9] Y.F. Su, R.R. Kotian, N. Lu, Energy harvesting potential of bendable concrete using polymer based piezoelectric generator, Compos. B Eng. 153 (2018) 124–129.

[10] S. Jana, S. Garain, S.K. Ghosh, S. Sen, D. Mandal, The preparation of γ-crystalline non-electrically poled photoluminescant ZnO−PVDF nanocomposite film for wearable nanogenerators, Nanotechnology 27 (44) (2016) 445403.
[11] A.C. Wang, C. Wu, D. Pisignano, Z.L. Wang, L. Persano, Polymer nanogenerators: Opportunities and challenges for large-scale applications, J. Appl. Polym. Sci. 135 (24) (2018) 45674.
[12] S. Crossley, R.A. Whiter, S. Kar-Narayan, Polymer-based nanopiezoelectric generators for energy harvesting applications, Mater. Sci. Technol. 30 (13) (2014) 1613−1624.
[13] Y. Yu, H. Sun, H. Orbay, F. Chen, C.G. England, W. Cai, et al., Biocompatibility and in vivo operation of implantable mesoporous PVDF-based nanogenerators, Nano Energy 27 (2016) 275−281.
[14] F. Hu, Q. Cai, F. Liao, M. Shao, S.T. Lee, Recent advancements in nanogenerators for energy harvesting, Small 11 (42) (2015) 5611−5628.
[15] J. Ryu, S. Eom, P. Li, C.H. Liow, S. Hong, Ferroelectric polymer PVDF-based nanogenerator, in: *Nanogenerator*. IntechOpen, 2019.
[16] C. Lee, J.A. Tarbutton, Electric poling-assisted additive manufacturing process for PVDF polymer-based piezoelectric device applications, Smart Mater. Struct. 23 (9) (2014) 095044.
[17] B. Bera, M.D. Sarkar, PVDF based piezoelectric nanogenerator as a new kind of device for generating power from renewable resources, IOSR J. Polym. Text. Eng. (IOSR-JPTE) 4 (2) (2017) 01−05.
[18] B. Zaarour, L. Zhu, X. Jin, Controlling the surface structure, mechanical properties, crystallinity, and piezoelectric properties of electrospun PVDF nanofibers by maneuvering molecular weight, Soft Mater. 17 (2) (2019) 181−189.
[19] T. Tansel, High beta-phase processing of polyvinylidenefluoride for pyroelectric applications, J. Polym. Res. 27 (4) (2020) 1−5.
[20] F. Liu, N.A. Hashim, Y. Liu, M.M. Abed, K. Li, Progress in the production and modification of PVDF membranes, J. Memb. Sci. 375 (1−2) (2011) 1−27.
[21] C. Chang, V.H. Tran, J. Wang, Y.K. Fuh, L. Lin, Direct-write piezoelectric polymeric nanogenerator with high energy conversion efficiency, Nano Lett. 10 (2) (2010) 726−731.
[22] F. Oliveira, Y. Leterrier, J.A. Månson, O. Sereda, A. Neels, A. Dommann, et al., Process influences on the structure, piezoelectric, and gas-barrier properties of PVDF-TrFE copolymer, J. Polym. Sci. B Polym. Phys. 52 (7) (2014) 496−506.
[23] E. Fukada, New piezoelectric polymers, Jpn. J. Appl. Phys. 37 (5S) (1998) 2775.
[24] T. Sekine, R. Sugano, T. Tashiro, J. Sato, Y. Takeda, H. Matsui, et al., Fully printed wearable vital sensor for human pulse rate monitoring using ferroelectric polymer, Sci. Rep. 8 (1) (2018) 1−10.
[25] C. Zhao, J. Niu, Y. Zhang, C. Li, P. Hu, Coaxially aligned MWCNTs improve performance of electrospun P (VDF-TrFE)-based fibrous membrane applied in wearable piezoelectric nanogenerator, Compos. B Eng. 178 (2019) 107447.
[26] P. Martins, A.C. Lopes, S. Lanceros-Mendez, Electroactive phases of poly (vinylidene fluoride): determination, processing and applications, Prog. Polym. Sci. 39 (4) (2014) 683−706.
[27] T. Liu, Z. Chang, Y. Yin, K. Chen, Y. Zhang, X. Zhang, The PVDF-HFP gel polymer electrolyte for Li-O2 battery, Solid State Ionics 318 (2018) 88−94.
[28] L. Yang, Q. Zhao, Y. Hou, L. Hong, H. Ji, L. Xu, et al., Flexible polyvinylidene fluoride based nanocomposites with high and stable piezoelectric performance over a wide temperature range utilizing the strong multi-interface effect, Compos. Sci. Technol. 174 (2019) 33−41.
[29] H. Li, W. Zhang, Q. Ding, X. Jin, Q. Ke, Z. Li, et al., Facile strategy for fabrication of flexible, breathable, and washable piezoelectric sensors via welding of nanofibers with multiwalled carbon nanotubes (MWCNTs), ACS Appl. Mater. Interfaces 11 (41) (2019) 38023−38030.

[30] S. Liu, C. Hu, F. Li, X.J. Li, W. Cui, C. Fan, Prevention of peritendinous adhesions with electrospun ibuprofen-loaded poly (L-lactic acid)-polyethylene glycol fibrous membranes, Tissue Eng. A 19 (3–4) (2013) 529–537.
[31] S. Mishra, L. Unnikrishnan, S.K. Nayak, S. Mohanty, Advances in piezoelectric polymer composites for energy harvesting applications: a systematic review, Macromol. Mater. Eng. 304 (1) (2019) 1800463.
[32] M. Smith, Y. Calahorra, Q. Jing, S. Kar-Narayan, Direct observation of shear piezoelectricity in poly-l-lactic acid nanowires, APL Mater. 5 (7) (2017) 074105.
[33] G. Zhao, B. Huang, J. Zhang, A. Wang, K. Ren, Z.L. Wang, Electrospun poly (l-lactic acid) nanofibers for nanogenerator and diagnostic sensor applications, Macromol. Mater. Eng. 302 (5) (2017) 1600476.
[34] S.J. Lee, A.P. Arun, K.J. Kim, Piezoelectric properties of electrospun poly (l-lactic acid) nanofiber web, Mater. Lett. 148 (2015) 58–62.
[35] B.Z. Mei, J.I. Scheinbeim, B.A. Newman, The ferroelectric behavior of odd-numbered nylons, Ferroelectrics 144 (1) (1993) 51–60.
[36] B.A. Newman, P. Chen, K.D. Pae, J.I. Scheinbeim, Piezoelectricity in nylon 11, J. Appl. Phys. 51 (10) (1980) 5161–5164.
[37] Y. Takase, J.W. Lee, J.I. Scheinbeim, B.A. Newman, High-temperature characteristics of nylon-11 and nylon-7 piezoelectrics, Macromolecules 24 (25) (1991) 6644–6652.
[38] A. Datta, Y.S. Choi, E. Chalmers, C. Ou, S. Kar-Narayan, Piezoelectric nylon-11 nanowire arrays grown by template wetting for vibrational energy harvesting applications, Adv. Funct. Mater. 27 (2) (2017) 1604262.
[39] F. Ram, S. Radhakrishnan, T. Ambone, K. Shanmuganathan, Highly flexible mechanical energy harvester based on nylon 11 ferroelectric nanocomposites, ACS Appl. Polym. Mater. 1 (8) (2019) 1998–2005.
[40] H.S. Kim, Y. Li, J. Kim, Electro-mechanical behavior and direct piezoelectricity of cellulose electro-active paper, Sens. Actuator A Phys. 147 (1) (2008) 304–309.
[41] L. Csoka, I.C. Hoeger, O.J. Rojas, I. Peszlen, J.J. Pawlak, P.N. Peralta, Piezoelectric effect of cellulose nanocrystals thin films, ACS Macro Lett. 1 (7) (2012) 867–870.
[42] S. Rajala, T. Siponkoski, E. Sarlin, M. Mettänen, M. Vuoriluoto, A. Pammo, et al., Cellulose nanofibril film as a piezoelectric sensor material, ACS Appl. Mater. Interfaces 8 (24) (2016) 15607–15614.
[43] M.M. Alam, D. Mandal, Native cellulose microfiber-based hybrid piezoelectric generator for mechanical energy harvesting utility, ACS Appl. Mater. Interfaces 8 (3) (2016) 1555–1558.
[44] S. Bairagi, S. Ghosh, S.W. Ali, A fully sustainable, self-poled, bio-waste based piezoelectric nanogenerator: electricity generation from pomelo fruit membrane, Sci. Rep. 10 (1) (2020) 1–13.
[45] M. Mousavi, Effect of gelatin on structural properties of lead-free piezoelectric $(K_{0.5}Na_{0.5})NbO_3$ nanopowders synthesized by a modified sol-gel route in gelatin media, Iran. J. Crystallogr. Mineral. 25 (3) (2017) 647–654.
[46] H. Ueda, E. Fukada, Piezoelectric relaxations in collagen gelatin systems, Polym. J. 11 (2) (1979) 103–111.
[47] Y. Han, Y. Han, X. Zhang, L. Li, C. Zhang, J. Liu, et al., Fish gelatin based triboelectric nanogenerator for harvesting biomechanical energy and self-powered sensing of human physiological signals, ACS Appl. Mater. Interf. 12 (14) (2020) 16442–16450.
[48] H. Yuan, T. Lei, Y. Qin, R. Yang, Flexible electronic skins based on piezoelectric nanogenerators and piezotronics, Nano Energy 59 (2019) 84–90.
[49] S.K. Ghosh, P. Adhikary, S. Jana, A. Biswas, V. Sencadas, S.D. Gupta, et al., Electrospun gelatin nanofiber based self-powered bio-e-skin for health care monitoring, Nano Energy 36 (2017) 166–175.

[50] A. Khan, Z. Abas, H. Soo Kim, I.-K. Oh, Piezoelectric thin films: an integrated review of transducers and energy harvesting, Smart Mater. Struct. 25 (5) (2016) 053002.
[51] K.S. Ramadan, D. Sameoto, S. Evoy, A review of piezoelectric polymers as functional materials for electromechanical transducers, Smart Mater. Struct. 23 (3) (2014) 033001.
[52] H. Li, C. Tian, Z.D. Deng, Energy harvesting from low frequency applications using piezoelectric materials, Appl. Phys. Rev. 1 (4) (2014) 041301.
[53] B. Hilczer, J. Kułek, E. Markiewicz, M. Kosec, B. Malič, Dielectric relaxation in ferroelectric PZT–PVDF nanocomposites, J. Non-Crystall. Solids 305 (1–3) (2002) 167–173.
[54] A.K. Zak, W.C. Gan, W.A. Majid, M. Darroudi, T.S. Velayutham, Experimental and theoretical dielectric studies of PVDF/PZT nanocomposite thin films, Ceram. Int. 37 (5) (2011) 1653–1660.
[55] S. Revathi, L.J. Kennedy, S.K. Basha, R. Padmanabhan, Synthesis, structural, optical and dielectric properties of nanostructured 0–3 PZT/PVDF composite films, J. Nanosci. Nanotechnol. 18 (7) (2018) 4953–4962.
[56] S.K. Pradhan, A. Kumar, A.N. Sinha, P. Kour, R. Pandey, P. Kumar, et al., Study of ferroelectric properties on PVDF-PZT nanocomposite, Ferroelectrics 516 (1) (2017) 18–27.
[57] S.H. Wankhade, S. Tiwari, A. Gaur, P. Maiti, PVDF–PZT nanohybrid based nanogenerator for energy harvesting applications, Energy Rep. 6 (2020) 358–364.
[58] Y. Zhang, Y. Zhang, X. Xue, C. Cui, B. He, Y. Nie, et al., PVDF–PZT nanocomposite film based self-charging power cell, Nanotechnology 25 (10) (2014) 105401.
[59] J.S. Yun, C.K. Park, Y.H. Jeong, J.H. Cho, J.H. Paik, S.H. Yoon, et al., The fabrication and characterization of piezoelectric PZT/PVDF electrospun nanofiber composites, Nanomater. Nanotechnol. 6 (2016) 20.
[60] N. Chamankar, R. Khajavi, A.A. Yousefi, A. Rashidi, F. Golestanifard, A flexible piezoelectric pressure sensor based on PVDF nanocomposite fibers doped with PZT particles for energy harvesting applications, Ceram. Int. 46 (2020) 19669–19681.
[61] M. Koç, L. Paralı, O. Şan, Fabrication and vibrational energy harvesting characterization of flexible piezoelectric nanogenerator (PEN) based on PVDF/PZT, Polym. Test. 90 (2020) 106695.
[62] K. Li, H. Wang, A. Ding, Polyamide-11/lead zirconate titanates—a novel composite material with high dielectric properties, Polym. Sci. Ser. B 52 (7–8) (2010) 438–442.
[63] A. Rjafallah, A. Hajjaji, F. Belhora, A. El Ballouti, S. Touhtouh, D. Guyomar, et al., PZT ceramic particles/polyurethane composites formalism for mechanical energy harvesting, Eur. Phys. J. Appl. Phys. 89 (3) (2020) 30901.
[64] K.T. Arul, M. Ramanjaneyulu, M.R. Rao, Energy harvesting of PZT/PMMA composite flexible films, Curr. Appl. Phys. 19 (4) (2019) 375–380.
[65] Y. Zhao, Q. Liao, G. Zhang, Z. Zhang, Q. Liang, X. Liao, et al., High output piezoelectric nanocomposite generators composed of oriented $BaTiO_3$ NPs@ PVDF, Nano Energy 11 (2015) 719–727.
[66] J. Nunes-Pereira, V. Sencadas, V. Correia, V.F. Cardoso, W. Han, J.G. Rocha, et al., Energy harvesting performance of BaTiO3/poly (vinylidene fluoride–trifluoroethylene) spin coated nanocomposites, Compos. B Eng. 72 (2015) 130–136.
[67] S. Siddiqui, D.I. Kim, M.T. Nguyen, S. Muhammad, W.S. Yoon, N.E. Lee, High-performance flexible lead-free nanocomposite piezoelectric nanogenerator for biomechanical energy harvesting and storage, Nano Energy 15 (2015) 177–185.
[68] U. Yaqoob, G.S. Chung, Effect of reduced graphene oxide on the energy harvesting performance of P (VDF-TrFE)-$BaTiO_3$ nanocomposite devices, Smart Mater. Struct. 26 (9) (2017) 095060.
[69] B. Dudem, D.H. Kim, L.K. Bharat, J.S. Yu, Highly-flexible piezoelectric nanogenerators with silver nanowires and barium titanate embedded composite films for mechanical energy harvesting, Appl. Energy 230 (2018) 865–874.

[70] D. Ponnamma, M.A.A. Al-Maadeed, Influence of $BaTiO_3$/white graphene filler synergy on the energy harvesting performance of a piezoelectric polymer nanocomposite, Sustain. Energy Fuels 3 (3) (2019) 774–785.
[71] S.C. Karumuthil, K. Singh, U. Valiyaneerilakkal, J. Akhtar, S. Varghese, Fabrication of poly (vinylidene fluoride-trifluoroethylene)–Zinc oxide based piezoelectric pressure sensor, Sensors Actuat. A Phys. 303 (2020) 111677.
[72] A. Anand, M.C. Bhatnagar, Role of vertically aligned and randomly placed zinc oxide (ZnO) nanorods in PVDF matrix: used for energy harvesting, Mater. Today Energy 13 (2019) 293–301.
[73] B. Sun, X. Li, R. Zhao, H. Ji, J. Qiu, N. Zhang, et al., Electrospun poly (vinylidene fluoride)-zinc oxide hierarchical composite fiber membrane as piezoelectric acousto-electric nanogenerator, J. Mater. Sci. 54 (3) (2019) 2754–2762.
[74] M. Aleksandrova, G. Kolev, Y. Vucheva, I. Pandiev, K. Denishev, Flexible oxide-polymeric composites for piezoelectric energy harvesting, in: Proceedings of the Thirty-first International Conference on Microelectronics (MIEL), 2019, September (pp. 141–144). IEEE.
[75] R. Sahoo, S. Mishra, L. Unnikrishnan, S. Mohanty, S. Mahapatra, S.K. Nayak, et al., Enhanced dielectric and piezoelectric properties of Fe-doped ZnO/PVDF-TrFE composite films, Mater. Sci. Semicond. Process. 117 (2020) 105173.
[76] M. Dietze, M. Es-Souni, Large area thick films of PVDF-TrFE and relaxor-ceramics for piezo-and pyroelectric applications, Macromol. Mater. Eng. 304 (12) (2019) 1900538.
[77] S. Das, A.K. Biswal, K. Parida, R.N.P. Choudhary, A. Roy, Electrical and mechanical behavior of PMN-PT/CNT based polymer composite film for energy harvesting, Appl. Surf. Sci. 428 (2018) 356–363.
[78] J. Kim, J.H. Ji, D.J. Shin, S. Yoon, Y.H. Ko, K.H. Cho, et al., 2-Dimensional rGO introduced PMN-PT and P (VDF-TrFE) flexible films for enhanced piezoelectric energy harvester, Appl. Surf. Sci. 494 (2019) 1000–1006.
[79] H.B. Kang, C.S. Han, J.C. Pyun, W.H. Ryu, C.-Y. Kang, Y.S. Cho, $(Na,K)NbO_3$ nanoparticle-embedded piezoelectric nanofiber composites for flexible nanogenerators, Compos. Sci. Technol. 111 (2015) 1–8.
[80] R.I. Mahdi, W.C. Gan, N.A. Halim, T.S. Velayutham, W.A. Majid, Ferroelectric and pyroelectric properties of novel lead-free polyvinylidenefluoride-trifluoroethylene–$Bi0.5Na0.5TiO_3$ nanocomposite thin films for sensing applications, Ceram. Int. 41 (10) (2015) 13836–13843.
[81] S. Saba, G.M. Mustafa, M. Saleem, S.M. Ramay, S. Atiq, Ferroelectric polymer/ceramic nanocomposites with low energy losses, Polym. Compos. 41 (8) (2020) 3271–3281.

CHAPTER

4

2D materials–polymer composites for developing piezoelectric energy-harvesting devices

Shivam Tiwari and Pralay Maiti

School of Materials Science and Technology, Indian Institute of Technology (BHU), Varanasi, India

4.1 Introduction

The ever increasing energy demand and the limited supply of energy to fulfill these demands do not go hand in hand. The depletion of the conventional energy sources leads to the need to refine an alternative method which could be sustainable and has the potential to meet energy demands. The advancement in every sector just leads to increased consumption of present energy sources. According to the world energy statistics for the year 2015, around 62% of the energy demands are fulfilled by the conventional sources, that is, fossil fuels. The dependency on the renewable sources is around 7%. From the data it is obvious that the world's energy demands are still dependent on nonrenewable sources [1]. Also the rising population of the world just makes the situation of energy demand even worse. Hence to overcome the dependency on the fossil fuels for our energy needs, alternative sources must be the pressing need of the hour which could be effective and sustainable for the coming generations. The shortage of energy supply for everyone is not the only concern; the adverse effects that befall the environment due to the continuous use of these fossil fuel-based energy sources is also a major apprehension. The rise in carbon dioxide and carbon monoxide in the last few years has led to an unexpected rise in global temperature and it will continue to rise if necessary

Engineered Polymer Nanocomposites for Energy Harvesting Applications
DOI: https://doi.org/10.1016/B978-0-12-824155-4.00001-6

replacements are not utilized. Hence the current world scenario is that there is an depleting nonrenewable source and rapidly increasing global warming due to the greenhouse gases. Certain alternative sources like solar, wind, geothermal, and hydroenergy have been continuously improved to overcome the demands for conventional energy sources and reduce the global pollution level. But these energy sources are dependent on the natural proforma and hence being totally dependent on them might not solve the energy demands from a wider perspective. Hence all small and large energies which could be fruitful in any manner should be considered and investigated. Extracting energies from waste is an alternative which can be very useful as it does not affect the environment and is economic. There are various types of useful and productive energies which go to waste and thus if such energies get converted and stored they could be of high importance [2,3]. Energy-harvesting (EH) from the waste mechanical sources is one such kind of alternative energy which can be generated from waste sources. Presently, most of the research community is trying to broaden this field as the energy produced is environment-friendly, no harmful by-products are generated, and less effort and maintenance are required.

4.1.1 Energy-harvesting

Energy-scavenging or EH is a way to convert the waste mechanical energy from different sources to useful electrical energy. It is a viable technique where the unwanted or waste energies from different sources are converted to the productive energies which are applicable to small or miniaturized devices. The different sources for such EH can be vibrational, based on human or animal body movements, acoustics, and other similar sources. EH is a process that has the potential to provide battery-less energy which reduces the dependency on batteries and prevents the after effects of used batteries. The other major merits of EH are that it is environmentally safe, requires less maintenance, and also is cost-effective [4,5]. EH can be procured from different sources and harmless means of energy can be extracted from this type of process. The need to create or broaden the renewable sources to meet the future energy demands and reduce the dependency on the conventional sources of energy is at its peak. Several potential renewable sources have been developed like solar energy, wind energy, hydrothermal energy, and vibration-based energy. Although they have higher efficiency they possess certain limitations in terms of continuous availability, not applicable to every location, and systematic distribution. Hence, modifications are continuously made in research and on an industrial scale to overcome those limitations and develop a continuous source of energy for a sustainable future [6,7].

TABLE 4.1 Common mechanical energy-harvesting sources [11].

Nature and human body-based sources	Infrastructure or transportation	Industrial operations
Wind, tides or waves of water, acoustic vibrations, walking, breathing, pulse rate, exercise	Tunnels, bridges, pipelines, vibration from the movement of the transport, aircraft, rail, turbines	Compressors, motors of instruments and machinery, chillers and condensers, boilers

Of the different EH techniques, mechanical EH (MEH) is one with wide applicability and it has the potential to harness viable energies from different mechanical sources which are wasted or unused. Mechanical energies are the most abundant energies because they are associated with each and every thing. Most mechanical energies go unused or wasted. The techniques to harness these unwanted energies are exponentially growing for a sustainable future. Common sources of the wasted energies are the movements of objects or bodies, vibrational sources from water, wind, instruments, human body movements, and stretching or bending of objects. Though the amount of energy generated from this might be low, it has all the potential for small-scale electronics or devices. The ability to store these energies can be fruitful for the longer term and be effective for many applications. Waste mechanical energies are mainly harvested using the vibration to electrical energy unit conversion technique [8–10]. Some common techniques to harvest energies from the mechanical sources are presented in Table 4.1. There are mainly three primary techniques to conduct the vibration-to-energy conversion: piezoelectric, electrostatic, and electromagnetic transductions. Electromagnetic transduction is a process where the output voltage is produced by the combined effect of the conductor and magnetic field. The electrostatic phenomenon is the process where an external voltage is applied to activate the relative vibratory movements of the capacitor materials which generate the alternating currents [12–18]. Piezoelectric materials are the materials where an external load leads to charge generation which produces an output signal. In piezoelectric materials no such external potential is required to trigger the active materials. Out of the three transduction mechanisms, the piezoelectric process is the most efficient and has been of higher interest among the researchers. The output generated from the electromagnetic and electrostatic mechanisms are of comparatively lower values compared to the piezoelectric ones. The output voltage is easily generated from the piezoelectric materials and has the potential to sense small vibrations for effective output generations [19]. One more added benefit of the piezoelectric materials are their scalability, that is, it is feasible to design the material at both the small and large scale [20].

4.1.2 Piezoelectricity

The term piezoelectricity is originated from the Greek word *"piezen,"* which means to apply pressure or to squeeze [21]. The piezoelectricity phenomenon was discovered by Jacques and Pierre Curie in 1880. The piezoelectric effect is the process where the application of mechanical stress leads to electrical energy generation or the input of electrical energy produces strain in the materials. There are mainly two types of piezoelectric process:

1. Direct piezoelectric effect an external mechanical load is applied onto the material and an external stimuli in the form of electrical energy is obtained. Most of the transducers or sensors are based on these effects [22].

$$D_i = e_{ij}{}^{\sigma}E_j + d_{in}{}^{d}\sigma_n \tag{4.1}$$

where D_i is the dielectric displacement, E_j is the electric field applied, σ_n denotes the mechanical stress vector, $d_{in}{}^{d}$ is the piezoelectric coefficient, and $e_{ij}{}^{\sigma}$ is the dielectric permittivity.

2. Indirect piezoelectric effect: an electrical input is given or applied to the material and an output strain is developed. Actuators and vibration damping are based on this type of piezoelectric effect [22].

$$S_x = d_{yx}{}^{a}E_j + s_{xn}{}^{E}\sigma_n \tag{4.2}$$

The term Sx denotes the mechanical strain vector, dyxa signifies piezoelectric coefficient, and sxnE is elastic compliance matrix [23].

The fundamentals of piezoelectricity can be best discussed using the crystalline solids and the arrangement of the ions within the unit cell. The crystalline structure consists of the equivalent positive and negative charges which remain neutralized along the polar axis. When an external mechanical stress or load is applied, the periodic arrangement within the crystal gets disturbed leading to improper arrangement of the dipoles, which results in the development of a dipole moment and consequently electric charge generation. The symmetry of the crystals is an important aspect in determining the piezoelectric property of the system. There exist about 32 classes of crystallographic units, of which 21 are noncentrosymmetric, and among them 20 exhibit piezoelectricity features. The deficiency of the symmetry among the ions arrangement in the materials leads to the development of permanent dipoles which correspond to the piezo response [3,4,24].

In terms of piezoelectric-based EH, a direct piezoelectric effect is considered. In this phenomenon, the application of external pressure generates electrical charges in the material. The absence of the external load does not allow the flow of charges within the material, and as a result any electrical output is not obtained. The origin of the phenomenon can be related to the nature of the surface charge, which varies accordingly with the mechanical stress applied. The polarization with the dipoles and domains develops surface charge at each surface of the material and the opposite charge gets attracted to the counter surface and the flow of ions or electrons is favored.

The amount of charge (Q_p) produced across the counter surface of the piezoelectric material of area A_p, on application of external stress ($\Delta\sigma$) is expressed as [4,5]:

$$Q_p = d_{33} \times A_p \times \Delta\sigma \tag{4.3}$$

where d_{33} denotes the piezoelectric coefficient (C/N) and the subscript 33 represents the application of load in the direction of the polarization.

When considering the piezoelectric materials, polarization plays an important role. Polarization is a process where the negative and positive charges get separated across the ends of the dielectric material under the influence of mechanical stress. The strength and the direction of the polarization depend mainly on three factors: symmetry of the crystal; orientation of the polarization, and application of mechanical stress on the system. When an external stress is applied on the piezoelectric material, the dipoles tend to align in the same direction providing a better piezoelectric property [25].

There are certain classes of materials which are effectively used as the potential piezoelectric materials. The ability of these materials to respond to the minute external load makes them next-generation materials for a sustainable future. The detailed discussion will be carried out in the next section.

4.1.3 Piezoelectric materials

Piezoelectric materials convey a solid-state conversion between mechanical and electrical energy at the small or large scale. When a mechanical load is applied on to the piezoelectric materials, a change in electric polarization occurs. This change in the polarization leads to a flow of electric current in the external circuit, and thus an output signal in the form of an electromechanical response is obtained. Hence to achieve the electrical signal from the effect of the external load, the device or materials must be able to generate the potential or charge. From the EH point of view, the performance of the material along with the applied mechanical stress is of primary importance. The coupling between the piezoelectric material and the applied mechanical source is the determining factor for efficient EH [26]. The electromechanical coupling factor (k^2) is the parameter which relates to the efficacy of the piezoelectric material, which is given as:

$$k^2 = (\text{generated mechanical energy})/(\text{input electrical energy})$$

Or,

$$k^2 = (\text{generated electrical energy})/(\text{input mechanical energy})$$

Piezoelectric coefficient (d_{xy}) is one of the important terms which explain the ability of the piezoelectric material. It is defined as the obtained polarization per unit of applied mechanical stress or the mechanical strain developed per unit of the applied electrical field. The subscript **x** denotes the direction of obtained polarization or field strength applied and the term **y** represents the direction of the applied external load or the induced strain.

There exists a large number of piezoelectric materials that are being used to harness energies from different sources. Table 4.2 presents the piezoelectric properties in terms of some coefficients values of some common piezoelectric materials.

Based on their properties and origin, there are mainly three classes of piezoelectric materials:

4.1.3.1 *Piezoceramics*

Piezoelectric ceramics have been the prominent materials in the field of piezoelectric EH. These are the inorganic polycrystalline materials

TABLE 4.2 Piezoelectric parameter values for some common piezoelectric materials [4,5].

Piezoelectric Material	d_{33}(pC/N)	ε_r	Electromechanical coupling factor (k_{33})
PZT	593	3400	0.75
PMN-PT	2820	8200	0.94
$BaTiO_3$	149	1200	0.49
ZnO	12.4	11	0.48
Quartz	− 2.3 (d_{31})	4.6	0.1
GaN	3.7	–	–
AlN	5	12	0.23
CdS	10.3	10.3	0.26
$LiNbO_3$	6	28.7	0.23
PVDF	− 33	7.6	0.19

consisting of several single crystals with specific chemical compositions. These materials are hard, stiff, and brittle with very high piezoelectric coefficients and dielectric constants. Piezoceramics consist of both ferroelectric materials, like lead zirconate titanate (PZT), barium titanate ($BaTiO_3$), and nonferroelectric materials, like zinc oxide (ZnO) and aluminum nitride (AlN). Among these materials, PZT possesses the superior properties in terms of piezoelectricity. PZT possesses a very high value of dielectric constant, d_{33} value, and electromechanical coupling factor. But the material is not considered to be suitable in some field of application due to its toxicity. The nonferroelectric materials are also a potential source of piezoelectricity. Though their piezoelectric values are comparatively less compared to their ferroelectric counterparts, these materials are suitable for semiconducting applications and biochemical sensors because of their high selectivity and sensitivity [10,27–31] (Fig. 4.1).

4.1.3.2 Piezo single crystals

The single crystal material consists of the periodic arrangement of the positive and negative ions over their structure which generates a dipole moment in the material due to the proper dipole alignment. These features of the single crystal lead to high piezoelectric properties and they are used in the field of sensors and actuators. Common single crystal piezomaterials are quartz, lithium tantalite, and lithium niobate. The presence of strong polarization among these materials makes them efficient material for piezoelectric-based applications. Apart from such merits of the piezoelectric single crystals, certain disadvantages also persist like complex fabrication methods and the high cost of synthesis, which limit their use in various applications. Single crystals are generally brittle in nature due to the unavailability of the ceramic grain boundaries. Also the single crystal tends to lose its piezoelectric properties in a higher electric field environment [3,32].

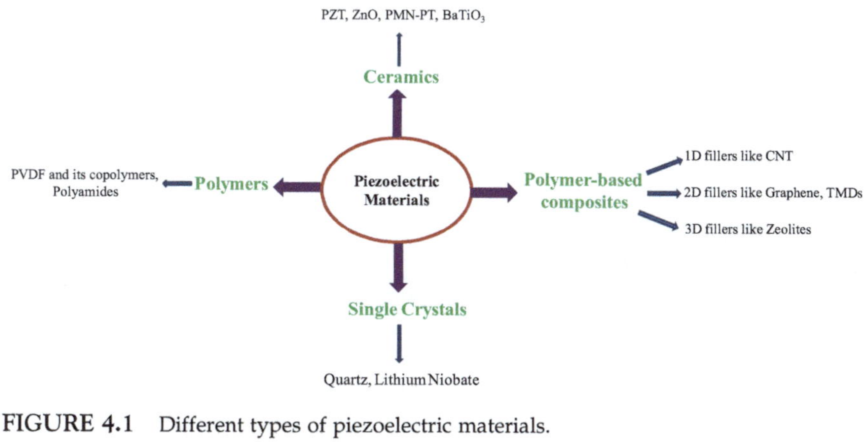

FIGURE 4.1 Different types of piezoelectric materials.

4.1.3.3 *Piezopolymers*

Piezoelectric polymers are organic compounds which consist of long-range order and high molecular weight. These are the materials with superior flexibility and toughness compared to other classes. Common polymers which are used for piezoelectric-based EH are poly(vinylidene difluoride) (PVDF) and its copolymers like PVDF–trifluoroethylene (PVDF-TrFE), PVDF–hexafluoropropylene (PVDF-HFP), and PVDF–chlorotrifluoroethylene (PVDF-CTFE), cellulose, polylactic acids (PLA), polyamides (PA), and polyurethanes (PU). The polymer-based EH has been a profound area of research due to their low density, cost-effectiveness, ease of processing, and better mechanical properties [3,23]. A detailed discussion of the piezopolymers will be given later in the chapter.

4.1.4 Composites

Composites may be defined as the mixture of two or more components with the same or different properties that combine together to provide improved physical and chemical properties of the system. The primary intention of this is to develop a material which incorporates the finest features of the individual constituents to provide sustainable and modified properties best suited for corresponding applications. Composite materials provide better mechanical, electrical, and thermal properties, along with low density and corrosion resistance. The composite finds its application in several fields like packaging materials, infrastructure, medical, aircraft, and many others. The composites consist of mainly two parts: matrix and reinforcer or filler. The matrix is the primary phase which acts as the base for the dispersed filler phase and provides stability to the material. The reinforcer phase is the part which remains dispersed in the matrix and provides strength to the system. Basically there are three kinds of composites based on the type of the matrix: metal–matrix composites, ceramic–matrix composites, and polymer–matrix composites. Polymer-based composites will be the primary discussion here. In the polymer composites, the polymer matrix is used for its flexibility and better toughness properties which add to the merits like abrasion resistance, chemical resistance, better impact properties, and corrosion resistance [33–35].

4.1.4.1 *Polymer-based composites*

Polymer composites are made of a polymer matrix with different fillers, which can be organic or inorganic depending upon the mode of applications. Polymer-based composites find use in different areas like EH, biomedical applications, fuel cells, structural applications, and many more [33,36–41]. The characteristic features of the polymer composites depend upon the nature of the reinforcer, its compatibility with the matrix, aspect ratio, and size of the filler [42,43]. The dimension of the phases plays an important role in attaining the specific properties for some applications. In the polymer-based composites, when the dimension of one of the components is in the nanometer scale, it is termed as a polymer nanocomposite. The nanocomposites have gained much wider attention due to their high surface to volume ratio and better electrical, mechanical, and thermal properties [6,7,11]. There are different types of nanocomposites based on the dimension of the filler incorporated into the polymer matrix, such as zero-dimensional (0D) like quantum dots; one-dimensional (1D) such as carbon nanotubes (CNT); two-dimensional (2D) reinforcers like graphene and its derivative, nanoclay; and three-dimensional (3D) like zeolites. Of the different dimensional fillers, 2D materials have attracted much attention due to their high surface area, better intercalation properties, and enhanced electrical and mechanical features. The detailed discussion will be carried out in the next unit.

4.1.5 2-dimensional materials

The development of the single-layer graphene by Novoselov et al. in 2004 led the scientific community to think about the use of the 2D materials for different technological and research advancements. Now the annals of the 2D materials have enormously expanded over a wide range to include the conductors, insulators, and semiconductors. The 2D materials possess some exceptional features like electrical, thermal, mechanical, and optical properties which provide an added advantage over their counterparts and thus are used in varying applications. EH is one of the premier fields where the incorporation of 2D materials has been preferred. The noncentrosymmetric 2D materials possess a greater advantage in electromechanical applications like piezoelectric-based EH [44]. In the 2D materials, there exists a van der Waals superlattice which arises from the dangling bond-free surface of the atomic layers, which allows the stacking integration of the 2D materials, avoiding the matching of lattice constraints. This property leads to the versatile applicability of the material in numerous fields. The crystal structure of the 2D materials is simpler in comparison with 1D and 3D material, which facilitates the better property–structure interaction in piezoelectricity [45].

The preparation of 2D materials is generally carried out in two major processes: exfoliation and vapor phase deposition. The exfoliation process is a top-down approach where the thin materials are developed from the bulk crystals. Mechanical exfoliation and chemical exfoliation are two important classes which are used to prepare these materials. The prepared materials from the mechanical exfoliation methods yield better crystallinity and neat surfaces which are applicable for many applications. Though the materials extracted from the chemical exfoliation process have the disadvantage of being degraded due to the change in their crystal lattice arrangement, this process produces better material compared to its counterparts [46]. Graphene, boron nitride, and molybdenum disulfide are some of the 2D materials that are prepared using the exfoliation process [47–49].

Vapor phase deposition is a method to prepare atomically thin films or flakes onto several layers on the required substrates. This process has mainly two approaches: chemical vapor deposition (CVD) and physical vapor deposition (PVD). The CVD process involves a chemical route to develop the material over some substrates. The 2D-materials developed from this process are more promising due to their scalability and controlled thickness features, which leads to it being a suitable process in electronic industries. PVD is the technique to develop 2D material using the recrystallization of the material through the solid–vapor process. In_2Se_3 is one of the 2D materials prepared using this process [50].

As the science progresses, the advancement of the materials at atomic and molecular level is following an upward trend. One such area is the piezoelectricity in the 2D materials which is progressively advancing due to the availability of the noncentrosymmetric point groups available in these materials [51,52]. Thus there exist different 2D materials which are used in piezoelectric-based EH applications and these will be discussed later in the chapter.

4.2 Role of 2-dimensional materials in polymer composites for piezoelectric-based energy-harvesting devices

Polymer-based piezoelectric materials have attracted wide attention in scientific research in different applications because of their flexibility, scalability, low density, biocompatibility, and better mechanical properties [53]. Polymer has the advantage of being able to be processed into fibers, films, foams, and coatings, which has led to its use in diverse applications like EH, sensors, storage devices, and biomedicine. To attain certain advanced purposes, the polymers are used in a matrix with a reinforcer to form polymer-based composites or nanocomposites, as discussed previously. For piezoelectric-based EH applications, the electroactive polymers find much wider attention because of their higher polarization. The electroactivity in these polymers mainly depends on their macromolecular structure which corresponds to their dipole orientation, crystalline structure, chain conformation, and processing parameters [54,55]. Table 4.3 provides the common piezoelectric polymers and their piezoelectric parameter values.

TABLE 4.3 Piezoelectric polymers and their corresponding d_{33} and dielectric constant values [55–60].

Polymer	d_{33}(pC/N)	ε_r
PLA	9.82	3–4
Polyamide-11	4	5
PVDF	− 24 to −33	6–12
P(VDF-TrFE)	− 38	18
P(VDF-HFP)	− 24	11
P(VDF-CTFE)	140	13
P(VDF-TrFE-CTFE)	–	65
Cellulose (wood)	0.1	–
Chitin (lobster apodeme)	1.5	–
Collagen (bone)	0.2	–
Polyacrylonitrile (PAN)	2	38
Polyurethane	–	6.8

Some commonly used piezoelectric polymers for EH are as follows:

PVDF and its copolymers: PVDF has emerged as one of the finest and most efficient piezopolymers. It has been a better replacement for many ceramic materials that are toxic, brittle, and more density. PVDF is a semicrystalline polymer composed of $-CH_2$-CF_2- units. The existence of fluorine atoms induces dipole moments in the polymer due to the higher electronegativity of the fluorine atoms [55]. PVDF consists primarily of five phases: α, β, γ, δ, and ε, based on their molecular arrangements. PVDF mainly possesses α phase, which is hexagonal in nature and where the molecules are arranged in TGTG′ type conformations. The antiparallel arrangement of the molecules makes the α phase nonpolar in nature [61]. The β phase consists of the all-trans conformations and has an orthorhombic crystal structure and possesses a net dipole moment which makes it polar in nature, and it is the most valuable phase in terms of piezoelectric applications. Semipolar γ phase has the T_3GT_3G' type conformations and has better electroactive properties than the α phase [62]. Although PVDF has a major part of α phase, it is of electroactive importance because it possesses the ability to get transformed from the nonpolar phase to the polar phase through some processing techniques. The transition of the nonpiezophase to the piezoactive phase is termed as Curie transition, which defines its potential in piezoelectric-based EH applications [4]. Certain methods have been used to transform the α phase to electroactive phases, such as copolymerization, mechanical stretching, poling, electrospinning, and the addition of fillers. The copolymerization method is an efficient way to tune the piezoelectric properties of PVDF. The addition of comonomer leads to a polar nature in the polymer, which facilitates its electroactive phases. Some of the copolymers of PVDF are PVDF-TrFE, PVDF–hexafluoropropylene (PVDF-HFP), and PVDF-CTFE. The main benefit of the copolymers are that they possess the β phase without any processing requirements, which is due to the steric hindrance of the added comonomer that leads to a reduction in activation energy in phase transformation and expands the interchain distance in the unit [55,63]. Mechanical stretching is an effective way to enhance the piezoelectric phase in the PVDF. The stretching leads to the alignment of the chain which results in the net dipole moment in the polymer and thus an increase in β-phase content is obtained. Poling is a process where the material is subjected to an electric field at higher temperature, which leads to the orientation of the polar axis and the nonpolar phase gets converted to the polar phase with a net dipole moment due to the aligned dipoles [4]. Poling develops an intrinsic polarization in the material that leads to the free movement of the molecules and thus gets aligned with respect to the electric field applied and heat. The dipoles remain oriented after the removal of the field and the lowering of the temperature which induces a piezoelectric behavior [23]. Gaur et al. processed the PVDF via stretching and poling to attain a maximum β phase content of ~75% [64]. Electrospinning is a promising method to achieve higher electroactive phases for better EH applications. In this technique, the material is subjected to higher potential which leads to the stretching and poling of the material that results in the self-alignment of the dipoles. The quality of the electrospun fibers can be tuned by varying the processing parameters like potential applied, flow rate, solvent concentration, and spinneret to collector distance. Andrew et al. [65] reported the increase in electroactive phase with an increase in the potential due to the better poling of the polymer. Lei et al. [66] used a different ratio of solvents (NMP and acetone) to prepare electrospun PVDF fibers and found that the maximum piezo-phase was ~95% at 5/5 vol./vol. ratio of solvents. The addition of different fillers also enhances the polar phase in the polymer. Certain electroactive fillers like ceramic fillers, biobased reinforcer, metal salts, and carbon-based materials are incorporated into the polymer matrix to transform the nonpolar part to a piezoactive phase. The addition of the fillers or nanofillers to the PVDF leads to enhanced piezoelectric properties like an increase in output voltage and current, and a rise in d_{33} and dielectric constant. Siddiqui et al. prepared a composite of PVDF-TrFE with barium titanate (BT) and it was seen that there was a rise in output voltage with BT loading compared with the pristine polymer [67]. Yu et al. [68] prepared electrospun nanofibers of PVDF-MWCNT and concluded that the addition of MWCNT led to a rise in output voltage to 5 V at 5% loading of filler. Kumar et al. [69] prepared a hybrid of PVDF and fish scales and the piezoelectric nature of fish scales enhanced the electroactive properties of the material, which was evident from the rise in output voltage and power density. The applicability and efficacy of PVDF and its copolymers has allowed tremendous growth in the field of EH due to its phase transformation ability and higher polarization values.

PA: The ability to generate piezoelectric behavior in nylon-11 has added one more feather to the cap of piezoelectric polymers [70]. PA exist as odd- and even-numbered polymers where the odd-numbered polymers, like nylon-5 or nylon-11, have piezoelectric behavior because of their effective dipole moment. In the case of even-numbered PA like nylon-6 the net dipole moment is zero due to the canceling of the dipoles [71]. Takase and coworkers found a change in values of piezoelectric stain coefficient with the change in nylon configuration (d_{31} for nylon-11 was around 14 pC/N and for Nylon -7 it was around 17 pC/N) [72]. PA have the potential to be effective piezoelectric polymers in spite of their lower piezoelectric constant (as given in Table 4.3). Hence work is being carried out to enhance its electroactive properties.

PLA: PLA is becoming one of the promising candidates for piezoelectric EH due to the presence of dipole moments in the moiety, which is attributed to the presence of the carbonyl group available in the polymeric chain. Apart from piezoelectric properties, PLA possesses good mechanical properties, biocompatibility, and scalability which enhances its merit for applications [73–75]. Varga and team prepared a composite of PLA and barium titanate via electrospinning and a piezoelectric coefficient was found to be around 0.5 nC/N, which explains its piezoelectric behavior [76].

PU is a polymer made of urethane groups, which are made from the monomers consisting of −NCO and −OH groups in a long polymeric chain. They are available in both thermosetting and thermoplastic forms. The isotropic form of PU possesses transverse piezoelectricity related to the bias DC potential and electrostriction. PU film shows a piezoelectric constant value of 38 pC/N at a DC bias field of 5MV per minute [77]. Sakamoto et al. [78] prepared a composite with PU and PZT along with carbon filler to analyze its electroactive properties. It was reported that the addition of carbon filler led to a rise in the piezoelectric coefficient value (d_{33}) to 25% as compared to the composite prepared in the absence of carbon filler.

Cellulose: Cellulose is a natural polymer obtained mainly from the plant's cell wall which is biocompatible and biodegradable in nature [79].

The presence of polar groups leads to a net dipole moment in the polymer which makes it a suitable material in the field of EH. Cellulose finds use in various applications like sensors, piezoelectric devices, and actuators. Cellulose has the advantage of being used in various ways, like nanofibrils and nanocrystals, which enhance its mode of applicability. Cellulosed-based electroactive papers are a new addition in the field of EH. These cellulose-based electroactive papers are prepared from the metal-electrode deposition on both sides of the film/paper, which act as a promising material for electromechanical actuators. These materials have advantages, like being low density, cheaper, and biocompatible, that facilitate their usage in various applications [23,80,81]. Rajala et al. [82] developed self-supporting films using the cellulose nanofibrils and measured their piezoelectric properties to find dielectric permittivity of 3.47 and 3.38 at 1 KHz and 9.97 GHz frequency.

4.2.1 Common device configuration for piezoelectric energy-harvesting

The performance of the material is best seen when it is fabricated into the productive and efficient device form which can sense the minimal force applied onto it. For a better piezoelectric based device, the coupling between the input mechanical source and the developed electroactive material becomes an important aspect for better electromechanical response. The output generated depends upon the capability of the piezomaterial to uphold the input load and undergo repeatedly deformation without much failure [5]. The common structural configurations available for the piezoelectric EH devices are shown in Fig. 4.2 [25]. In the cantilever beam type devices, maximum amount of stain is obtained over a mechanical vibration due to its better and simple geometry and ease in application. The device comprises of a piezoelectric active layer bonded to a nonpiezoelectric elastic layer fixed at one end or by joining the both piezoelectric layers in opposite directions. These are often termed as unimorph or bimorph respectively. When an external load is applied on to the cantilever, it produces an oscillation of the device which results in the AC voltage from the piezoelectric layer. This type of devices is best suited for low impact forces [3,23,83].

Cymbal type devices are mainly used for the purpose where high impact pressure is needed. The device is made up of a piezoelectric layer positioned between the two metal cap ends on both sides of the configuration. When an axial pressure is subjected to the metal caps, the axial stress gets converted to radial stress over the piezoelectric layer which increases the charge generation ability and thus its piezoelectric properties gets enhanced [3].

Stack type configuration based devices are made up of multiple layers of the piezoelectric materials placed one over the other in a way

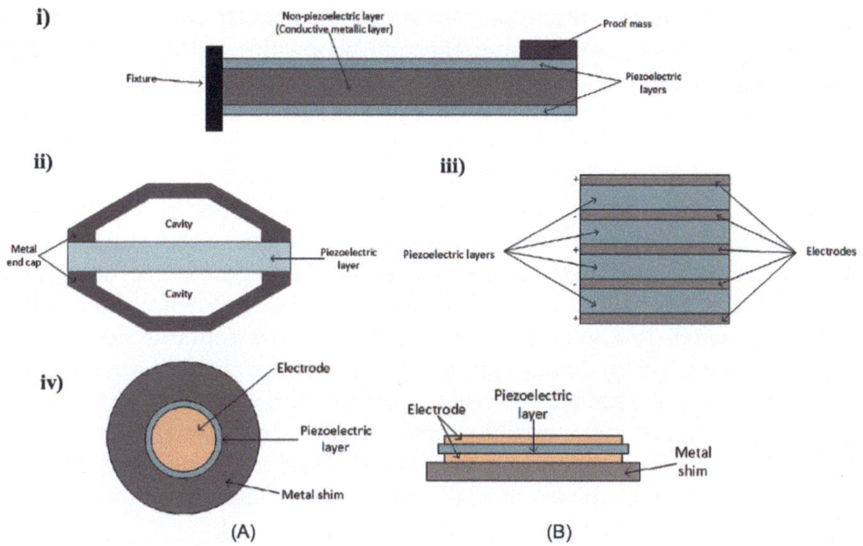

FIGURE 4.2 Different structural configurations for piezoelectric EH: (A) cantilever beam type; (B) cymbal type; (C) stacked type; and (D) circular diaphragm; (i) front view; and (ii) side view [25].

that the poling directions of the piezoelectric layers get aligned with the external applied pressure. The larger stiffness of the prepared stack devices leads to lowering of the mechanical energy generated from the applied stress. To avoid such situations external mechanical amplifiers are attached to the configuration [83].

A circular diaphragm is one of the other types of devices employed for piezoelectric EH applications. It consists of a thin disk-type layered piezoelectric material attached to the metal shim which is held onto a clamp ring. Prestress is preapplied to the device to raise its operational efficiency and to generate better output. These types of devices are capable of harvesting energy from the low-frequency, periodic acoustic waves or pressure fields and vibrations. The structure of the circular diaphragm is stiffer, which results in higher resonating frequency under the vibration type of operation [23].

However, the type of the device structure and configuration is developed and modified based on the mode of the applications needed.

4.2.2 2-dimensional-materials and polymer composites for piezoelectric energy-harvesting devices

The introduction of the 2D materials has led to some remarkable works in different fields of applications. The incorporation of the 2D materials with the polymer has even broadened its characteristic values. The better electrical, optical, and thermal properties of the 2D materials and the high mechanical stability, low density, and biocompatibility features of the polymers have led to some efficient composites that are applicable to several diverse areas. EH is one such field where the 2D materials have made some significant contributions.

Graphene and its derivatives are the most common 2D material explored for wider application. Graphene is a one atom thick planar sheet of densely arranged sp^2 hybridized carbon atoms in a hexagonal crystal lattice. The 2D π to π conjugation and better crystallinity of the graphene makes it a versatile material with some remarkable features [84]. Graphene, being a conductive filler, possesses a high surface area, extraordinary mechanical, thermal, optical, and electrical properties, and a high aspect ratio, which makes it a widely used filler for EH applications. These outstanding properties make it a good candidate for polymer-based applications by enhancing the mechanical, thermal, and electrical properties of the composites [85,86].

PVDF and its copolymers are the most used polymers for the polymer-based composites for piezoelectric EH. When an electroactive filler or reinforcer is added to PVDF or its copolymers, it increases the piezoelectric properties of the materials and also acts as a nucleating agent that leads to stretching of the molecules, enhancing its crystallinity through electrostatic interactions. Conducting fillers like graphene and its derivatives provide better electrostatic interactions that synergize with the polymers resulting in a better piezoelectric material for EH applications. Electrospun nanofibers of PVDF–graphene are produced by Abolhasani and coworkers using an electrospinning process with varying graphene content from 0–5 wt.%. On evaluating the content of electroactive phase formation, 0.1 wt.% graphene and PVDF composite yielded maximum β phase content. As the content of the graphene was increased the electroactive phase gets reduced. A piezoelectric nanogenerator was prepared from the electrospun fiber which generated an output voltage of 7.9 V and current of 4.5 μA for 0.1 wt.% graphene-based PVDF nanohybrid under 0.2 MPa pressure with a constant frequency of 1 Hz. Under similar conditions, pure PVDF fiber produced 3.8 V and 2 μA of output voltage and current, respectively. Also the developed nanogenerator was capable of charging a capacitor and lighting up an LED for around 30 seconds [87]. To enhance the interfacial interactions and better dispersion with the polymers, the graphene is surface modified to achieve higher performance at lower filler loadings. Wang et al. [88] used thermally exfoliated graphene (EG) with different oxygen-containing groups like ozone-EG, PMMA-g-EG, and fluorination-EG, added to PVDF via the solution route. The PMMA-g-EG generated the maximum β phase, which was attributed to the

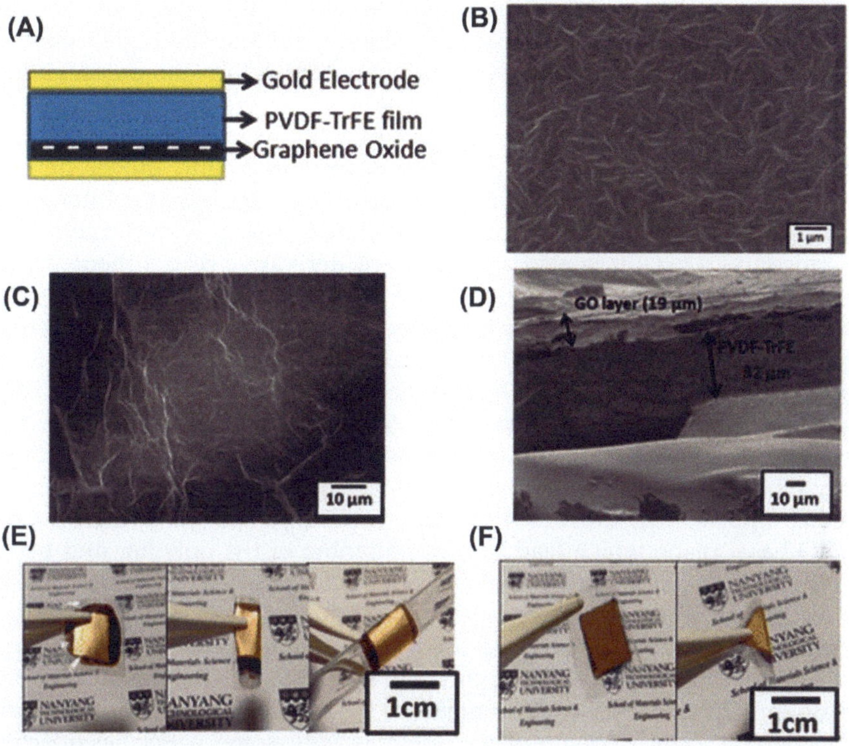

FIGURE 4.3 (A) Diagrammatic representation of the fabricated device from PVDF-TrFE and GO; surface morphology of (B) PVDF-TrFE film and (C) graphene oxide layer. (D) Cross-sectional micrograph of the designed bilayer films; prepared device of (E) bilayer and (F) PVDF-TrFE film. *Reprinted with permission from V. Bhavanasi, V. Kumar, K. Parida, J. Wang, P.S. Lee Enhanced piezoelectric energy harvesting performance of flexible PVDF-TrFE bilayer films with graphene oxide, ACS Appl. Mater. & Interfaces 8 (1) (2016) 521–529 Copyright (2016) American Chemical Society.*

interaction between the CF_2 of PVDF and carbonyl group of PMMA. Also the PVDF and PMMA-g-EG produced the maximum dielectric permittivity at 100 Hz. Graphene oxide (GO) consists of negative charges because of the availability of functional groups, which can be used in the polymers to design piezoelectric-based nanogenerators to harvest energy from different mechanical sources. PVDF-TrFE and GO was used by Bhavanasi et al. to prepare a bilayer film to measure the piezoelectric response of the composite. The device configuration and surface morphology of the prepared film is represented in Fig. 4.3. The effect of poling on the performance for EH was clearly demonstrated. The bilayer film of poled PVDF-TrFE and GO generated a maximum potential of 4.3 V as compared to the nonpoled bilayer film of voltage 0.3 V against a mechanical stress of 0.32 MPa. Hence the poling of the film leads to better alignment of dipoles which results in the higher electromechanical response. The maximum power output for the poled bilayer film was around 4.41 $\mu W/cm^2$ which was better than the pristine polymer PVDF-TrFE film with power density 1.77 $\mu W/cm^2$ at the load resistance of 1 MΩ. The increase in the output voltage and power density was attributed to the combined effect of the electrostatic interaction from the GO and effective polarization from the polymer layers [89]. Reduced GO (RGO) is also an important derivative of graphene acting as a filler in the field of piezoelectric EH. The PVDF-TrFE-RGO-based piezoelectric nanogenerator is fabricated by Habibur et al. [90]. The effect of RGO loading into the polymer matrix was seen without the use of any poling process. The 0.1 wt.% RGO content of the electroactive filler generated the maximum output voltage (2.4 V) and power (3.2 μW). The maximum output current of 0.8 μA was also obtained from the fabricated device with the filler content of 0.1 wt.%. The rise in the performance of the composite was attributed to the electrostatic charges present in the RGO basal plane which lead to better aligning of the polymer and hence enhance the efficiency of the device. Stretchable and flexible devices have attracted much wider attention in the field of EH applications. Highly flexible devices are being used to prepare wearable nanogenerators, biological implantations, and portable electronic devices. However the main limitations in these fabrications are the durability and efficient power generations. Hence polymer-based devices with some better electroactive fillers like 2D materials provide better insight and performance in piezoelectric EH.

A highly flexible and stretchable device is fabricated from PVDF-TrFE, PDMS-CNT composite, and graphene. Here graphene acts as a top electrode and the polymer PVDF-TrFE acts as the piezoelectric–pyroelectric material. A maximum output voltage of 0.7 V was obtained from stretching the device. The device was stable after stretching or compressing, which explains its robustness as a device [91]. Maity et al. [92] prepared a nanostructure composite of polyaniline grafted GO on the PVDF matrix at different variations. The filler content of 5 wt.% resulted in the maximum electroactive phase (91%) formation. The DC conductivity and the dielectric permittivity increased at similar filler content explaining the role of the 2D electroactive filler in EH applications. Graphene nitride ($g\text{-}C_3N_4$) is a 2D piezoelectric material as the graphitic form of carbon nitride. The tri-s-triazine layer of the material consists of intrinsic noncentrosymmetric holes over the structure which leads to in-plane piezoelectricity [93]. Khalifa et al. [94] prepared an electrospun fiber from PVDF and graphitic carbon nitride and it was found that the addition of the carbon-based 2D filler led to a rise in the β phase content. Along with the rise in piezoelectric phase content, the graphene carbon nitride enhanced the mechanical and thermal stability of the fibers. A device was being developed to analyze the EH performance of the electrospun fibers. The nanofiber mat was placed between two interdigitated copper electrodes of specific dimension. The electrode was then covered with polyvinyl chloride sheet which is nonconducting in nature. Different modes of mechanical stress like finger tapping, palm tapping, thumb imparting, and bending were applied on to the fabricated device which resulted in better output performance. A maximum voltage generated from the composite fiber was around 7.5 V which was nearly eight times higher than pure PVDF fibers. The maximum output current and power density achieved from the nanogenerator were 0.23 μA and 0.22 $\mu W/cm^2$, respectively. In other work, a facile approach to prepare highly flexible PVDF and PANI- graphene carbon nitride nanofibers (PPBF) from an electrospinning process was demonstrated. The device was fabricated using copper electrode and PVC sheet and a piezoelectric EH response was observed. Fig. 4.4 shows the different motions of the human body like arm, toe, palm, and heel leading to the generation of output voltage and subsequent LEDs glowing from the fabricated device. The prepared nanogenerator generated output voltage around 30 V and maximum current of 3.7 μA. The device was used to light up commercial LEDs and around 70 LEDs were able

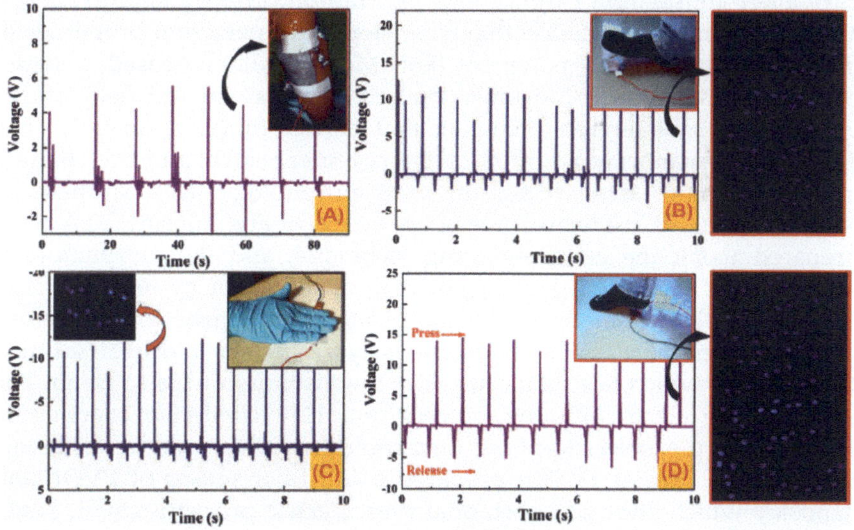

FIGURE 4.4 Output voltage generation from PPBF-7.5 device using different motions of human body (A) arm movement; (B) toe tapping (C) palm, and (D) heel pressure. *Reprinted with permission from M. Khalifa, S. Anandhan, PVDF nanofibers with embedded polyaniline–graphitic carbon nitride nanosheet composites for piezoelectric energy conversion, ACS Appl. Nano Mater. 2 (11) (2019) 7328–7339. Copyright (2019) American Chemical Society.*

to light up on the application of external stress. Hence the effect of the 2D materials on the polymer-based composites was effective and a better piezoelectric response was generated on application of the external pressure [95].

Two-dimensional nanoclay is a finely grained layered crystalline material. The basic structuring of these hydrous silicates or layered silicates involves individual layers of tetrahedral and/or octahedral sheets placed one over the other. Each layer is bonded through van der Waals force, hydrogen bonding, and electrostatic interactions forming a stack or group of lamellae. Nanoclay is generally organically modified to interact better with the polymer matrix forming a template base structure. Organically modified nanoclay or organoclay enhances the electroactive phase formation which is very negligible in the case of unmodified clays. Nanoclays are rigid type fillers which when combined with the soft matrices of the polymers, lead to free movement of the polymeric chain molecules and depending upon the interaction, enhance mechanical properties, thus providing better stability to the composite structure. There are different types of nanoclay like montmorillonite, saponite, hectorite, laponite, and halloysite. The surface modifications of the nanoclay are done to reduce its surface energies and enhance its better compatibility and interaction with the polymer substrates. Different modifiers like primary, secondary, tertiary, and quaternary alkyl ammonium cations, and pyridinium orphosphonium cations are implemented to increase the dispersion and interaction properties of the nanoclay with the polymers [96]. Montmorillonite-based polymer composites have been the most studied nanoclay in the field of EH because of its better reinforcement and barrier functions to polymers. The effect of nanoclay addition to the polymer matrix and the changes in its polymorphism, structure, and morphology was studied by Neppalli et al. [97]. Electrospun fibers of PVDF and Cloisite 20 A were prepared using the electrospinning technique, and the morphological investigation suggested that the addition of the nanoclay resulted in an increase in fiber diameter of the composite as compared to the pristine polymer fibers. The addition of nanoclay results in a rise in solution conductivity which has its influence in the viscosity of the solution. The effect of the electrospinning process and clay combined together to enhance the piezoelectric phase formation providing a better electroactive material. Xin et al. [98] prepared a full fiber sensor of PVDF and nanoclay which showed substantial piezoelectric properties with better response speed and reproducibility. The addition of nanoclay leads to a rise in the electroactive phase as confirmed from the XRD and FTIR analysis. The nanoclay-based composite generated the maximum output voltage of 2.76 V as compared to 0.78 V for PVDF fibers. Also impacts from finger tapping generated an average output voltage of 0.83 V, which shows its potential as a device for sensor-based applications. Leveque et al. [99] incorporated different layered silicates to the polyamide 11 (PA 11) to analyze its vibrational EH applications. Addition of the layered silicates resulted in the rise in mechanical properties of the developed nanocomposites. The Cloisite Na^+ nanoclay induced better piezoelectric properties in the polymer composites compared to the other silicates like Cloisite 20 A and Cloisite 10 A. The addition of 5 wt.% of Cloisite Na^+ resulted in a piezo response of about 2.2 which was similar to the PA 11. However, the EH enactment of the PA 11 was better than the Cloisite Na^+-based composite due to the higher stiffness of the nanoclay. Nanoclay acts as a nucleating agent which decreases the nucleation energy barrier and hence increases the crystallization of the system. The positive part of the organically modified nanoclay interacts with the $-CF_2$ part of the PVDF or its copolymers, thus facilitating the piezoelectric phase formation for better performance. Tiwari et al. [100] prepared the electrospun nanofiber of PVDF and nanoclay (Cloisite 30B) using the electrospinning process. The different electrospun fibers were prepared at optimized electrospinning condition by varying the nanoclay content. The addition of nanoclay leads to a rise in the electroactive phase content which was then analyzed using the electromechanical response. A device was fabricated to measure the piezoelectric performance of the material using different human body movements. An electrospun scaffold was placed between two aluminum foil electrodes using copper wires. The whole system was encapsulated using poly (dimethyl siloxane) to provide rigidity and stability to the device. The maximum output voltage and power density obtained were about 70 V and 68 $\mu W/cm^2$, respectively.

Two-dimensional transition metal dichalcogenides (TMDs) have attracted great interest due to their excellent physical, chemical, electrical, and optical properties. TMDs possess a layered type structure similar to that of graphite, and the thickness of each monolayer is around 0.6–0.7 nm, consisting of hexagonally packed layers of metal atoms placed between two chalcogen atoms. The intralayer bonds of the TMDs are generally covalent in nature, while the layered part consists of the weak van der Waals force, which gets easily cleaved. The quantum size effect and the nanolevel thickness enable them to be used in different applications. TMDs have a bandgap ranging from 0–2 eV for semimetal and semiconducting statuses, respectively. This bandgap is dependent on the number of layers, doping, and element combinations, which make it a suitable candidate for electronics and energy-based applications. Molybdenum disulfide (MoS_2) is the most studied TMDs in the field of EH. MoS_2 possesses a crystalline nature with one Mo atom sandwiched between the two S layers enclosed in a hexagonal geometry. MoS_2 is found to be piezoelectric due to the broken inversion symmetry in its crystal structure. The rhombic prismatic cell consists of asymmetrically arranged S and Mo atoms such that the hexagonal lattice when stretched leads to deformation of the Mo-S bonds which leads to in-plane piezoelectricity. The mechanical flexibility and stretchability and the piezoelectric properties of the 2D- MoS_2 explains its potential in various applications like EH, sensors, actuators, and wearable materials [101–103]. Wu et al. [104] was the first to report the mechanical EH in 2D MoS_2 flakes. The odd layer of the MoS_2 flakes on repeated stretching and releasing generated electrical output due to the development of strain-induced polarization. A flexible device was fabricated of MoS_2 flakes on polyethylene terephthalate (PET) flexible substrate electrode using Cr/Pd/Au deposition. When the PET substrate was subjected to a uniaxial strain, strain-induced polarization charges flow to the external circuit and on releasing the strain; the electrons flow back in the opposite direction. For a single-layer MoS_2 flake, the maximum output current and voltage generated were around 27 pA and 18 mV, respectively. Han et al. [105] developed a novel piezoelectric nanogenerator from MoS_2 shells, Al_2O_3 thin layer, and polystyrene (PS) beads. The shell size was controlled using the PS bead size and the effect of the size of the shell was investigated on output power generation. The piezoelectric material developed generated output voltage of 1.2 V on application of 4.2 kPa pressure. Maity et al. [106] prepared an electrospun fiber of PVDF and 2D-MoS_2. Addition of MoS_2 leads to enhancement of the electroactive phase and crystallinity of the composite nanofiber. The piezoelectric device was used to analyze the electromechanical response generated from different mechanical sources like falling leaf and match stick from different heights substantial output was being generated which explains the efficacy of the material. On application of stress from human finger motion produced 14 V of output voltage from the composite nanofiber as compared to 2.5 V from the neat PVDF fiber based nanogenerator. The device was able to light up LEDs which hence approves the materials potential as a piezoelectric material. Regenerated cellulose and MoS_2 based composite films were used to prepare a flexible piezoelectric nanogenerator. The films were poled under the electric field of 15 MV/m and the poled films were sandwiched between two aluminum foils connected through copper wires encapsulated using polyimide sheet. The fabricated device was subjected to external stress which leads to generation of output voltage and current of 2 V of 150 nA respectively which was higher as compared to the pure PVDF film (0.4 V and 20 nA). The d_{33} values for the composite and pristine polymer was 19 pC/N and 3 pC/N respectively [107]. Two-dimensional hexagonal boron nitride (hBN) is another 2D material of utmost importance in the field of EH. The structure comprises of alternating boron and nitrogen atoms arranged over the honeycomb lattices of the nanosheets which results in the in-plane piezoelectricity in these materials. The intrinsic piezoelectric properties of the HBN make it suitable filler for polymer based composites for piezoelectric based EH applications. Zhang et al. [108] prepared an electrospun fiber from the PVDF and boron nitride nanosheets (BNNs) through electrospinning process at different concentration of the BNNs. The maximum output voltage and current generated was around −0.9 V to + 0.7 V and −0.85 μA to + 0.7 μA respectively for PVDF-1.5 BNNs. A highly flexible and efficient piezoelectric nanogenerator was prepared from PVDF and hBN nanoflakes. The free standing nanofiber was fabricated to a device form using the copper foils acting as electrode and Kapton tape. The tape was used to keep the fiber and the electrode intact. The device was structured to avoid any external contacts and short circuit. The maximum open circuit voltage generated from different type of motion like tapping hard, compression and flapping was around 68 V, 16 V and 20 V respectively. The maximum power density and output current was around 53.2 $\mu W/cm^2$ and 0.1 μA, respectively. The prepared device was the subjected to human motion like finger folding and walking which led to generation of high output voltage. Walking generated potential of nearly 98 V which demonstrated the performance of the device as potential pedometer [109]. Synergism plays an important role in bringing out the better properties of the system. Synergy is the term used to indicate the interaction between the two or more materials leading to enhanced effective properties which were better than the individual material. Polymer based composites with the hybrid fillers to achieve improved efficacy has been of importance in recent years [110]. The concentration of the fillers in the polymer matrix needs to optimize for achieving better synergistic effect based properties. Ponnamma et al. [111] fabricated PVDF-HFP based nanocomposites with two electroactive fillers $BaTiO_3$ and hBN and studied the piezoelectric properties. The h-BN nanolayers provide the better dielectric strength and tensile properties while the $BaTiO_3$ provides the better piezoelectric properties. Due to the synergistic effect of the combined fillers with the polymer leads to a flexible device with enhanced piezoelectric output at the optimized concentration of the two fillers that is 3 wt.% $BaTiO_3$ and 1 wt.% h-BN, maximum output voltage of 2.4 V and power density of 0.95 $\mu W/cm^2$ was achieved which was due to the synergistic effect of the hybrid fillers.

4.3 Applications

Polymer-based piezoelectric composites have attracted much attention due to their durability, reliability, low cost, and better accuracy in various fields like actuation, sensing, biomedicine, transduction, robotics, and others. Piezoelectric-based EH devices or techniques have been implemented in various sectors to harvest energies from different mechanical sources. The movement of vehicles on the road generates a much higher load of mechanical energy which can be used to power the traffic lights and vehicle weighing systems. The use of piezoelectric generators to harvest energy from the road can reduce the cost and also it is a clean energy source [23,112]. Song et al. [113] prepared a piezoelectric generator composed of 48 piezoelectric cantilever beams under asphalt. The maximum power output generated was around 736 μW. A cantilever type piezoelectric device was formulated by Karimi et al. to harvest energy from the vibrations produced from the bridges. The maximum power obtained at optimal load resistance of 200 kΩ was around 35 μW [114]. Wearable and implantable nanogenerator have also gained much interest based on piezoelectric materials due to the ability to sense small changes produced from the mechanical sources. Human body movements are a rich source of different mechanical energies which remain unused. 2D BNNs were used to prepare a piezoelectric nanogenerator to convert the produced mechanical energy from the human body into usable electrical energy. The maximum output of voltage and power generated were around 22 V and 40 μW, respectively. The developed device was placed on various parts of the human body like neck, elbow, wrist, knee, and foot and a substantial amount of energy was produced [115]. In the field of medical science, implantable devices such as pacemakers, stimulators for neurons, tissues, artificial retinas, or drug delivery systems are of great technological importance. But these devices have a limited lifetime and they need to be replaced often. Hence to avoid such barriers piezoelectric-based EH devices are one of the potential alternatives [109]. A piezoelectric-based EH device prepared from a porous thin film of helical type structure was developed to harvest mechanical energy from the cardioverter-defibrillator or from the lead of a pacemaker to produce usable electrical energy [116]. The migration of the cells also produces convertible mechanical energy to electrical energy using the piezoelectric materials. Hence the piezoelectric-based devices are promising technology that can be used for tissue engineering or tissue repair applications to avoid the need to implant batteries or electrodes [112]. These devices are the potential materials for biomedical-based applications because of their scalability and better performance [117].

4.4 Conclusion

Piezoelectric materials are functional materials that possess the potential to be used in various fields. EH based on the piezoelectric materials has been the most accomplished field of applications. 2D materials possessing noncentrosymmetry are used as piezoelectric materials and their higher surface to volume ratio, and better electrical and mechanical properties make them versatile materials for EH application. Polymer composites using two-dimensional piezoelectric materials provide the combined advantages of the polymer and 2D materials, and as a result the performance of the device-based system is enhanced. Hence, the incorporation of the 2D materials into the polymer matrix is of great technological importance in the field of nanoscience and nanotechnology. This chapter covers some of the important piezoelectric materials and their role in EH applications, as well as providing an insight into the importance of these materials in different fields.

References

[1] International Energy Agency (IEA) (2017) Key World Energy Statistics, Archived (PDF) from the original on 20 May 2018, p. 30.

[2] L.B. Kong, T. Li, H.H. Hng, F. Boey, T. Zhang, S. Li, Waste energy harvesting, Lect. Notes Energy 24 (2014) 263–403.

[3] H. Li, C. Tian, Z.D. Deng, Energy harvesting from low frequency applications using piezoelectric materials, Appl. Phys. Rev. 1 (4) (2014) 041301.

[4] C. Wan, C.R. Bowen, Multiscale-structuring of polyvinylidene fluoride for energy harvesting: the impact of molecular-, micro-and macro-structure, J. Mater. Chem. A 5 (7) (2017) 3091–3128.

[5] C.R. Bowen, H.A. Kim, P.M. Weaver, S. Dunn, Piezoelectric and ferroelectric materials and structures for energy harvesting applications, Energy Environ. Sci. 7 (1) (2014) 25–44.

[6] J.P. Thomas, M.A. Qidwai, J.C. Kellogg, Energy scavenging for small-scale unmanned systems, J. Power Sources 159 (2) (2006) 1494–1509.

[7] J.A. Paradiso, T. Starner, Energy scavenging for mobile and wireless electronics, IEEE Pervasive Comput. 4 (1) (2005) 18–27.

[8] M. Deterre, B. Boutaud, R. Dalmolin, S. Boisseau, J.J. Chaillout, E. Lefeuvre et al. Energy harvesting system for cardiac implant applications. Symposium on Design, Test, Integration & Packaging of MEMS/MOEMS (DTIP) IEEE, 2011, pp. 387–391.

[9] S.P. Beeby, M.J. Tudor, N.M. White, Energy harvesting vibration sources for microsystems applications, Meas. Sci. Technol. 17 (12) (2006) R175.

[10] S.R. Anton, H.A. Sodano, A review of power harvesting using piezoelectric materials (2003–2006), Smart Mater. Struct. 16 (3) (2007) R1.

[11] S. Roundy, P.K. Wright, J. &Rabaey, A study of low level vibrations as a power source for wireless sensor nodes, Computer Commun. 26 (11) (2003) 1131–1144.

[12] S. Priya, Advances in energy harvesting using low profile piezoelectric transducers, J. Electroceram. 19 (1) (2007) 167–184.

[13] C. Williams, R.B. Yates, Analysis of a micro-electric generator for microsystems, Sens. Actuators A: Phys. 52 (1–3) (1996) 8–11.

[14] P. Glynne-Jones, M.J. Tudor, S.P. Beeby, N.M. White, An electromagnetic, vibration-powered generator for intelligent sensor systems, Sens. Actuators A: Phys. 110 (1–3) (2004) 344–349.

[15] D.P. Arnold, Review of microscale magnetic power generation, IEEE Trans. Magnetics 43 (11) (2007) 3940–3951.

[16] P.D. Mitcheson, P. Miao, B.H. Stark, E.M. Yeatman, A.S. Holmes, T.C. Green, MEMS electrostatic micropower generator for low frequency operation, Sens. Actuators A: Phys. 115 (2–3) (2004) 523–529.

[17] Y.B. Jeon, R. Sood, J.H. Jeong, S.G. Kim, MEMS power generator with transverse mode thin film PZT, Sens. Actuators A: Phys. 122 (1) (2005) 16–22.
[18] L. Wang, F.G. Yuan, Vibration energy harvesting by magnetostrictive material, Smart Mater. Struct. 17 (4) (2008) 045009.
[19] K.A. Cook-Chennault, N. Thambi, A.M. Sastry, Powering MEMS portable devices—a review of non-regenerative and regenerative power supply systems with special emphasis on piezoelectric energy harvesting systems, Smart Mater. Struct. 17 (4) (2008) 043001.
[20] W.J. Choi, Y. Jeon, J.H. Jeong, R. Sood, S.G. Kim, Energy harvesting MEMS device based on thin film piezoelectric cantilevers, J. Electroceram. 17 (2–4) (2006) 543–548.
[21] G.H. Haertling, Ferroelectric ceramics: history and technology, J. Am. Ceram. Soc. 82 (4) (1999) 797–818.
[22] H.S. Kim, J.H. Kim, J. Kim, A review of piezoelectric energy harvesting based on vibration, Int. J. Precis. Eng. Manuf. 12 (6) (2011) 1129–1141.
[23] S. Mishra, L. Unnikrishnan, S.K. Nayak, S. Mohanty, Advances in piezoelectric polymer composites for energy harvesting applications: a systematic review, Macromol. Mater. Eng. 304 (1) (2019) 1800463.
[24] R. Caliò, U.B. Rongala, D. Camboni, M. Milazzo, C. Stefanini, G. De Petris, et al., Piezoelectric energy harvesting solutions, Sensors 14 (3) (2014) 4755–4790.
[25] C. Covaci, A. Gontean, Piezoelectric energy harvesting solutions: a review, Sensors 20 (12) (2020) 3512.
[26] S.G. Kim, S. Priya, I. Kanno, Piezoelectric MEMS for energy harvesting, MRS Bull. 37 (11) (2012) 1039–1050.
[27] J. Fang, H. Niu, H. Wang, X. Wang, T. Lin, Enhanced mechanical energy harvesting using needleless electrospun poly (vinylidene fluoride) nanofibre webs, Energy Environ. Sci. 6 (7) (2013) 2196–2202.
[28] P. Muralt, Ferroelectric thin films for micro-sensors and actuators: a review, J. Micromech. Microeng. 10 (2) (2000) 136.
[29] S. Tadigadapa, K. Mateti, Piezoelectric MEMS sensors: state-of-the-art and perspectives, Meas. Sci. Technol. 20 (9) (2009) 092001.
[30] T. Funasaka, M. Furuhata, Y. Hashimoto, & K. Nakamura, Piezoelectric generator using a LiNbO/sub 3/plate with an inverted domain. 1998 IEEE Ultrasonics Symp. Proc. (Cat. No. 98CH36102) 1 (1998) 959–962.
[31] N. Setter, D. Damjanovic, L. Eng, G. Fox, S. Gevorgian, S. Hong, et al., Ferroelectric thin films: Review of materials, properties, and applications, J. Appl. Phys. 100 (5) (2006) 051606.
[32] M.A. Karami, O. Bilgen, D.J. Inman, M.I. Friswell, Experimental and analytical parametric study of single-crystal unimorph beams for vibration energy harvesting, IEEE Trans. Ultrason. Ferroelectr. Freq. Control 58 (7) (2011) 1508–1520.
[33] A. Shaw, S. Sriramula, P.D. Gosling, M.K. Chryssanthopoulos, A critical reliability evaluation of fibre reinforced composite materials based on probabilistic micro and macro-mechanical analysis, Compos. Part. B: Eng. 41 (6) (2010) 446–453.
[34] A. Gaur, A. Shukla, D. Saxena, P. Maiti, Polymer composites for structural, device, and biomedical applications, Mater. Sci. Technol. (2006) 1–34.
[35] A.F. Avila, P.C. Rodrigues, D.B. Santos, A.C. Faria, A dual analysis for recycled particulate composites: linking micro-and macro-mechanics, Mater. Charact. 50 (4–5) (2003) 281–291.
[36] S.H. Wankhade, S. Tiwari, A. Gaur, P. Maiti, PVDF–PZT nanohybrid based nanogenerator for energy harvesting applications, Energy Rep. 6 (2020) 358–364.
[37] A. Gaur, S. Tiwari, C. Kumar, P. Maiti, Flexible, lead-free nanogenerators using poly (vinylidene fluoride) nanocomposites, Energy Fuels 34 (5) (2020) 6239–6244.
[38] A. Gaur, S. Tiwari, C. Kumar, P. Maiti, Polymer biowaste hybrid for enhanced piezoelectric energy harvesting, ACS Appl. Electron. Mater. 2 (5) (2020) 1426–1432.

[39] A. Gaur, S. Tiwari, C. Kumar, P. Maiti, Bio-waste orange peel and polymer hybrid for efficient energy harvesting, Energy Rep. 6 (2020) 490–496.

[40] D.K. Patel, S. Senapati, P. Mourya, M.M. Singh, V.K. Aswal, B. Ray, et al., Functionalized graphene tagged polyurethanes for corrosion inhibitor and sustained drug delivery, ACS Biomater. Sci. & Eng. 3 (12) (2017) 3351–3363.

[41] O. Prakash, K.K. Jana, R. Jain, P. Shah, M. Manohar, V.K. Shahi, et al., Functionalized poly(vinylidene fluoride-co-hexafluoro propylene) membrane for fuel cell, Polym. (Guildf). 151 (2018) 261–268.

[42] R.X. Zhang, Q.Q. Ni, T. Natsuki, M. Iwamoto, Mechanical properties of composites filled with SMA particles and short fibers, Compos. Struct. 79 (1) (2007) 90–96.

[43] G. Kickelbick (Ed.), Hybrid Materials: Synthesis, Characterization, and Applications, John Wiley & Sons, 2007.

[44] C. Cui, F. Xue, W.J. Hu, L.J. Li, Two-dimensional Mater. piezoelectric ferroelectric functionalities, npj 2D Mater. Appl. 2 (1) (2018) 1–14.

[45] P. Lin, C. Pan, Z.L. Wang, Two-dimensional nanomaterials for novel piezotronics and piezophototronics, Mater. Today Nano 4 (2018) 17–31.

[46] X. Rocquefelte, F. Boucher, P. Gressier, G. Ouvrard, P. Blaha, K. Schwarz, Mo cluster formation in the intercalation compound LiMoS2, Phys. Rev. B 62 (4) (2000) 2397.

[47] K.S. Novoselov, D. Jiang, F. Schedin, T.J. Booth, V.V. Khotkevich, S.V. Morozov, et al., Two-dimensional atomic crystals, Proc. Natl. Acad. Sci. 102 (30) (2005) 10451–10453.

[48] J.N. Coleman, M. Lotya, A. O'Neill, S.D. Bergin, P.J. King, U. Khan, et al., Two-dimensional nanosheets produced by liquid exfoliation of layered materials, Science 331 (6017) (2011) 568–571.

[49] R.J. Smith, P.J. King, M. Lotya, C. Wirtz, U. Khan, S. De, et al., Large-scale exfoliation of inorganic layered compounds in aqueous surfactant solutions, Adv. Mater. 23 (34) (2011) 3944–3948.

[50] M. Lin, D. Wu, Y. Zhou, W. Huang, W. Jiang, W. Zheng, et al., Controlled growth of atomically thin In2Se3 flakes by van der Waals epitaxy, J. Am. Chem. Soc. 135 (36) (2013) 13274–13277.

[51] J. Zhang, S.A. Meguid, Piezoelectricity of 2D nanomaterials: characterization, properties, and applications, Semicond. Sci. Technol. 32 (4) (2017) 043006.

[52] R. Hinchet, U. Khan, C. Falconi, S.W. Kim, Piezoelectric properties in two-dimensional materials: simulations and experiments, Mater. Today 21 (6) (2018) 611–630.

[53] Y. Yu, H. Sun, H. Orbay, F. Chen, C.G. England, W. Cai, et al., Biocompatibility and in vivo operation of implantable mesoporous PVDF-based nanogenerators, Nano Energy 27 (2016) 275–281.

[54] H.S. Nalwa (Ed.), Ferroelectric Polymers: Chemistry: Physics, and Applications, CRC Press, 1995.

[55] P. Martins, A.C. Lopes, S. Lanceros-Mendez, Electroactive phases of poly (vinylidene fluoride): Determination, processing and applications, Prog. Polym. Sci. 39 (4) (2014) 683–706.

[56] E. Fukada, History and recent progress in piezoelectric polymers, IEEE Trans. Ultrason. Ferroelectr. Freq. Control 47 (6) (2000) 1277–1290.

[57] P. Frübing, A. Kremmer, R. Gerhard-Multhaupt, A. Spanoudaki, P. Pissis, Relaxation processes at the glass transition in polyamide 11: from rigidity to viscoelasticity, J. Chem. Phys. 125 (21) (2006) 214701.

[58] E. Fukada, Y. Ando, Piezoelectric properties of poly-β-hydroxybutyrate and copolymers of β-hydroxybutyrate and β-hydroxyvalerate, Int. J. Biol. Macromol. 8 (6) (1986) 361–366.

[59] S. Hikosaka, H. Ishikawa, Y. Ohki, Effects of crystallinity on dielectric properties of poly (L-lactide), IJTFM 129 (4) (2009) 217–222.

[60] P. Costa, J. Nunes-Pereira, N. Pereira, N. Castro, S. Gonçalves, S. Lanceros-Mendez, Recent progress on piezoelectric, pyroelectric, and magnetoelectric polymer-based energy-harvesting devices, Energy Technol. 7 (7) (2019) 1800852.

[61] B. Ameduri, From vinylidene fluoride (VDF) to the applications of VDF-containing polymers and copolymers: recent developments and future trends, Chem. Rev. 109 (12) (2009) 6632–6686.
[62] H. Song, Fabrication and characterisation of electrospun polyvinylidene fluoride (PVDF) nanocomposites for energy harvesting applications, Doctoral dissertation, Brunel University London, 2016.
[63] T. Furukawa, Ferroelectric properties of vinylidene fluoride copolymers, Phase Transit.: Multinat. J. 18 (3–4) (1989) 143–211.
[64] A. Gaur, C. Kumar, S. Tiwari, P. Maiti, Efficient energy harvesting using processed poly (vinylidene fluoride) nanogenerator, ACS Appl. Energy Mater. 1 (7) (2018) 3019–3024.
[65] J.S. Andrew, D.R. Clarke, Effect of electrospinning on the ferroelectric phase content of polyvinylidene difluoride fibers, Langmuir 24 (3) (2008) 670–672.
[66] T. Lei, L. Yu, G. Zheng, L. Wang, D. Wu, D. Sun, Electrospinning-induced preferred dipole orientation in PVDF fibers, J. Mater. Sci. 50 (12) (2015) 4342–4347.
[67] S. Siddiqui, D.I. Kim, E. Roh, T.Q. Trung, M.T. Nguyen, N.E. Lee, A durable and stable piezoelectric nanogenerator with nanocomposite nanofibers embedded in an elastomer under high loading for a self-powered sensor system, Nano Energy 30 (2016) 434–442.
[68] H. Yu, T. Huang, M. Lu, M. Mao, Q. Zhang, H. Wang, Enhanced power output of an electrospun PVDF/MWCNTs-based nanogenerator by tuning its conductivity, Nanotechnology 24 (40) (2013) 405401.
[69] C. Kumar, A. Gaur, S. Tiwari, A. Biswas, S.K. Rai, P. Maiti, Bio-waste polymer hybrid as induced piezoelectric material with high energy harvesting efficiency, Compos. Commun. 11 (2019) 56–61.
[70] Q. Gao, J.I. Scheinbeim, Dipolar intermolecular interactions, structural development, and electromechanical properties in ferroelectric polymer blends of nylon-11 and poly (vinylidene fluoride), Macromolecules 33 (20) (2000) 7564–7572.
[71] Z. Ounaies, J.S. Harrison, Piezoelectric polymers, NASA/CR-2001- 211422, ICASE Report No. 2001–43, USA, 2001, 1.
[72] Y. Takase, J.W. Lee, J.I. Scheinbeim, B.A. Newman, High-temperature characteristics of nylon-11 and nylon-7 piezoelectrics, Macromolecules 24 (25) (1991) 6644–6652.
[73] Y. Chen, L.M. Geever, J.A. Killion, J.G. Lyons, C.L. Higginbotham, D.M. Devine, Review of multifarious applications of poly (lactic acid), Polym. Technol. Eng. 55 (10) (2016) 1057–1075.
[74] C. Zhao, J. Zhang, Z.L. Wang, K. Ren, A poly (l-lactic acid) polymer-based thermally stable cantilever for vibration energy harvesting applications, Adv. Sustain. Syst. 1 (9) (2017) 1700068.
[75] J. Morvan, E. Buyuktanir, J.L. West, A. Jakli, Highly piezoelectric biocompatible and soft composite fibers, Appl. Phys. Lett. 100 (6) (2012) 063901.
[76] M. Varga, J. Morvan, N. Diorio, E. Buyuktanir, J. Harden, J.L. West, et al., Direct piezoelectric responses of soft composite fiber mats, Appl. Phys. Lett. 102 (15) (2013) 153903.
[77] S.B. Lang, Guide to the Literature of Piezoelectricity and Pyroelectricity, 19, 2003, pp. 71–266.
[78] W.K. Sakamoto, E.D. Souza, D.K. Das-Gupta, Electroactive properties of flexible piezoelectric composites, Mater. Res. 4 (3) (2001) 201–204.
[79] J. Plackner, The converse piezoelectric effect in wood and cellulose materials, 2009.
[80] J. Kim, Y.B. Seo, Electro-active paper actuators, Smart Mater. Struct. 11 (3) (2002) 355.
[81] Z. Abas, H.S. Kim, J. Kim, J.H. Kim, Cellulose electro-active paper: From discovery to technology applications, Front. Materials 1 (2014) 17.
[82] S. Rajala, T. Siponkoski, E. Sarlin, M. Mettänen, M. Vuoriluoto, A. Pammo, et al., Cellulose nanofibril film as a piezoelectric sensor material, ACS Appl. Mater. Interfaces 8 (24) (2016) 15607–15614.

[83] A. Toprak, O. Tigli, Piezoelectric energy harvesting: State-of-the-art and challenges, Appl. Phys. Rev. 1 (3) (2014) 031104.
[84] Z. Li, L. Wang, Y. Li, Y. Feng, W. Feng, Carbon-based functional nanomaterials: Preparation, properties and applications, Compos. Sci. Technol. 179 (2019) 10–40.
[85] J. Yan, M. Liu, Y.G. Jeong, W. Kang, L. Li, Y. Zhao, et al., Performance enhancements in poly (vinylidene fluoride)-based piezoelectric nanogenerators for efficient energy harvesting, Nano Energy 56 (2019) 662–692.
[86] T. Kuilla, S. Bhadra, D. Yao, N.H. Kim, S. Bose, J.H. Lee, Recent advances in graphene based polymer composites, Prog. Polym. Sci. 35 (11) (2010) 1350–1375.
[87] M.M. Abolhasani, K. Shirvanimoghaddam, M. Naebe, PVDF/graphene composite nanofibers with enhanced piezoelectric performance for development of robust nanogenerators, Compos. Sci. Technol. 138 (2017) 49–56.
[88] S. Wang, L. Liu, Y. Zeng, B. Zhou, K. Teng, M. Ma, et al., Improving dielectric properties of poly (vinylidene fluoride) composites: effects of surface functionalization of exfoliated graphene, J. Adhes. Sci. Technol. 29 (7) (2015) 678–690.
[89] V. Bhavanasi, V. Kumar, K. Parida, J. Wang, P.S. Lee, Enhanced piezoelectric energy harvesting performance of flexible PVDF-TrFE bilayer films with graphene oxide, ACS Appl. Mater. Interfaces 8 (1) (2016) 521–529.
[90] R.M. Habibur, U. Yaqoob, S. Muhammad, A.I. Uddin, H.C. Kim, The effect of RGO on dielectric and energy harvesting properties of P (VDF-TrFE) matrix by optimizing electroactive β phase without traditional polling process, Mater. Chem. Phys. 215 (2018) 46–55.
[91] J.H. Lee, K.Y. Lee, M.K. Gupta, T.Y. Kim, D.Y. Lee, J. Oh, et al., Highly stretchable piezoelectric-pyroelectric hybrid nanogenerator, Adv. Mater. 26 (5) (2014) 765–769.
[92] N. Maity, A. Mandal, A.K. Nandi, Hierarchical nanostructured polyaniline functionalized graphene/poly (vinylidene fluoride) composites for improved dielectric performances, Polymer 103 (2016) 83–97.
[93] M. Zelisko, Y. Hanlumyuang, S. Yang, Y. Liu, C. Lei, J. Li, et al., Anomalous piezoelectricity in two-dimensional graphene nitride nanosheets, Nat. Commun. 5 (1) (2014) 1–7.
[94] M. Khalifa, A. Mahendran, S. Anandhan, Synergism of graphitic-carbon nitride and electrospinning on the physico-chemical characteristics and piezoelectric properties of flexible poly (vinylidene fluoride) based nanogenerator, J. Polym. Res. 26 (3) (2019) 1–13.
[95] M. Khalifa, S. Anandhan, PVDF Nanofibers with Embedded Polyaniline–Graphitic Carbon Nitride Nanosheet Composites for Piezoelectric Energy Conversion, ACS Appl. Nano Mater. 2 (11) (2019) 7328–7339.
[96] M. Jawaid, A. Qaiss, R. Bouhfid, Nanoclay reinforced polymer composites, 2016.
[97] R. Neppalli, S. Wanjale, M. Birajdar, V. Causin, The effect of clay and of electrospinning on the polymorphism, structure and morphology of poly (vinylidene fluoride), Eur. Polym. J. 49 (1) (2013) 90–99.
[98] Y. Xin, X. Qi, H. Tian, C. Guo, X. Li, J. Lin, et al., Full-fiber piezoelectric sensor by straight PVDF/nanoclay nanofibers, Mater. Lett. 164 (2016) 136–139.
[99] M. Leveque, C. Douchain, M. Rguiti, K. Prashantha, C. Courtois, M.F. Lacrampe, et al., Vibrational energy-harvesting performance of bio-sourced flexible polyamide 11/layered silicate nanocomposite films, Int. J. Polym. Anal. Charact. 22 (1) (2017) 72–82.
[100] S. Tiwari, A. Gaur, C. Kumar, P. Maiti, Enhanced piezoelectric response in nanoclay induced electrospun PVDF nanofibers for energy harvesting, Energy 171 (2019) 485–492.
[101] M. Ahmadi, O. Zabihi, S. Jeon, M. Yoonessi, A. Dasari, S. Ramakrishna, et al., 2D transition metal dichalcogenide nanomaterials: advances, opportunities, and challenges in multi-functional polymer nanocomposites, J. Mater. Chem. A 8 (3) (2020) 845–883.

[102] S.K. Ghosh, D. Mandal, Piezoelectricity of 2D materials and its applications toward mechanical energy harvesting, 2D Nanomaterials Energy Appl. (2020) 1–38.
[103] F.R. Fan, W. Tang, Z.L. Wang, Flexible nanogenerators for energy harvesting and self-powered electronics, Adv. Mater. 28 (22) (2016) 4283–4305.
[104] W. Wu, L. Wang, Y. Li, F. Zhang, L. Lin, S. Niu, et al., Piezoelectricity of single-atomic-layer MoS2 for energy conversion and piezotronics, Nature 514 (7523) (2014) 470–474.
[105] J.K. Han, S. Kim, S. Jang, Y.R. Lim, S.W. Kim, H. Chang, et al., Tunable piezoelectric nanogenerators using flexoelectricity of well-ordered hollow 2D MoS_2 shells arrays for energy harvesting, Nano Energy 61 (2019) 471–477.
[106] K. Maity, B. Mahanty, T.K. Sinha, S. Garain, A. Biswas, S.K. Ghosh, et al., Two-Dimensional piezoelectric MoS_2-modulated nanogenerator and nanosensor made of poly(vinlydine Fluoride) nanofiber webs for self-powered electronics and robotics, Energy Technol. 5 (2) (2017) 234–243.
[107] S. Chen, J. Li, Y. Song, Q. Yang, Z. Shi, C. Xiong, Flexible and environment-friendly regenerated cellulose/MoS_2 nanosheet nanogenerators with high piezoelectricity and output performance, Cellulose (2021) 1–10.
[108] J. Zhang, D. Liu, Q. Han, L. Jiang, H. Shao, B. Tang, et al., Mechanically stretchable piezoelectric polyvinylidene fluoride (PVDF)/Boron nitride nanosheets (BNNSs) polymer nanocomposites, Compos. Part. B: Eng. 175 (2019) 107157.
[109] P. Yadav, T.D. Raju, S. Badhulika, Self-poled hBN-PVDF nanofiber mat-based low-cost, ultrahigh-performance piezoelectric nanogenerator for biomechanical energy harvesting, ACS Appl. Electron. Mater. 2 (7) (2020) 1970–1980.
[110] U. Szeluga, B. Kumanek, B. Trzebicka, Synergy in hybrid polymer/nanocarbon composites. A review, Compos. Part. A: Appl. Sci. Manuf. 73 (2015) 204–231.
[111] D. Ponnamma, M.A.A. Al-Maadeed, Influence of BaTiO3/white graphene filler synergy on the energy harvesting performance of a piezoelectric polymer nanocomposite, Sustain. Energy Fuels 3 (3) (2019) 774–785.
[112] N. Sezer, M. Koç, A comprehensive review on the state-of-the-art of piezoelectric energy harvesting, Nano Energy (2020) 105567.
[113] Y. Song, C.H. Yang, S.K. Hong, S.J. Hwang, J.H. Kim, J.Y. Choi, et al., Road energy harvester designed as a macro-power source using the piezoelectric effect, Int. J. Hydrog. Energy 41 (29) (2016) 12563–12568.
[114] M. Karimi, A.H. Karimi, R. Tikani, S. Ziaei-Rad, Experimental and theoretical investigations on piezoelectric-based energy harvesting from bridge vibrations under travelling vehicles, Int. J. Mech. Sci. 119 (2016) 1–11.
[115] K.B. Kim, W. Jang, J.Y. Cho, S.B. Woo, D.H. Jeon, J.H. Ahn, et al., Transparent and flexible piezoelectric sensor for detecting human movement with a boron nitride nanosheet (BNNS), Nano Energy 54 (2018) 91–98.
[116] L. Dong, A.B. Closson, M. Oglesby, D. Escobedo, X. Han, Y. Nie, et al., In vivo cardiac power generation enabled by an integrated helical piezoelectric pacemaker lead, Nano Energy 66 (2019) 104085.
[117] G. Ciofani, A. Menciassi (Eds.), Piezoelectric Nanomaterials for Biomedical Applications, Springer, Berlin, 2012, pp. 213–238.

CHAPTER

5

Non-fluorinated piezoelectric polymers and their composites for energy harvesting applications

Farsa Ram[1,2] *and Kadhiravan Shanmuganathan*[1,2]

[1]Polymer Science and Engineering Division, CSIR-National Chemical Laboratory, Pune, India, [2]Academy of Scientific and Innovative Research (AcSIR), Ghaziabad, India

5.1 Introduction

Piezoelectric materials generate electrical polarization/electricity (direct piezoelectric effect) in response to applied mechanical deformations. This phenomenon is also reversible, wherein the materials show mechanical deformation on the application of external field (converse piezoelectric effect). The direct piezoelectric effect is used in energy-harvesting and sensing applications, while the converse piezoelectric effect is used in transducer and inkjet printers [1]. Bulk piezoelectric materials have randomly oriented dipoles (Fig. 5.1A) and hence yield lower piezoelectric output. Maximum piezoelectric output can be achieved by aligning these dipoles in the thickness direction using an electrical poling process, wherein direct current (DC) bias is applied across the thickness direction of the material at a specific temperature (Fig. 5.1B) [1]. Piezoelectric materials have two subclasses, pyroelectric (materials exhibiting electrical polarization due to change in temperature) and ferroelectric (materials with a permanent dipole moment, the direction of which can be reversed by external field) [1]. The extent of piezoelectric charges generated per unit force is defined as the piezoelectric charge constant, and it determines the performance of

Engineered Polymer Nanocomposites for Energy Harvesting Applications
DOI: https://doi.org/10.1016/B978-0-12-824155-4.00005-3

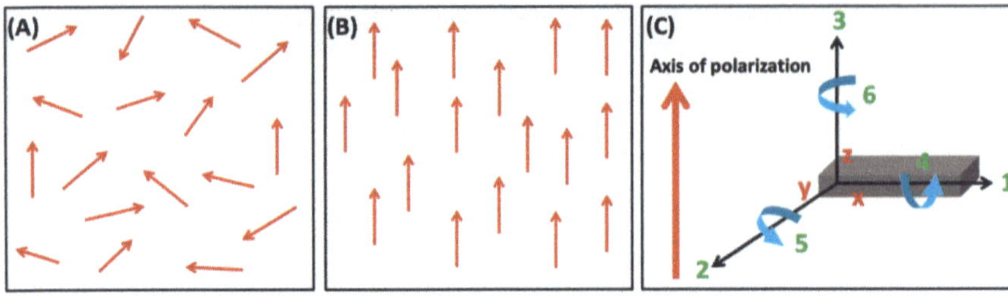

(D)

$$d_{mn} = \begin{pmatrix} d_{11} & d_{12} & d_{13} & d_{14} & d_{15} & d_{16} \\ d_{21} & d_{22} & d_{23} & d_{24} & d_{25} & d_{26} \\ d_{31} & d_{32} & d_{33} & d_{34} & d_{35} & d_{36} \end{pmatrix}$$

FIGURE 5.1 Dipole orientation in the materials (A) before poling and (B) after poling; (C) schematic representation of piezoelectric charge constant and (D) matrix for piezoelectric coefficient. [2] Source: *Author's original figure, reproduced from reference F. Ram, Nanocellulose reinforced polymer composites for energy application. 2020. Thesis submitted to Academy of Scientific and Innovative Research at Council of Scientific and Industrial Research-National Chemical Laboratory, Pune. http://dspace.ncl.res.in:8080/xmlui/handle/20.500.12252/5920.*

piezoelectric materials. The x, y, and z-axes are electrical axes; the z-axis is considered as the axis of polarization and denoted by 3, while the y- and x-axes are denoted by 2 and 1, respectively. The polarization due to shear along x, y, and z-axes is denoted as 4, 5, and 6, respectively (Fig. 5.1C). In piezoelectric charge constant (d_{mn}), m defines the direction in which charges are being collected/generated, and n defines the direction in which mechanical force is applied. Based on the direction of applied force and the direction in which charges are being collected, the piezoelectric constants can be denoted by a 3×6 matrix (Fig. 5.1D) [3]. d_{33}, d_{31}, and d_{15} have higher values compared to other piezoelectric charge constants and are mostly used in practical applications.

- d_{33} (applied stress in direction 3 and induced polarization in direction 3)
- d_{31} (applied stress in direction 1 and induced polarization in direction 1)
- d_{15} (applied shear stress along the direction 2 and induced polarization in direction 1)

In general, ceramic materials show a high piezoelectric coefficient. For example, lead zirconate titanate (PZT) [4] and lead magnesium niobate-lead titanate (PMN-PT) [5,6] have piezoelectric charge constants in the range of 100–2800 pC/N. These ceramic materials are widely used in transducers, microphones, accelerometers, etc. However, ceramic materials have high density and are brittle in nature. This drawback is mitigated to some extent by fabricating ceramic/polymer composites, wherein the polymer can be piezoelectric or nonpiezoelectric. Among the piezoelectric polymers, poly(vinylidene fluoride) (PVDF) and their copolymers show the highest piezoelectric properties. On other hand, nonfluorinated polymers, such as poly(acrylonitrile), nylons, cellulose, collagen, and poly(lactic acid) (PLA), have significant piezoelectric properties and some of them are bio-based (e.g., nylon 11, PLA) or naturally produced (e.g., collagen, cellulose). In this chapter, we discuss the piezoelectric properties of nonfluorinated semicrystalline and amorphous polymers of natural and synthetic origin. Further, we also discuss the utilization of these nonfluorinated polymers for energy-harvesting applications.

5.2 Piezoelectricity in semicrystalline polymers

In general, a high piezoelectric output is provided by crystalline materials that exhibit a net dipole moment within the unit cell (found in asymmetric crystal types, for example, trigonal, triclinic, etc.). In semicrystalline polymers, crystallites are dispersed and embedded in the amorphous matrix. Most of the polymers show polymorphism, and only polar phases are electroactive. The nonpolar phase can be converted into polar phases by mechanical stretching, thermal annealing, and high voltage treatment [1]. Electrical poling of polymeric materials is performed to orient all dipoles in one direction for maximized piezoelectric output. Poling can be performed by contact poling (with an applied field of $\sim$50 MV/m) or by corona discharge [1]. PVDF and its copolymers show the highest piezoelectric coefficient among polymers. Odd nylons are another set of polymers that show significant piezoelectric behavior. Bio-derived polymers like PLA, nylon 11, and poly(caprolactone) (PCL) are also known to exhibit some piezoelectric characteristics. Aliphatic and aromatic polyureas have been explored for their piezoelectric properties, yet are barely utilized in practical applications.

5.3 Piezoelectricity in natural polymers

Piezoelectricity in natural polymers was first reported in 1941 in keratin (in a bundle of hair) [7]. Natural polymers like cellulose, collagen, chitin, deoxyribonucleic acid (DNA), etc. have a low piezoelectric coefficient compared to other polymers like PVDF and nylons [1]. These polymers are semicrystalline in nature, and piezoelectricity originates from various mechanisms, as shown in Table 5.1. Most of the polymers possess D

TABLE 5.1 Origin of piezoelectricity in nonfluorinated polymers.

S. no.	Polymer	Piezoelectricity due to	References
1.	Nylon 11	Amide dipoles in polar crystallite phases	[8]
2.	Nylon 6, 12	Enlarge interchain spacing and disordered hydrogen bonding in mesophase	[9]
3.	PLA	$-C=O$ dipole in chiral and helical chain packing	[10]
4.	Cellulose	Intermolecular hydrogen bonding and uniaxially oriented dipoles	[11]
5.	PAN	Nitrile dipoles in zigzag planar polymeric chains	[12]
6.	PCL	$-C=O$ dipoles in electrospun fiber	[13]
7.	Chitin	Intrinsic molecular polarization arising from the noncentrosymmetric crystal structure of both α- and β-chitin polymorphs	[14,15]
8.	Collagen	Assembly of helical collagen molecule in fibrils and axial oriented arrangement of these fibrils in collagen fiber	[16]
9.	Silk	Hydrogen bonding between polypeptide chains	[17]
10.	Polyurea	Amide dipoles and hydrogen bonding	[18,19]

symmetry, which makes the material show shear piezoelectricity. The piezoelectricity in these polymers is attributed to the rotation of polar atomic groups linked to the asymmetric carbon. Electric displacement is produced perpendicular to applied shear stress in the polarization axis [20]. Collagen, gelatin, chitin, chitosan, cellulose, polypeptides, DNA, and PLA are some examples of piezoelectric biopolymers [20].

5.4 Piezoelectricity in amorphous polymers

Traditionally, it was believed that crystallinity is a prerequisite for piezoelectricity. However, it has been shown that piezoelectric-like behavior can also be evoked in amorphous polymers. Piezoelectricity in amorphous polymers can be explained with the help of dielectric and electrical double layer theories. According to dielectric theory, piezoelectricity arises from frozen molecular dipole orientation in a quasistable state [21]. Amorphous polymers have a glass transition temperature (T_g) above which the polymer chains are sufficiently mobile to align the dipoles in the presence of an applied electrical field (80–100 MV/m) [20]. This orientation is partially retained by lowering the temperature below T_g in the presence of an applied electric field, which leads to a piezoelectric-like effect arising from oriented dipoles and space charge injection or doped ions (if any). Interestingly, a few recent reports [22] state that the contribution from space charge injection is negligible as compared to oriented dipoles.

Piezoelectricity in amorphous polymers can also be induced and further enhanced by adding small molecules with a high dipole moment. Contrary to the semicrystalline polymers, where the physical origin of piezoelectricity is due to the polar crystalline phases; in amorphous polymers it is due to the formation of an electrical double layer interface within different phases of polymers in the presence of an external electrical field. This direct interfacial origin of piezoelectricity is further governed by the electrical field strength and duration of poling time [23]. A specific case of polyamide is further discussed in the respective section. Mechanical stretching was also found to increase the piezoelectric output of these polymers [20]. Poly(acrylonitrile) (PAN), poly(vinylidenecyanide vinylacetate) (PVDCN/VAc), polyphenylethernitrile (PPEN), and poly(L-bicyclobutanecarbonitrile), polyvinyl chloride (PVC), and polyvinyl acetate (PVAc) are some examples of amorphous piezoelectric polymers [20]. Table 5.1 lists various polymers and thier origin of piezoelectricity.

5.5 Energy-harvesting applications

In this section, we focus on the energy-harvesting application of nonfluorinated polymers, including all subclasses, viz. semicrystalline, amorphous, and biopolymers. The piezoelectric output is optimized by poling a material in the presence of a high electrical field, where the dipole orientation in the direction of the applied field improves the piezoelectric output. Contact poling and corona poling are two commonly used poling methods. In a typical contact poling procedure, the piezoelectric material is sandwiched between two electrodes. Then the sample is submerged in a high dielectric material, for example, silicon oil, to avoid arcing during application of an external electrical field. The dielectric material is then heated around the glass transition temperature (T_g) of the polymer so the chains can relax and reorient themselves in the applied field direction. The electric field is applied (based on the thickness of materials, usually $\sim$100 MV/m) when the polymers are around T_g and removed once they are cooled to room temperature to arrest the dipole orientation in the applied field direction [1,24]. In the corona poling procedure, the material is exposed to corona discharge (produced by the breakdown of gaseous molecules in the high electric field between two asymmetric electrodes for example, needle and plate). Upon applying a high electric field (2–20 kV), the gaseous molecule around the tip results in ionization (corona discharge) and travel toward the plate electrode. A grid is placed (applied field ~0.2–4 kV) between needle and sample to homogenously distribute the corona discharge. The corona discharge treatment results in the polarization of the polymers. Similar to contact poling, the samples are heated for better control of polarization and the electric field is only removed when the samples are in a cooled state [25,26].

A device fabricated for piezoelectric energy-harvesting is known as a piezoelectric nanogenerator (PENG), which converts small-scale mechanical deformations into electricity. For the fabrication of nanogenerators, piezoelectric material is sandwiched between two conductive electrodes, as shown in Fig. 5.2A, and capsulated in another nonpiezoelectric polymer matrix. The additional wires are attached to the electrodes for connecting an external load to draw the

current. The application of compressive force in the thickness direction generates piezoelectric polarization charges in the material at the electrode–material interface (Fig. 5.2B). The polarization charge density is dependent on applied force and increases with increased force (Fig. 5.2C) [27]. A piezoelectric nanogenerator has the potential to harvest energy from all types of mechanical motions, such as human physiological motions, wind, water wave, vibrations, airflow, sound, traffic, and noise [28]. In this section, energy-harvesting applications of nonfluorinated semicrystalline, amorphous, and biopolymers are discussed.

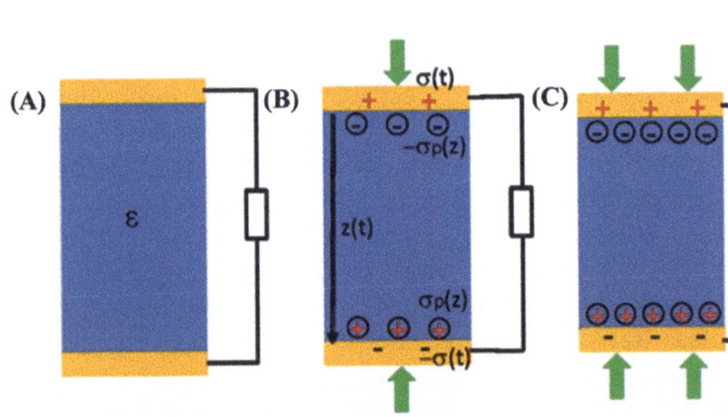

FIGURE 5.2 Working principle of piezoelectric nanogenerator: (A) PENG when no strain is applied; (B) PENG when small amount of force is applied. The piezoelectric polarization charges are generated in the materials, which creates the piezopotential. The piezopotential is equalized with the flow of electrons from one electrode to the other electrode. The same process continues with the next application of force; (C) piezoelectric polarization charge density increases with increased force or frequency. [27] Source: *Reproduced with permission from reference Z.L. Wang, On Maxwell's displacement current for energy and sensors: the origin of nanogenerators. Mater. Today 20 (2) (2017) 74–82, copyright (2017), Elsevier.*

5.5.1 Polyamides

Polyamides, also known as nylons, have a repeating unit of -NH $(CH_2)_nCO$-. When n is an even number, it results in an odd number of carbon atoms between two amides groups and these are called odd nylons. When n is an odd number, it results in an even number of carbon atoms between two amide groups and these are called even nylons. In odd nylons, the carbonyl groups and amine groups pack to maximize the hydrogen bonding and alignment of amide dipoles, which results in maximum net polarization [1]. Odd nylons usually crystallize in various polar polymorphs and the common stable forms are triclinic α form, pseudohexagonal γ form, and hexagonal δ form. These polymorphs can be obtained or interconverted by suitable treatments, like vapor phase deposition [29], uniaxial stretching [30], melt quenching [31], cold drawing [32], and spin coating [33]. In even nylons, the carbonyl groups and amine groups of adjacent chains pack in such a way that the amide dipoles cancel each other and crystallize in nonpolar polymorphs [1]. A recent report by Zhang et al. established that the requisite of polar phase is not necessary to show the piezoelectric effect in even nylons. The electric displacement-electric field (D-E) hysteresis measurements show a loop like nylon 11, which confirms their ferroelectric nature. The study, which was supported by X-ray diffraction (XRD) and infrared spectroscopy, showed that melt-quenched nylon 6 and nylon 12 films result in metastable mesophase, which is different from crystalline and amorphous phases. The mesophase is a highly defective crystalline phase and disordered in the perpendicular direction to the polymer chains. The chains are poorly packed, and chain conformation contains multiple twists per unit. The poorly organized hydrogen bonding among amide groups and enlarged interchain spacing is attributed to be the critical factor for ferroelectricity in these films. However, there is no literature on the energy-harvesting properties of even nylons [9].

Aliphatic polyamides are hydrophilic in nature, which results in difficulty in handling under normal conditions. This problem was addressed by increasing methylene groups between the amides, which increases the distance between the amide groups and hence reduced the piezoelectricity. Although nylon 5, 7, and 9 have high piezoelectricity, due to ease of handling under normal conditions, nylon 11 is preferred for practical applications [34]. It has been used as nanowires and films for mechanical energy-harvesting.

Datta et al. [8] prepared nylon 11 nanowires in an anodized aluminum oxide (AAO) template via capillary wetting technique. Nylon 11 nanowires embedded in AAO were used for energy-harvesting from low frequency and small amplitude vibrations. The nanogenerator resulted in an output voltage of 1 V and a short circuit current of 100 nA. Ram et al. [35] prepared flexible nylon 11 films with the addition of glycerol as a plasticizer. These nylon 11 (NYG) films crystallized predominantly in less polar α-phase. The addition of 5 wt.% carboxylated cellulose nanocrystals (CNC) in nylon 11 (5NYCG) resulted in the highly polar γ-phase in nylon 11 films (Fig. 5.3). Strong hydrogen bonding interaction between carboxyl groups of CNC and amide groups of nylon 11 led to enhanced γ-phase crystals (Fig. 5.4). Nonpoled nylon 11 and 5 wt.% nylon11/CNC nanogenerators resulted in an output voltage of 0.7 and 2.3 V at ~23 N impact force. The corona poled nylon 11, and 5 wt.% nylon11/CNC nanogenerators resulted in 2.8 and

6.95 V at similar conditions (Fig. 5.5). The addition of 5 wt.% CNC in nylon 11 led to a two to threefold increase in piezoelectric output due to enhanced polar γ-phase. The 5 wt.% nylon 11/CNC flexible nanogenerators were highly durable, yielding similar output voltage over 800 cycles and after 3 months of storage. The maximum output power density of nylon 11 and 5 wt.% nylon/CNC nanogenerators were found to be ~90 and 500 $\mu W/cm^3$ (with 820 kΩ resistors), respectively. The resultant power density could be utilized to run small sensors and charge a dielectric capacitor. These nanogenerators were used to charge a dielectric capacitor, where 5 wt.% nylon 11/CNC resulted in 2.8 times higher voltage in the capacitor in 90 s, as shown in Fig. 5.6.

5.5.2 Poly(L-lactic acid)

Lactic acid monomer has a chiral center, and its polymerization results in PLA with a helical structure. PLA crystallizes in α, β, and γ-phases, where β-phase is the most piezoelectric phase [36]. α-Phase has 10_3 types of helical structure, and β-phase has a 3_1 type helical structure. In normal conditions, all polarization from $-C=O$ dipoles along a PLA chain cancel out but shear stresses direct and align this polarization in one direction.

This results in net polarization out of plane of a polymeric chain [10]. Zhao et al. [37] reported piezoelectricity in electrospun PLA nanofibers. CO_2-treated fibers exhibited increased piezoelectric output due to enhanced crystallinity. Single fiber nanogenerator of poly(L-lactic acid) (PLLA) nanofibers resulted in open-circuit voltage (V_{oc}) and short circuit current (I_{sc}) of 25 mV and 8 pA during a simple push−release mechanism. Lee et al. [10] electrospun PLA nanofiber web (Fig. 5.7A). PLA chains were aligned along the axis of nanofibers with a preferential orientation of the $-C=O$ along the chain-helix. The piezoelectric voltage output increased with the number of stacked layers and reached 7.2 V at eight layers of PLA, as shown in Fig. 5.7B. Zhu et al. [38] prepared porous electrospun PLLA nanofibers for strain sensing and energy-harvesting applications. PLLA nanofibers produced 250 mV output voltage and 96 pA short circuit current on a bending setup with a

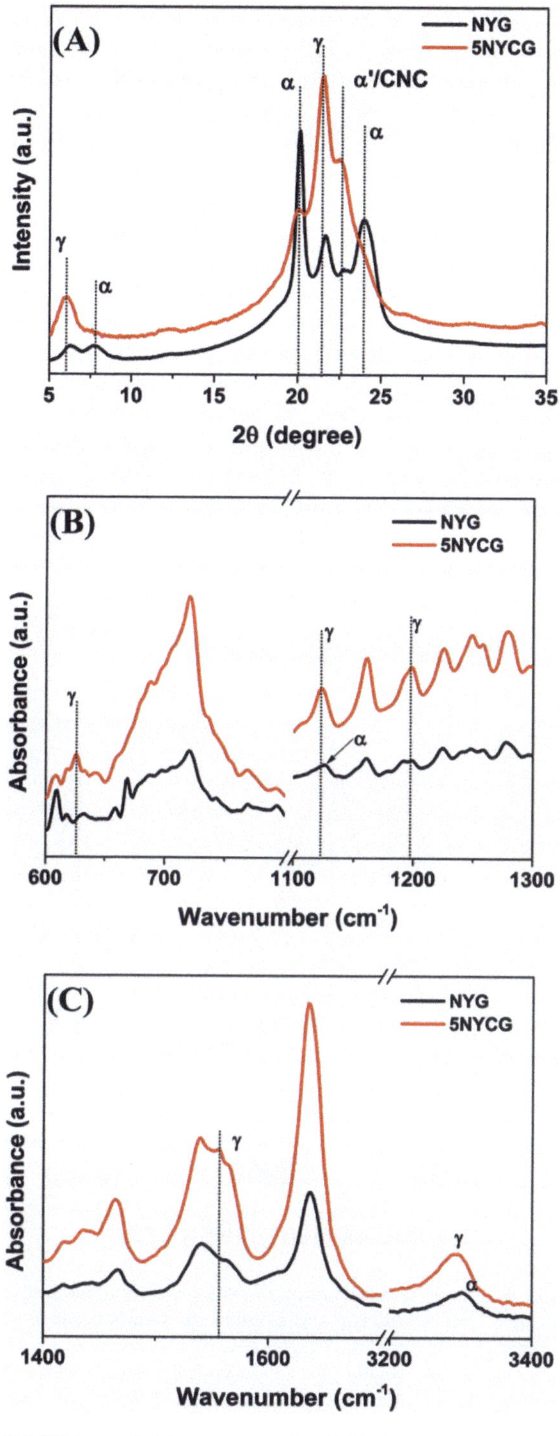

FIGURE 5.3 (A) Wide-angle X-ray scattering; (B,C) Fourier-transform infrared analysis of solution drawn nylon 11 and its composite films. Source: *Reproduced with permission from reference F. Ram, S. Radhakrishnan, T. Ambone, K. Shanmuganathan, Highly flexible mechanical energy harvester based on nylon 11 ferroelectric nanocomposites. ACS Appl. Polym. Mater. 1 (8) (2019) 1998–2005, copyright (2019), American Chemical Society.*

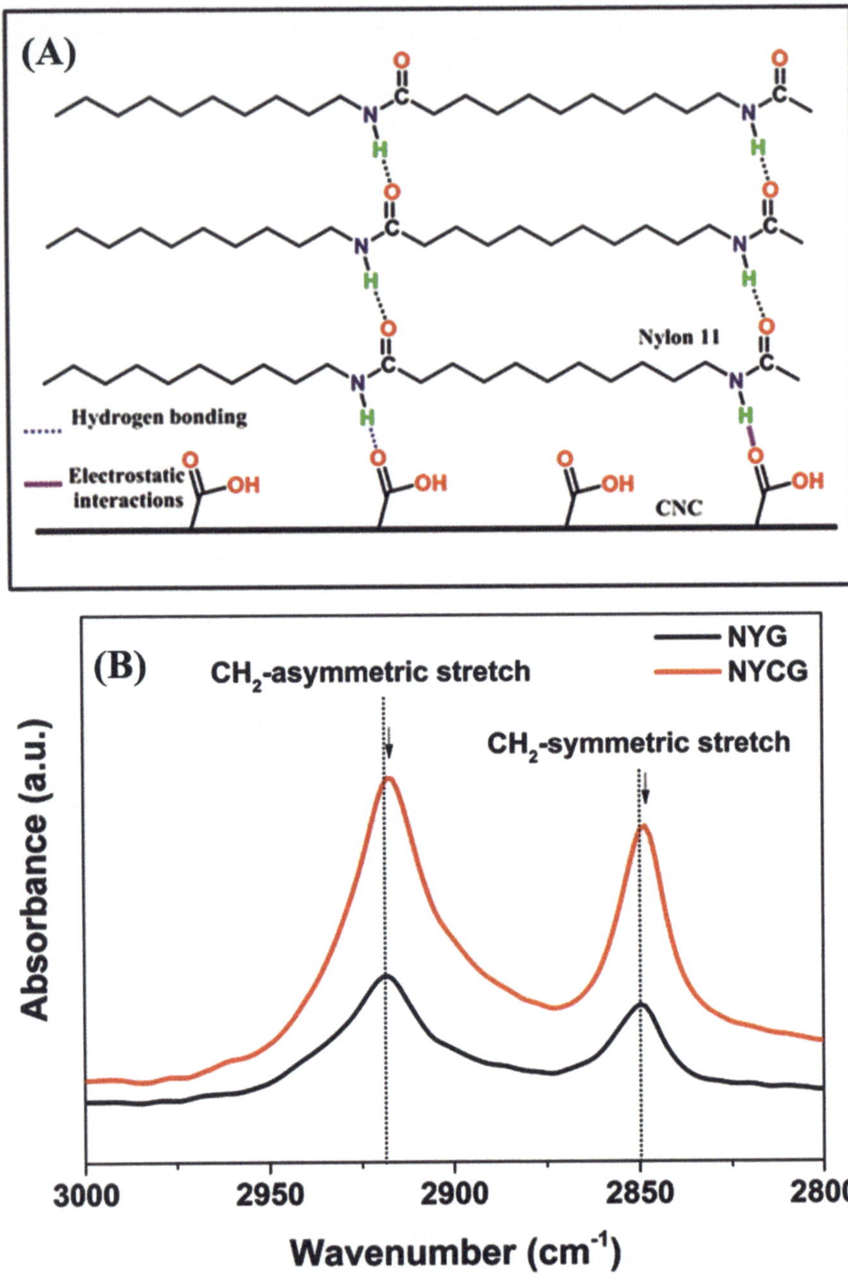

FIGURE 5.4 (A) Schematic representation of possible interaction between nylon 11 and CNC and (B) shift in $-CH_2$ symmetric and asymmetric bands because of these interfacial interactions. [35] Source: *Reproduced with permission from reference F. Ram, S. Radhakrishnan, T. Ambone, K. Shanmuganathan, Highly flexible mechanical energy harvester based on nylon 11 ferroelectric nanocomposites. ACS Appl. Polym. Mater. 1 (8) (2019) 1998–2005, copyright (2019) American Chemical Society.*

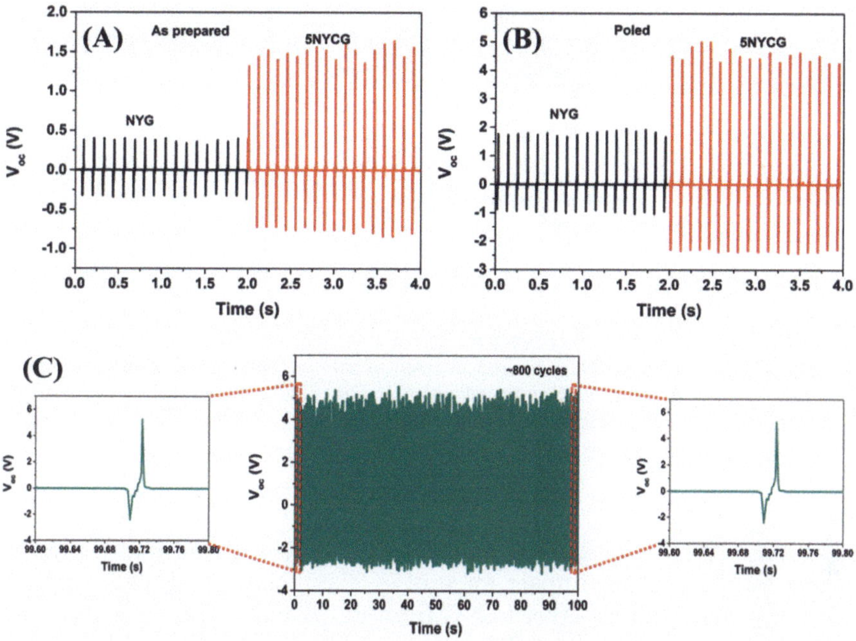

FIGURE 5.5 Voc of flexible-film-based PENGs: (A) NYG; (B) 5NYCG; and (C) durability test of 5NYCG PENG. Source: *Reproduced with permission from reference F. Ram, S. Radhakrishnan, T. Ambone, K. Shanmuganathan, Highly flexible mechanical energy harvester based on nylon 11 ferroelectric nanocomposites. ACS Appl. Polym. Mater. 1 (8) (2019) 1998–2005, copyright (2019) American Chemical Society.*

pushing displacement of 1 mm. The V_{oc} and I_{sc} increased to 0.55 V and 230 pA at a displacement of 5 mm, and it was found to be 2 V and 25 nA from human joint motion.

Curry et al. [39] fabricated a double-layer biodegradable and biocompatible sensor for energy-harvesting and physiological pressure monitoring. A thin film of PLLA was prepared by hot pressing and then mechanically stretched at an annealing temperature of 90°C. Mechanical stretching helped in the alignment of PLLA chains, and the annealing process helped to increase crystallinity. These processes helped to achieve enhanced piezoelectric properties. Further, it was found that piezoelectric output was optimum when films were cut at a 45° angle to the stretching direction. These films were labeled as treated films, and films with only stretching and annealing labeled as untreated or processed films. The maximum crystallinity and orientation were found at a draw ratio (DR) of ~5. The sensors were fabricated from various films having the same area and tested for energy-harvesting from vibrations and impact. Films with a DR of 2.5 resulted in the highest piezoelectric output from

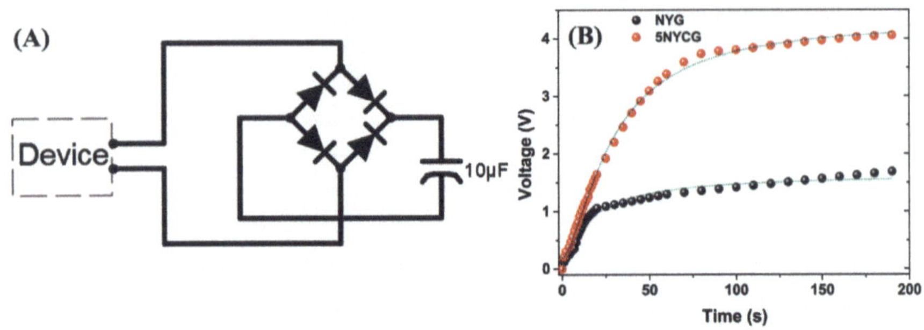

FIGURE 5.6 (A) Bridge rectifier circuit and (B) capacitor charging curves for 10 μF using NYG and 5NYCG PENGs. Source: *Reproduced with permission from reference F. Ram, S. Radhakrishnan, T. Ambone, K. Shanmuganathan, Highly flexible mechanical energy harvester based on nylon 11 ferroelectric nanocomposites. ACS Appl. Polym. Mater. 1 (8) (2019) 1998–2005, copyright (2019) American Chemical Society.*

vibrations. Films with a draw ratio of 6.5 resulted in the highest output voltage of 0.9 V at an impact force of ~23 N. Zhao et al. [40] prepared single-layer and double-layer PENG from PLLA films. The films were prepared by solution casting in dichloromethane (DCM), followed by annealing (135°C, 4 h), stretching at DR of 4, and cutting into rectangular shapes at 45 degrees relative to the stretching direction. The devices were constructed of two active layers of PLLA and one passive layer of PLA.

Active and passive layers having the same Young's modulus helped to realize the enhanced output in energy-harvesting from a bending setup by improving the input mechanical energy to the active piezo-layer. The device was tested by dropping a certain amount of weight at varied frequencies and resistances. The maximum performance was achieved at 19.7 Hz frequency and 0.55 MΩ resistance at a dropped weight of 4.71 g. The weight dropped deflected the cantilever by 15.5 mm, and the cantilever had a radius of 27.42 cm with 3.65 cm^{-1} curvature. At these conditions, V_{oc}, I_{sc}, and power were found to be 16.9 V, 26.7 μA, and 0.28 mW, respectively. The device was then integrated as a hybrid tribo and piezo device for e-skin applications. Zhao et al. previously reported V_{oc} and power of 9.05 V and 14.17 μW, respectively, from a double-layer PLLA device with poly(ethylene terephthalate) passive layer at 1.8 g weight drop on cantilever [41].

5.5.3 Poly(caprolactone)

It is a semicrystalline polymer and its crystallinity can be as high as 70% depending on processing conditions. Sencadas et al. [13] prepared PCL nanofibers via electrospinning. PCL crystallizes in a

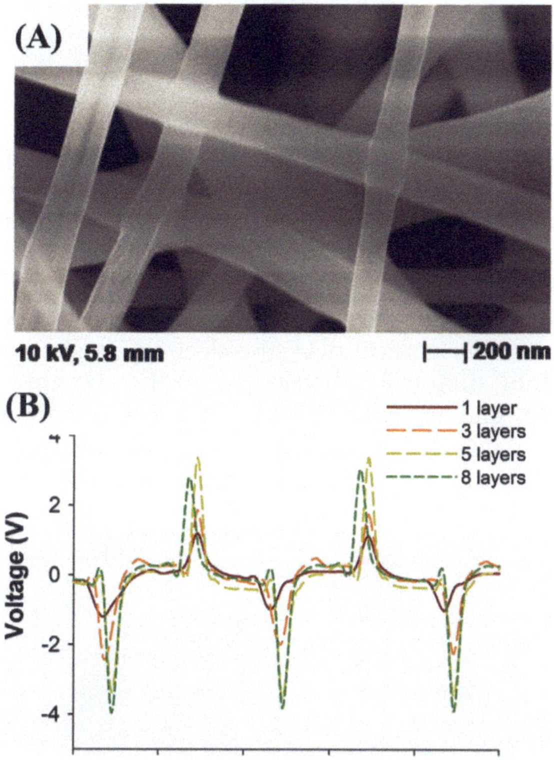

FIGURE 5.7 (A) PLLA electrospun nanofibers and (B) piezoelectric voltage output of PLLA nanofibers web with increasing layers. Source: *Reproduced with permission from reference S.J. Lee, A.P. Arun, K.J. Kim, Piezoelectric properties of electrospun poly (L-lactic acid) nanofiber web. Mater. Lett. 148 (2015) 58–62., copyright (2015) Elsevier.*

noncentrosymmetric orthorhombic crystal structure, where –C = O dipoles get polarized and aligned in the applied field direction due to mechanical stretching and high electrical field. During the electrospinning process, fast evaporation of solvents prevents the reorganization of polymer chains in a more stable conformation. A single PCL fiber was coated with an Au electrode to examine the piezoelectric properties. Piezoresponse force microscopy (PFM) measurement showed that with the application of tip voltage of ± 5 V the –C = O dipole polarity can be reversed by 180°. Amplitude–dc bias voltage measurements showed a butterfly-shaped hysteresis loop, which confirms the electromechanical behavior of PCL nanofibers. A device fabricated from PCL nanofibers was used for mechanical energy-harvesting. The device resulted in V_{oc} of 6 V at 10 Hz frequency and power density of 13μW/cm^2 at an applied force of 10 N.

5.5.4 Poly(acrylonitrile)

PAN is an amorphous vinyl-like polymer with a –CN pendant group. Wang et al. [12] used the electrospinning technique (Fig. 5.8A) to prepare PAN nanofibers (Fig. 5.8B). PAN has two conformations: planar zigzag or saw-tooth, and 3_1 helical. These conformations appear at 1250 and 1230 cm^{-1} in a Fourier-transform infrared spectrum. The zigzag conformation has an all trans-form structure with a dipole moment of 3 Debye. The fraction of zigzag conformation was found to be 79.7% in electrospun PAN fibers. Compared to 3_1 helical, planar zigzag has a shorter C-C bond distance and shorter interplanar spacing in the c-axis direction. Under the electrical field, the electron cloud shifts toward carbon atoms from uniformly distributed C-H• •N; this weakens the hydrogen bond and eventually leads to a conformational change in PAN chains. A device fabricated from a nonwoven PAN nanofibers mat

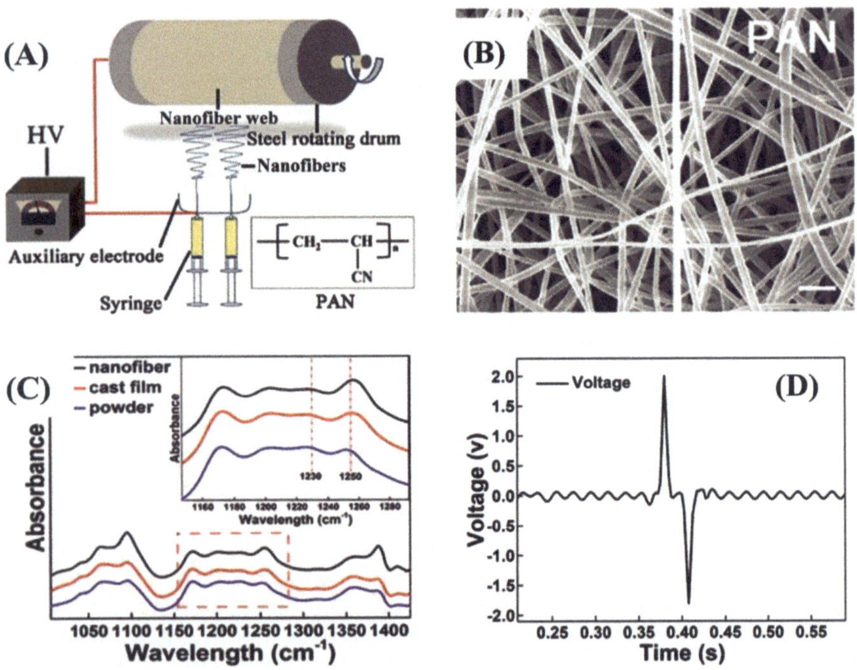

FIGURE 5.8 (A) Schematic illustration of preparation of nanofiber wed via electrospinning; (B) SEM image of as-spun PAN nanofibers; (C) Fourier-transform infrared spectra of PAN powder, PAN cast film, and electrospun PAN nanofiber membrane; and (D) piezoelectric outputs of the nanofiber device. Source: *Reproduced with permission from reference W. Wang, Y. Zheng, X. Jin, Y. Sun, B. Lu, H. Wang, et al. Unexpectedly high piezoelectricity of electrospun polyacrylonitrile nanofiber membranes. Nano Energy 56 (2019) 588–594, copyright (2019) Elsevier.*

resulted in V_{oc}, I_{sc}, and power of 2 V, 1.2 μA, and 0.91 μW (Fig. 5.8D). The piezoelectric output was further enhanced to 6 V in aligned fiber (zigzag content ~85%). Shao et al. [42] used electrospun PAN fibers for energy-harvesting from sound. The device was fabricated from PAN nanofibers having holes in it. These holes absorb the sound vibrations and produce electrical signals. The piezoelectric device resulted in V_{oc}, I_{sc}, and power of 58 V, 12 μA, and 210.3 μW, respectively, under 117 dB sound. Lower pressure and higher frequency of sound resulted in lower output. Cai et al. [43] prepared thin films of PAN by electric field induction technology. The best piezoelectric output was obtained when films were polarized under an applied electric field of 4 kV/cm. The piezoelectric device fabricated from polarized film resulted in V_{oc} and I_{sc} of 1.2 V and 0.13 μA at 20 N applied force. In another study [44], they optimized the poling temperature and found that poling at 80°C resulted in the best voltage output of ~0.9 V at an applied force of 15 N.

5.5.5 Cellulose

Bazhenov discovered the direct piezoelectric effect in wood in 1950 [45], and in 1955 Fukuda showed that wood also exhibits a converse piezoelectric effect [46]. Native crystalline cellulose is known as cellulose I and occurs in two crystalline forms: I_α and I_β, which consist of triclinic and monoclinic crystal units, respectively. These two noncentrosymmetric crystalline units are responsible for the piezoelectricity of cellulose-I [46]. García et al. [11] predicted that the piezoelectricity of cellulose-I_β 2D crystal is due to inter- and intralayer hydrogen bonds. Frka-Petesic et al. [47] reported the first experimental proof of permanent dipole moment of 4400 ± 400 Debye along the long axis of CNCs. Csoka et al. reported the shear piezoelectric constant (d25) of a polarized thin film of aligned CNC to be around 211 pm/V [48]. Tuukkanen et al. [49] fabricated a thin film of cellulose nanofiber (CNF) and measured its polarization-voltage (P-V) loop. A characteristic of P-V loop was observed between 40–50 V/μm, and maximum polarization of 0.15 μC/cm^2 was achieved at 50 V/μm field. They also measured the piezo sensitivity (charges per unit force) of these films at various places, which was 4.7 ± 0.9 pC/N. Further, they fabricated the CNC [50] and bacterial cellulose (BC) [51] film-based piezoelectric sensors and piezo sensitivity of these sensors was found to be 7.3 ± 2.5 and 5–20 pC/N, respectively. Ram et al. [52] prepared composite aerogels from isotropic and directionally ice-templated carboxylated CNCs and polyethyleneimine. CNCs were dispersed in water with the addition of polyethyleneimine and a crosslinker (poly(ethylene glycol) diglycidyl ether). The dispersions were frozen isotropically to yield isotropic aerogels and directionally frozen to prepare anisotropic aerogels. The schematic of setup and mechanism of preparation of anisotropic aerogel is illustrated in Fig. 5.9A. The 3-D microtomographic images of isotropic and anisotropic aerogels are shown in Fig. 5.9B and Fig. 5.9C. These aerogels were highly flexible and porous (~ 85%) and exhibited greater than 90% recovery at 50% compressive strain even after 100 compression–decompression cycles (Fig. 5.9D and Fig. 5.9E). Further, these aerogels were used in mechanical energy-harvesting applications.

Anisotropic aerogels led to an enhanced output voltage of 840 mV (at ~ 8 N applied force), which is over 2.6 times higher than isotropic aerogels (320 mV), as shown in Figs. 5.9F and 5.9G. The anisotropic aerogel

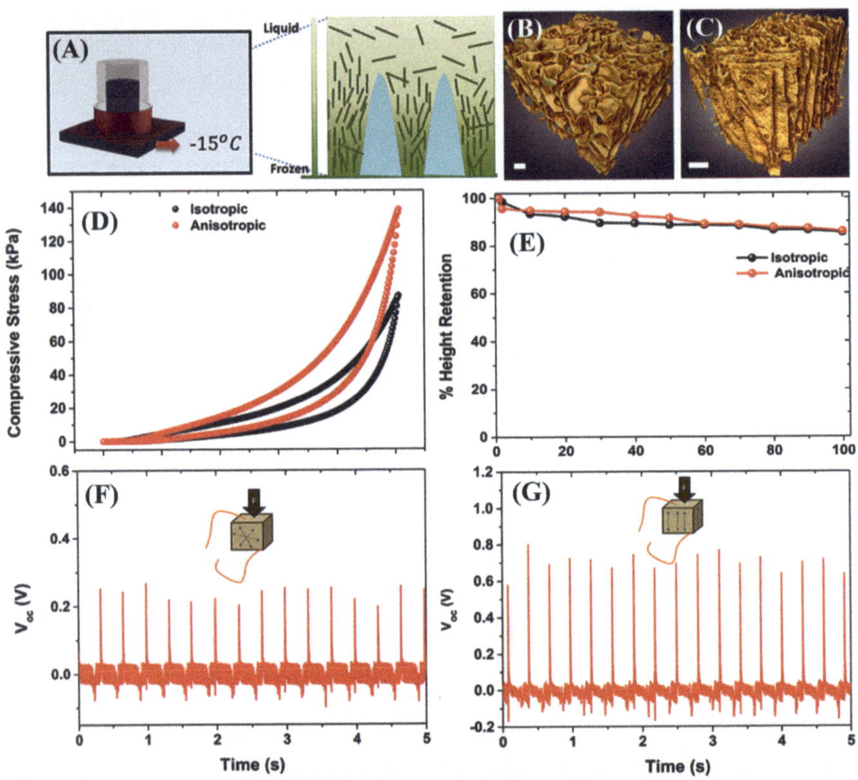

FIGURE 5.9 (A) Schematic of setup and mechanism of preparation of ice-templated anisotropic CNC aerogel. 3-D X-ray microtomographic images of CNC aerogels; (B) isotropic; and (C) anisotropic (scale bars = 45 μm). (D) Stress–strain curve of isotropic and anisotropic aerogels and (E) their percentage height retention at 50% strain. Voltage output from CNC aerogel PENG; (F) isotropic and (G) anisotropic aerogel. Source: *Reproduced with permission from reference S. Mun, H.U. Ko, L. Zhai, S.K. Min, H.C. Kim, J. Kim, Enhanced electromechanical behavior of cellulose film by zinc oxide nanocoating and its vibration energy harvesting. Acta Mater. 114 (2016) 1–6, copyright (2021) Elsevier.*

also led to an energy density of 52 nW/cm^2. The higher output performance of anisotropic aerogels was accredited to the higher active piezoelectric material and their alignment in the aerogels walls in the direction of the applied force.

Several cellulose-based composites have been reported as piezoelectric energy harvesters, where primarily cellulose acted as a flexible support to the piezoelectric particles, for example, ZnO–cellulose [54], [55], vanadium-doped ZnO-BC [56], $BaTiO_3$-BC [57], and $BaTiO_3$–cellulose [58]. Kim et al. fabricated cellulose-based flexible electroactive papers (EApap) for actuation and energy-harvesting applications. An EApap fabricated from regenerated cellulose films with a gold electrode (400 mm^2) resulted in peak to peak voltage output (V_{pp}) of 66.2 mV [59]. Further, to study the electrode effect, various EApap were fabricated by the dissolution of cellulose in LiCl/dimethylacetamide with Au, Ag, and Al electrodes. Single electrode (400 mm^2) EApap with gold, silver, and Al coating resulted in V_{pp} of 59, 92, and 113 mV. The output was further enhanced by increased electrode area (1200 mm^2), where it reached 131, 196, and 266 mV for Au, Ag, and Al electrodes, respectively. The higher output voltage with Al electrode was attributed to the lower work function of the electrode [60]. Further, ZnO nanocoated cellulose film (ZONCE) based on the EApap concept was utilized for energy-harvesting from shaker vibration (~14.2 N). The origin of the electroactive nature of EApap was attributed to the ordered and disordered region (Fig. 5.10A) with doped ions and hydroxyl-bound water in cellulose paper. Under applied electrical potential, these ions with free water move toward the anode. Further, the addition of ZnO increases the induced charge density upon application of mechanical force, as shown in Fig. 5.10B, which leads to higher output. The device resulted in V_{pp}, I_{sc}, and power output of 908 mV, 3.21 μA, and 2.91 μW, respectively (Fig. 5.10C) [53].

In yet another approach, cellulose was incorporated into a poly (dimethylsiloxane) (PDMS) matrix, where the output voltage is improved due to homolytic dissociation of Si-O-Si bonds in PDMS and their interaction with cellulose. Further, a conductive filler is added to improve the output power density. Alam et al. [61] fabricated a piezoelectric nanogenerator by embedding native cellulose microfibers (NCMF) in the PDMS matrix. At 5 wt.% NCMF and 0.5 wt.% multiwalled carbon nanotube (MWCNT) in PDMS, the nanogenerator resulted in optimum performance, where V_{pp}, I_{sc}, and power density of 30 V, 500 nA, and 9 μW/cm^3, respectively, was achieved at 40 kPa of applied pressure. The output performance of the nanogenerator was attributed to many factors, such as increased dispersion and conduction by MWCNT, strain-induced polarization in dispersed NCMF, and increased dipole density by hydrogen bond-forming under applied

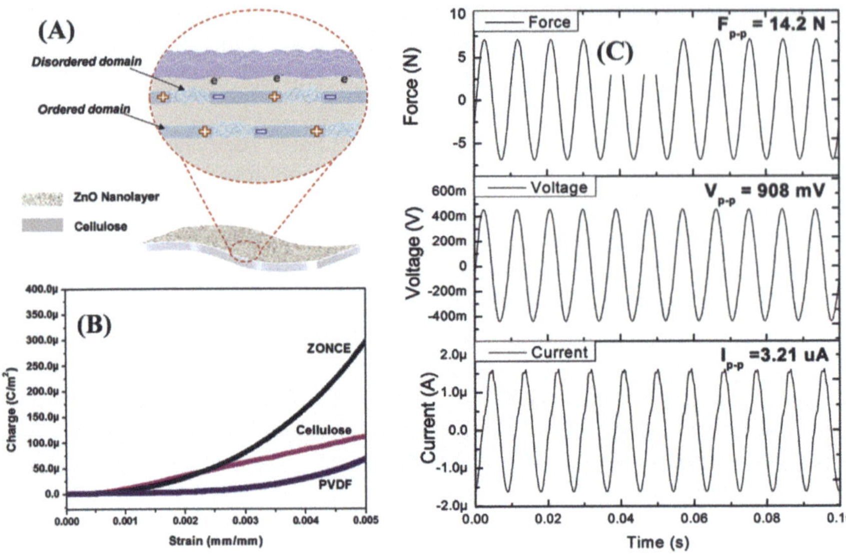

FIGURE 5.10 (A) Illustration showing mechanism of EApap; (B) enhancement in induced charge density under applied strain in presence of ZnO; and (C) piezoelectric output with applied external vibrations. Source: *Reproduced with permission from reference S. Mun, H.U. Ko, L. Zhai, S.K. Min, H.C. Kim, J. Kim, Enhanced electromechanical behavior of cellulose film by zinc oxide nanocoating and its vibration energy harvesting. Acta Mater. 114 (2016) 1–6, copyright (2016) Elsevier.*

stress due to homolytic cleavage of Si-O-Si. Zheng et al. [62] reported CNF/PDMS composite film-based PENG by applying PDMS on the porous surface of compressed freeze-dried CNF aerogel for energy-harvesting application. The nanogenerator resulted in V_{oc}, I_{sc}, and power density of 60.2 V, 10.1 μA, and 6.3 mW/cm^3, respectively. A pure CNF aerogel film resulted in an output voltage of 9.7 V due to displacement/relocation of electrical dipoles in cellulose under applied stress. The improved output voltage of the PDMS/CNF device was attributed to the above discussed homolytic dissociation of Si-O-Si bonds and larger mechanical deformation due to its porous nature [61,62]. Pusty et al. [63] reported Au-CNF/PDMS nanogenerators for energy-harvesting. The device resulted in V_{oc}, I_{sc}, and power density of 6 V, 700 nA, and 8.34 mW/m^2, respectively.

5.5.6 Chitin/chitosan

Chitin is a polymer of N-acetyl-D-glucoseamine having β(1→4) linkage and crystallizes in three polymorphs α, β, and γ. Fukuda reported shear piezoelectricity in α-chitin film in 1975 [14], which originates from oriented long-chain fibrous molecules. Li et al. [64] reported piezoelectric and ferroelectric properties in chitosan (from seashell) using various PFM measurements. Kim et al. [15] prepared β-phase transparent chitin nanofibers film for loudspeaker application. The film resulted in V_{oc} and power density of 1.04 V and 184 nW/cm^2 at 100 kPa. Street et al. [65] reported piezoelectricity in electrospun chitin nanofibers. A piezoelectric device fabricated from aligned nanofibers resulted in 300% improved output voltage (~0.26 V) than random fiber. Ghosh et al. [66] utilized prawn shell (chitin nanofibers) to fabricate a piezoelectric energy-harvesting device. Prawn shell contains protein (20%–40%), $CaCO_3$ (20%–50%), and chitin (15%–40%), where after demineralization only chitin acts as an active piezoelectric component. Field emission scanning electron microscope (FESEM) analysis of prawn shells revealed the hierarchical structure of chitin nanofibers, where chitin nanofibers had a diameter of 25 nm. 2D- fast Fourier-transform (FFT) of FESEM image suggested that the chitin nanofibers are highly oriented within a range of 90–118 degrees. The degree of fiber alignment was 25%, which was the origin of the self-polarization of chitin nanofibers in the prawn shell, and there was a piezoelectric charge coefficient (d_{33}) of ~2 pC/N. Under periodic applied stress of 5.2 kPa, the device resulted in V_{oc} and I_{sc} of 1 V and 1 nA. The device resulted in V_{oc}, I_{sc}, and power density of 4.4 V, 3.7 nA, and 0.76 μW/cm^2 when impacted with human finger (stress ~ 12kPa). Hoque et al. [67] utilized chitin nanofibers from crab shells to fabricate a mechanical energy-harvesting device. XRD characterization revealed that the thin film is predominately crystallized in α-form. The ferroelectric nature was examined by P-E loop measurement, where it showed a remnant polarization of 0.178 μ/cm^2. The film resulted in d_{33} value of 9.49 pCN^{-1}. A nanogenerator-based on this film was examined for mechanical energy-harvesting from human finger tapping (~27.5 N). With an increase in the frequency of finger tapping, the output performance also increased, and maximum V_{oc} and I_{sc} of 22 V and 0.12 μA were found at a frequency of 6 Hz. The device resulted in V_{oc} of 8 and 1.5 V when subject to a ball drop test and bath sonication vibrations. The device was also tested after 1, 2, 3, 6, 9, and 12 months and it was found that the device significantly retained output performances. The devices were biodegradable and biocompatible and could be used to harness energy from any mechanical deformation.

Deacetylated chitin, known as chitosan, is another semicrystalline biopolymer known to show a piezoelectric effect. In a hydrated state, chitosan shows eight-fold helical conformation, and in a dry state, it shows two-fold helical conformation. Praveen et al. [68] prepared chitosan films in different solvents and confirmed their noncentrosymmetric nature by second harmonic generation measurements. The films prepared in formic acid resulted in d_{33} of 18 pCN^{-1} and were then used for vibration sensors. The device fabricated from the film was mechanically excited at a vibration frequency of 80 Hz, which resulted in an output voltage of 3 mV. Films prepared in succinic, acetic, and adipic acid resulted in a lower voltage amplitude upon vibration excitation. The increased performance in the chitosan–formic acid film was due to the formation of formate, which led to increased crystallinity and hence increased piezoelectric properties. Hosseini et al. [69] reported a β-glycin/chitosan-based flexible nanogenerator, where chitosan's role was to provide support to crystalline β-glycin. The device fabricated from this resulted in an output voltage of ~32 mV.

5.5.7 Collagen

It is the most abundant animal protein, and a major constituent of bones. Piezoelectricity of type-I collagen in bone and tendon was first reported by Fukada and Yusuda [16,70]. Type-I collagen fibrils are formed by a quarter staggered arrangement of collagen molecules with periodic axial structure. Further, the collagen molecule comprises three polypeptide chains, two α1(I) chains and one α2(I) chain, in a unique triple-helical structure. These chains usually consist of a glycine–proline–hydroxyproline sequence [71]. The piezoelectricity in collagen originates from hydrogen bond formation between these polypeptide chains, and alignment of collagen fibrils causes polarization in collagen fiber (Fig. 5.11A) [72]. Ghosh et al. [72] fabricated a biopiezoelectric nanogenerator from fish swim bladder (FSB). FSB consists of well-aligned natural collagen fibrils, as seen in the FESEM image (Fig. 5.11B), that have a D-periodicity of ~62 nm. It means that the collagens are packed in a defined and ordered manner, conserving their original helical state. The device resulted in V_{oc}, I_{sc}, and power density of 10 V, 51 nA, and 4.15 μW/cm^2 respectively, when stressed by normal human finger pressure (1.4 MPa) as shown in Fig. 5.11C. In another report, they used fish scale (collagen nanofibrils) to prepare a piezoelectric energy harvester. The device generated V_{oc}, I_{sc}, and power density of 4 V, 1.5 μA, and 1.14 μW/cm^2 respectively, under compressive stress of 0.17 MPa [73]. A fish skin-based piezoelectric nanogenerator was used to harness mechanical energy from human physiological motions. Physiological motions by speaking, coughing, movement of Adam's apple, and wrist bending resulted in an output current of ~0.5–4 nA [74].

Wang et al. extracted collagen fiber aggregates from waste leather. A multilayered film was fabricated via a spin-coating method with doping of polyaniline-acidified MWCNT, which resulted in a 3D network structure of collagen fibers used for strain and pressure sensing applications [75]. Karan et al. utilized eggshell membrane (located between egg

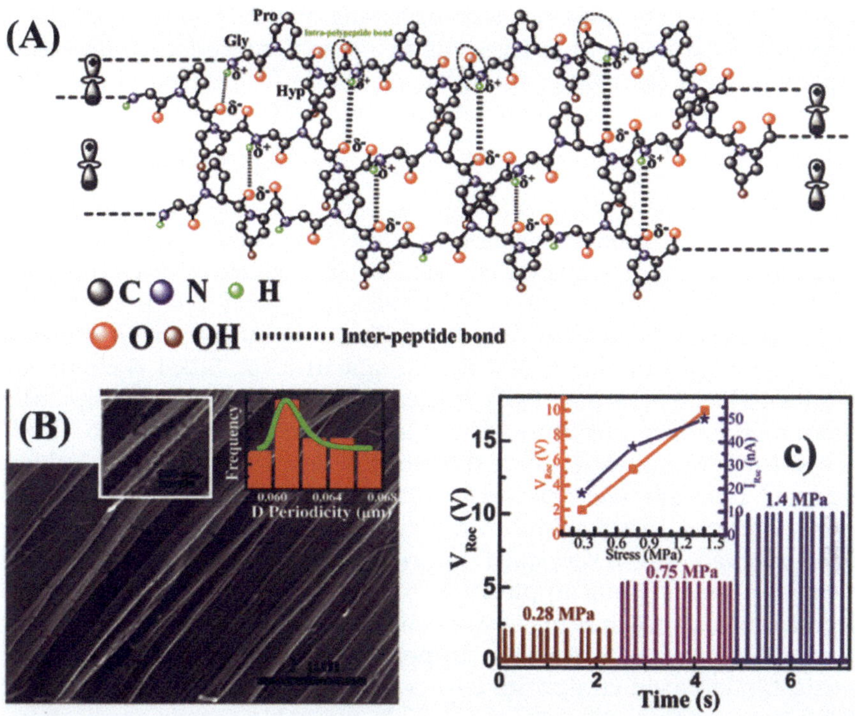

FIGURE 5.11 (A) The schematic illustration of intra- and inter-peptide chains of collagen with Gly-Pro-Hyp amino acids motif sequence to understand the origin of polarization in the collagen. (B) FESEM image with left inset representing a single collagen fiber to calculate the D-periodicity indicated by the histogram profile in the right inset; (C) rectified output voltages by different stresses with stress-dependent output voltages and currents in inset. Source: *Reproduced with permission from reference S.K. Ghosh, D. Mandal, Efficient natural piezoelectric nanogenerator: electricity generation from fish swim bladder. Nano Energy 28 (2016) 356–365, copyright (2016) Elsevier.*

cuticle and albumin) to fabricate a piezoelectric generator. Egg shell membrane (ESM) consists of proteins, where collagen type-I, V, and X account for 88%–93% by dry weight. The ESM-based piezoelectric device resulted in V_{oc}, I_{sc}, and power density of 26.4 V, 1.45 μA, and 2.59 mW/cm^3 at an applied stress of ~81.6 kPa. Further, the device was used for energy-harvesting from various mechanical inputs, such as human physiological motions, bath sonicator vibration, sound, sewing machine, and coin falling [76].

Vivekananthan et al. [77] fabricated a piezoelectric device from commercial collagen-coated cellulose fabric for energy-harvesting and humidity sensing. The electrically poled device resulted in V_{pp} and I_{sc} of 45 V and 250 nA under an applied force of 5 N. The resultant voltage decreased with increased relative humidity (RH), from 45 V @45% RH to 4 V @ 90% RH, due to a decrease in resistance with increasing humidity. Gelatin, produced by partial denaturation of collagen has also been used for energy-harvesting applications. An electrospun fish gelatin-based device resulted in an output voltage of 10 V at an applied pressure of 20 kPa [78].

5.5.8 Silk

Silk has been used in textiles for centuries, and it is being widely explored in medical applications. Silk fibroin consists of low- and high-molecular-weight chains crosslinked by disulfide bond. Silk is a semicrystalline polymer, which exhibits crystalline polymorphism in the form of metastable silk-I, stable silk-II, and silk-III form. Silk-I is composed of an α-helix structure and silk-II is composed of a pleated antiparallel β-sheet secondary structure. Shear stress converts silk-I into silk-II. The silk-III form consists of a threefold helical structure [17]. Fukuda reported inherent shear piezoelectricity in a fiber bundle of silk [79]. Piezoelectricity in silk fibroin originates from net polarization of uniaxial aligned silk-II crystals: inter- and intrachain sliding within pleated β-sheets upon application of stress could also rotate amide dipoles and cause internal polarization in silk [17]. Kim et al. [80] used silk fibroin as a flexible support to prepare piezoelectric nanoparticle/silk-based composite energy-harvesting devices. Joseph et al. [81] prepared thin silk film for flexible wearable sensor applications. The sensor resulted in V_{pp} of 2.1 V under 3 bar pressure of compressed dry air. Pan et al. [82] fabricated a nanogenerator from spider silk (SS), which resulted in output voltage and current of 40.7 mV and 105 nA at an applied strain rate of 0.02 s^{-1}. Karan et al. [83] established, using PFM, that SS has piezoelectricity at the molecular level, where the out of plane piezoelectric coefficient was 0.36 pm/V. Further, a piezoelectric nanogenerator was fabricated for mechanical energy-harvesting, which resulted in V_{oc}, I_{sc} and power density of 21.3 V, 0.68 μA, and 4.56 $\mu W/cm^2$, respectively at 82.2 kPa pressure. The performance of SS PENG was attributed to the inherent piezoelectricity of SS and dipoles inside the crystal lattice in the direction of the applied mechanical force. A wire-like nanogenerator fabricated from SS resulted in an output voltage of 5.7 and 3.2 V under bending and stretching conditions. The device was also explored for energy-harvesting/sensing of human physiological motions.

Sencadas et al. [84] reported electrospun silk fibroin nanofibers for energy-harvesting applications. PFM measurements confirmed their piezoelectric/ferroelectric nature, where a single fiber resulted in a piezoelectric coefficient of 38 ± 2 pm/V. Amplitude vs. voltage and phase vs. voltage loop exhibited butterfly and rectangular loops, which are characteristic of ferroelectricity. The device fabricated from these membranes resulted in an output voltage of 8 V at 24 Hz frequency, which was stable up to 10,000 cycles.

5.5.9 Other polymers

A few other nonfluorinated polymers have also shown piezoelectric characteristics, and they are listed in Sections 5.2, 5.3, and 5.4. These polymers exhibit weak to significant piezoelectricity. However, due to very low response, these polymers are barely explored for energy-harvesting applications. Koyama et al. [18] used polyurea thin film for vibrational energy-harvesting. Polyurea thin film was formed by vapor deposition polymerization of 4,4′-diphenylmethane diisocyanate (MDI) and 4,4′-diamino diphenyl ether (ODA). The vapor deposition polymerization eliminates the crosslinking in polyurea chains [19]. The thin-film base was mounted on Be-Cu cantilever of various thicknesses. Lower thickness and shorter cantilever length achieved a higher energy conversion efficiency maximum of 0.233% with an output voltage of ~100 mV. A comprehensive summary of mechanical energy-harvesting from various nonfluorinated polymers is shown in Table 5.2.

5.6 Summary and future outlook

Nonfluorinated polymers are economically viable and environment-friendly piezoelectric materials. They are widely utilized for mechanical energy-harvesting from various inputs such as human physiological motion, impact force, and vibrations. Solution-drawn films of bio-derived Nylon 11/CNC-based PENGs resulted in an output voltage of ~7 V with high durability. The voltage output from PLA electrospun fibers and films ranged from ~7–17 V. Natural polymers-based PENGs surprisingly have yielded a wider range of voltage output, for example, regenerated cellulose EApap (~66.2 mV), CNC aerogels (840 mV), chitin nanofiber film from prawn shell (1 V), chitin nanofibers from crab shell (22 V), collagen from FSB (10 V), SS (21.3 V), collagen coated cellulose fabric (45 V) and cellulose/PDMS aerogel films (60.2 V). PAN, which is commonly utilized in fabrics, is considered as a weak piezoelectric material but an electrospun nanofiber mat of PAN resulted in a high output voltage of 58 V. Utilization of these alternative polymers can reduce the cost and adverse impact on the environment. Among these, biopolymers have the highest piezoelectric output voltage and have the advantages of renewability and abundant availability. High piezoelectric output from commonly used polymers, such as cellulose,

TABLE 5.2 Overview of mechanical energy-harvesting using nonfluorinated polymers.

S. no.	Polymeric material	V_{oc} or Voltage output	I_{sc} or Current output	Power or Power density	Ref.
1.	Nylon nanowires in AAO	1 V	100 nA	–	[8]
2.	Plasticized nylon 11 and 5 wt.% nylon 11/CNC	2.8 and 6.95 V	–	90 and 500 μW/cm^3	[35]
3.	Electrospun PLA	7.2 V	–		[10]
4.	Double-layer PLLA film	9.05 V	–	14.17 μW	[41]
5.	Electrospun PCL	6 V	–	13 μW/cm^2	[13]
6.	Electrospun PAN	2 V	1.2 μA	0.91 μW	[12]
7.	Electroactive paper from regenerated cellulose	66.2 mV	–	–	[59]
8.	5 wt.% Native cellulose microfibrils and 0.5 wt.% MWCNT with PDMS	30 V	500 nA	9 μW/cm^3	[61]
9.	CNF aerogel film	9.7 V	–	–	[62]
10.	Transparent β-chitin nanofiber film	1.04 V	–	184 nW/cm^2	[15]
11.	Chitin nanofiber film derived from prawn shell	4.4 V	3.7 nA	0.76 μW/cm^2	[66]
12.	Chitosan formate (chitosan prepared in formic acid)	3 mV	–	–	[68]
13.	Collagen fibrils from fish swim bladder	10 V	51 nA	4.15 μW/cm^2	[72]
14.	Collagen coated cellulose fabric	45 V	250 nA	–	[77]
15.	Electrospun fish gelatin	10 V	–	–	[78]
16.	Silk film	2.1 V	–	–	[81]
17.	Spider silk	21.3 V	0.68 μA	4.56 μW/cm^2	[82]
18.	Vapor deposited polyurea film	100 mV	–	–	[19]

silk, and PAN, can revolutionize the field of wearable and intelligent clothing. Since materials manufacturing processes are already established for these polymers, transforming these materials into mechanical energy-harvesting devices, smart clothing, and tactile sensors would be low-hanging fruit.

References

[1] A.J. Lovinger, Ferroelectric polymers, Science 220 (4602) (1983) 1115–1121.

[2] F. Ram, Nanocellulose reinforced polymer composites for energy application. 2020. Thesis submitted to Academy of Scientific and Innovative Research at Council of Scientific and Industrial Research—National Chemical Laboratory, Pune. <http://dspace.ncl.res.in:8080/xmlui/handle/20.500.12252/5920>.

[3] J.S. Harrison, Z. Ounaies, Piezoelectric polymers, Encyclopedia of Polymer Science and Technology, 3, John Wiley & Sons, Inc, Hoboken, NJ, 2002, pp. 474–498.

[4] Q. Guo, G.Z. Cao, I.Y. Shen, Measurements of piezoelectric coefficient d_{33} of lead zirconatetitanate thin films using a mini force hammer, J. Vib. Acoust. 135 (2013) 011003.

[5] T.R. Shrout, Z.P. Chang, N. Kim, S. Markgraf, Dielectric behavior of single crystals near the $(1-X)Pb(Mg_{1/3}Nb_{2/3})O_3$-$(x)PbTiO_3$ morphotropic phase boundary, Ferroelectr. Lett. 12 (1990) 63–69.

[6] R. Zhang, B. Jiang, W. Cao, Elastic, piezoelectric, and dielectric properties of multidomain $0.67Pb(Mg_{1/3}Nb_{2/3})O_{3\text{-}0.33}PbTiO_3$ single crystals, J. Appl. Phys. 90 (2001) 3471–3475.

[7] A.J.P. Martin, Tribo-electricity in wool and hair, Proc. Phys. Soc. 53 (1941) 183.

[8] A. Datta, Y.S. Choi, E. Chalmers, C. Ou, S. Kar-Narayan, Piezoelectric nylon-11 nanowire arrays grown by template wetting for vibrational energy harvesting applications, Adv. Funct. Mater. 27 (2) (2017) 1604262.

[9] Z. Zhang, M.H. Litt, L. Zhu, Unified understanding of ferroelectricity in n-nylons: is the polar crystalline structure a prerequisite? Macromolecules 49 (8) (2016) 3070–3082.

[10] S.J. Lee, A.P. Arun, K.J. Kim, Piezoelectric properties of electrospun poly (L-lactic acid) nanofiber web, Mater. Lett. 148 (2015) 58–62.

[11] Y. García, Y.B. Ruiz-Blanco, Y. Marrero-Ponce, C.M. Sotomayor-Torres, Orthotropic piezoelectricity in 2D nanocellulose, Sci. Rep. 6 (2016) 34616.

[12] W. Wang, Y. Zheng, X. Jin, Y. Sun, B. Lu, H. Wang, et al., Unexpectedly high piezoelectricity of electrospun polyacrylonitrile nanofiber membranes, Nano Energy 56 (2019) 588–594.

[13] V. Sencadas, Energy harvesting applications from Poly(ε-caprolactone) electrospun membranes, ACS Appl. Polym. Mater. 2 (6) (2020) 2105–2110.

[14] E. Fukada, S. Sasaki, Piezoelectricity of α-chitin, J. Polym. Sci.: Polym. Phys. (Ed.) 13 (9) (1975) 1845–1847.

[15] K. Kim, M. Ha, B. Choi, S.H. Joo, H.S. Kang, J.H. Park, et al., Biodegradable, electroactive chitin nanofiber films for flexible piezoelectric transducers, Nano Energy 48 (1) (2018) 275–283.

[16] E. Fukada, I. Yasuda, Piezoelectric effects in collagen, Jpn. J. Appl. Phys. 3 (1964) 117–121.

[17] T. Yucel, P. Cebe, D.L. Kaplan, Structural origins of silk piezoelectricity, Adv. Funct. Mater. 21 (4) (2011) 779–785.

[18] D. Koyama, K. Nakamura, Electric power generation using vibration of a polyurea piezoelectric thin film, Appl. Acoust. 71 (5) (2010) 439–445.

[19] Y. Takahashi, M. Iijima, E. Fukada, Pyroelectricity in poled thin films of aromatic polyurea prepared by vapor deposition polymerization, Jpn. J. Appl. Phys. 28 (12A) (1989) L2245.

[20] J.S. Harrison, Z. Ounaies, Piezoelectric polymers. National Aeronautics and Space Administration-Langley Research Center. Report no. CR-2001–211422. <https://ntrs.nasa.gov/citations/20020044745> (accessed 11.03.21).
[21] M.G. Broadhurst, C.G. Malmberg, F.I. Mopsik, W.P. Harris, Piezo-and pyroelectricity in polymer electrets, in: M.M. Perlman (Ed.), Electrets: Charge Storage and Transport in Dielectrics, The Electrochemical Society, Inc, Princeton, NJ, 1973, pp. 492–504.
[22] F.I. Mopsik, M.G. Broadhurst, Molecular dipole electrets, J. Appl. Phys. 46 (10) (1975) 4204–4208.
[23] Lewis T.J. The origin of piezoelectricity in polymeric materials, in: Annual Report Conference on Electrical Insulation and Dielectric Phenomena, IEEE, 2002 Oct 20, pp. 291–294.
[24] S. Sukumaran, S. Chatbouri, D. Rouxel, E. Tisserand, F. Thiebaud, Ben, et al., Recent advances in flexible PVDF based piezoelectric polymer devices for energy harvesting applications, J. Intell. Mater. Syst. Struct. 32 (2021) 746–780.
[25] Q. Pan, C. Fang, Z. Qin, Q. Gu, X. Cheng, D. Xu, et al., Nonlinear optical properties and thermal stability of the poled polymer NTAB/PEK-c, Mater. Lett. 57 (2003) 2612–2615.
[26] J.A. Giacometti, O.N. Oliveira, Corona charging of polymers, IEEE Trans. Electr. Insul. 27 (1992) 924–943.
[27] Z.L. Wang, On Maxwell's displacement current for energy and sensors: the origin of nanogenerators, Mater. Today 20 (2) (2017) 74–82.
[28] Z. Wu, T. Cheng, Z.L. Wang, Self-powered sensors and systems based on nanogenerators, Sensors 20 (10) (2020) 2925.
[29] K. Meurisch, B. Gojdka, T. Strunskus, V. Zaporojtchenko, F. Faupel, Vapour phase deposition of highly crystalline self-poled piezoelectric nylon-11, J. Phys. D: Appl. Phys. 45 (5) (2012) 055304.
[30] M.H. Litt, C.H. Hsu, P. Basu, Pyroelectricity and piezoelectricity in nylon 11, J. Appl. Phys. 48 (1977) 2208–2212.
[31] J.I. Scheinbeim, Piezoelectricity in γ-form nylon 11, J. Appl. Phys. 52 (1981) 5939–5942.
[32] J.I. Scheinbeim, J.W. Lee, B.A. Newman, Ferroelectric polarization mechanisms in nylon 11, Macromolecules 25 (14) (1992) 3729–3732.
[33] N. Tsutsumi, T. Yamaoka, Ferroelectric properties of ultrathin films of Nylon 11, Thin Solid. Films 518 (2) (2009) 814–818.
[34] Dunn P.E., Carr S.H. Piezoelectricity in nylon 5, 7, in: 6th International Symposium on Electrets, (ISE 6) Proceedings, IEEE. 1988, pp. 256–260.
[35] F. Ram, S. Radhakrishnan, T. Ambone, K. Shanmuganathan, Highly flexible mechanical energy harvester based on nylon 11 ferroelectric nanocomposites, ACS Appl. Polym. Mater. 1 (8) (2019) 1998–2005.
[36] Chakhchaoui N., Frahan R., Eddiai A., Meddad M., Cherkaoui O., Boughaleb Y., et al. Improved piezoelectric properties of PLA/PZT hybrid composite films, IOP Conf. Ser.: Mater. Sci. Eng. 827 (1) (2020) 012012.
[37] G. Zhao, B. Huang, J. Zhang, A. Wang, K. Ren, Z.L. Wang, Electrospun poly(L-Lactic Acid) nanofibers for nanogenerator and diagnostic sensor applications, Macromol. Mater. Eng. 302 (5) (2017) 1600476.
[38] J. Zhu, L. Jia, R. Huang, Electrospinning poly(L-lactic acid) piezoelectric ordered porous nanofibers for strain sensing and energy harvesting, J. Mater. Sci.-Mater. Electron. 28 (16) (2017) 12080–12085.
[39] E.J. Curry, K. Ke, M.T. Chorsi, K.S. Wrobel, A.N. Miller, A. Patel, et al., Biodegradable piezoelectric force sensor, Proc. Natl. Acad. Sci. 115 (5) (2018) 909–914.

[40] S. Gong, B. Zhang, J. Zhang, Z.L. Wang, K. Ren, Biocompatible poly(lactic acid)-based hybrid piezoelectric and electret nanogenerator for electronic skin applications, Adv. Funct. Mater. 30 (14) (2020) 1908724.
[41] C. Zhao, J. Zhang, Z.L. Wang, K. Ren, A poly (L-lactic acid) polymer-based thermally stable cantilever for vibration energy harvesting applications, Adv. Sustain. Syst. 1 (9) (2017) 1700068.
[42] H. Shao, H. Wang, Y. Cao, X. Ding, J. Fang, H. Niu, et al., Efficient conversion of sound noise into electric energy using electrospun polyacrylonitrile membranes, Nano Energy (2020) 104956.
[43] T. Cai, T. Bi, Y. Yang, E. Bi, S. Xue, Preparation of high-performance polymer piezoelectric thin film by electric field induction technology, Polymer 20 (2019) 122486.
[44] T. Cai, Y. Yang, E. Bi, Preparation of high-performance polyacrylonitrile piezoelectric thin film by temperature control, React. Funct. Polym. 27 (2020) 104638.
[45] V.A. Bazhenov, V.P. Kanstantinovo, Daklady Akad. Nauk. U.S.S.R. 71 (1950) 283–286. (from Chem. Abst. 1951;45:2747).
[46] E. Fukada, Piezoelectricity of wood, J. Phys. Soc. Jpn. 10 (2) (1955) 149–154.
[47] B. Frka-Petesic, B. Jean, L. Heux, First experimental evidence of a giant permanent electric-dipole moment in cellulose nanocrystals, Europhys. Lett. 107 (2) (2014) 28006.
[48] L. Csoka, I.C. Hoeger, O.J. Rojas, I. Peszlen, J.J. Pawlak, P.N. Peralta, Piezoelectric effect of cellulose nanocrystals thin films, ACS Macro Lett. 1 (7) (2012) 867–870.
[49] S. Rajala, T. Siponkoski, E. Sarlin, M. Mettänen, M. Vuoriluoto, A. Pammo, et al., Cellulose nanofibril film as a piezoelectric sensor material, ACS Appl. Mater. Interfaces 8 (24) (2016) 15607–15614.
[50] S. Tuukkanen, S. Rajala, A survey of printable piezoelectric sensors, IEEE SENSORS, Busan (2015) 1–4. Available from: https://doi.org/10.1109/ICSENS.2015.7370542.
[51] R. Mangayil, S. Rajala, A. Pammo, E. Sarlin, J. Luo, V. Santala, et al., Engineering and characterization of bacterial nanocellulose films as low cost and flexible sensor material, ACS Appl. Mater. Interfaces 9 (2017) 19048–19056.
[52] F. Ram, B. Biswas, A. Torris, G. Kumaraswamy, K. Shanmuganathan, Elastic piezoelectric aerogels from isotropic and directionally ice-templated cellulose nanocrystals: comparison of structure and energy harvesting, Cellulose 281 (2021) 6323–6337.
[53] S. Mun, H.U. Ko, L. Zhai, S.K. Min, H.C. Kim, J. Kim, Enhanced electromechanical behavior of cellulose film by zinc oxide nanocoating and its vibration energy harvesting, Acta Mater. 114 (2016) 1–6.
[54] A. Kumar, H. Gullapalli, K. Balakrishnan, A. Botello-Mendez, R. Vajtai, M. Terrones, et al., Flexible ZnO–cellulose nanocomposite for multisource energy conversion, Small 7 (15) (2011) 2173–2178.
[55] H.U. Ko, S. Mun, S.K. Min, G.W. Kim, J. Kim, Fabrication of cellulose ZnO hybrid nanocomposite and its strain sensing behavior, Materials 7 (10) (2014) 7000–7009.
[56] G. Zhang, Q. Liao, M. Ma, F. Gao, Z. Zhang, Z. Kang, et al., Uniformly assembled vanadium doped ZnO microflowers/bacterial cellulose hybrid paper for flexible piezoelectric nanogenerators and self-powered sensors, Nano Energy 52 (2018) 501–509.
[57] G. Zhang, Q. Liao, Z. Zhang, Q. Liang, Y. Zhao, X. Zheng, et al., Novel piezoelectric paper-based flexible nanogenerators composed of $BaTiO_3$ nanoparticles and bacterial cellulose, Adv. Sci. 3 (2) (2016) 1500257.
[58] K. Shi, X. Huang, B. Sun, Z. Wu, J. He, P. Jiang, Cellulose/$BaTiO_3$ aerogel paper based flexible piezoelectric nanogenerators and the electric coupling with triboelectricity, Nano Energy (57)(2019) 450–458.
[59] Z. Abas, H.S. Kim, L. Zhai, J. Kim, J.H. Kim, Possibility of cellulose-based electro-active paper energy scavenging transducer, J. Nanosci. Nanotechnol. 14 (10) (2014) 7458–7462.

[60] Z. Abas, H.S. Kim, L. Zhai, J. Kim, J.H. Kim, Electrode effects of a cellulose-based electro-active paper energy harvester, Smart Mater. Struct. 23 (7) (2014) 074003.
[61] M.M. Alam, D. Mandal, Native cellulose microfiber-based hybrid piezoelectric generator for mechanical energy harvesting utility, ACS Appl. Mater. Interfaces 8 (3) (2016) 1555–1558.
[62] Q. Zheng, H. Zhang, H. Mi, Z. Cai, Z. Ma, S. Gong, High-performance flexible piezoelectric nanogenerators consisting of porous cellulose nanofibril (CNF)/poly (dimethylsiloxane)(PDMS) aerogel films, Nano Energy 26 (2016) 504–512.
[63] M. Pusty, P.M. Shirage, Gold nanoparticle–cellulose/PDMS nanocomposite: a flexible dielectric material for harvesting mechanical energy, RSC Adv. 10 (17) (2020) 10097–10112.
[64] T. Li, K. Zeng, Nanoscale piezoelectric and ferroelectric behaviors of seashell by piezoresponse force microscopy, J. Appl. Phys. 113 (18) (2013) 187202.
[65] R.M. Street, T. Huseynova, X. Xu, P. Chandrasekaran, L. Han, W.Y. Shih, et al., Variable piezoelectricity of electrospun chitin, Carbohydr. Polym. 195 (2018) 218–224.
[66] S.K. Ghosh, D. Mandal, Bio-assembled, piezoelectric prawn shell made self-powered wearable sensor for non-invasive physiological signal monitoring, Appl. Phys. Lett. 110 (12) (2017) 123701.
[67] N.A. Hoque, P. Thakur, P. Biswas, M.M. Saikh, S. Roy, B. Bagchi, et al., Biowaste crab shell-extracted chitin nanofiber-based superior piezoelectric nanogenerator, J. Mater. Chem. A 6 (28) (2018) 13848–13858.
[68] E. Praveen, S. Murugan, K. Jayakumar, Investigations on the existence of piezoelectric property of a bio-polymer–chitosan and its application in vibration sensors, RSC Adv. 7 (56) (2017) 35490–35495.
[69] E.S. Hosseini, L. Manjakkal, D. Shakthivel, R. Dahiya, Glycine–chitosan-based flexible biodegradable piezoelectric pressure sensor, ACS Appl. Mater. Interfaces 12 (8) (2020) 9008–9016.
[70] E. Fukada, I. Yasuda, On the piezoelectric effect of bone, J. Phys. Soc. Jpn. 12 (10) (1957) 1158–1162.
[71] K.A. Piez, J. Gross, The amino acid composition of some fish collagens: the relation between composition and structure, J. Biol. Chem. 235 (4) (1960) 995–998.
[72] S.K. Ghosh, D. Mandal, Efficient natural piezoelectric nanogenerator: electricity generation from fish swim bladder, Nano Energy 28 (2016) 356–365.
[73] S.K. Ghosh, D. Mandal, High-performance bio-piezoelectric nanogenerator made with fish scale, Appl. Phys. Lett. 109 (10) (2016) 103701.
[74] S.K. Ghosh, D. Mandal, Sustainable energy generation from piezoelectric biomaterial for noninvasive physiological signal monitoring, ACS Sustain. Chem. Eng. 5 (10) (2017) 8836–8843.
[75] X. Wang, O. Yue, X. Liu, M. Hou, M. Zheng, A novel bio-inspired multi-functional collagen aggregate based flexible sensor with multi-layer and internal 3D network structure, Chem. Eng. J. 392 (2020) 123672.
[76] S.K. Karan, S. Maiti, S. Paria, A. Maitra, S.K. Si, J.K. Kim, et al., A new insight towards eggshell membrane as high energy conversion efficient bio-piezoelectric energy harvester. Mater, Today Energy 9 (2018) 114–125.
[77] V. Vivekananthan, N.R. Alluri, Y. Purusothaman, A. Chandrasekhar, S. Selvarajan, S. J. Kim, Biocompatible collagen nanofibrils: an approach for sustainable energy harvesting and battery-free humidity sensor applications, ACS Appl. Mater. interfaces 10 (22) (2018) 18650–18656.
[78] S.K. Ghosh, P. Adhikary, S. Jana, A. Biswas, V. Sencadas, S.D. Gupta, et al., Electrospun gelatin nanofiber based self-powered bio-e-skin for health care monitoring, Nano Energy 36 (2017) 166–175.

[79] E. Fukada, On the piezoelectric effect of silk fibers, J. Phys. Soc. Jpn. 11 (12) (1956) 1301A.

[80] K.N. Kim, J. Chun, S.A. Chae, C.W. Ahn, I.W. Kim, S.W. Kim, et al., Silk fibroin-based biodegradable piezoelectric composite nanogenerators using lead-free ferroelectric nanoparticles, Nano Energy 14 (2015) 87–94.

[81] J. Joseph, S.G. Singh, S.R. Vanjari, Leveraging innate piezoelectricity of ultra-smooth silk thin films for flexible and wearable sensor applications, IEEE Sens. J. 17 (24) (2017) 8306–8313.

[82] Pan C.T., Yen C.K., Hsieh M.C., Zeng S.W., Kuo S.W., Shiue Y.L., et al. Energy harvester made of Taiwan local Nephila pilipes spider silk. In 19th International Conference on Solid-State Sensors, Actuators and Microsystems (Transducers) 2017:2220-2222. IEEE.

[83] S.K. Karan, S. Maiti, O. Kwon, S. Paria, A. Maitra, S.K. Si, et al., Nature driven spider silk as high energy conversion efficient bio-piezoelectric nanogenerator, Nano Energy 49 (2018) 655–666.

[84] V. Sencadas, C. Garvey, S. Mudie, J.J. Kirkensgaard, G. Gouadec, S. Hauser, Electroactive properties of electrospun silk fibroin for energy harvesting applications, Nano Energy 66 (2019) 104106.

CHAPTER

6

Polysaccharide-based nanocomposites for energy-harvesting nanogenerators

Avinash R. Pai[1], Sharin Maria Thomas[3] and Sabu Thomas[1,2]

[1]International and Inter University Centre for Nanoscience and Nanotechnology, Mahatma Gandhi University, Kottayam, India, [2]School of Energy Materials, Mahatma Gandhi University, Kottayam, India, [3]Centre for Nanotechnology Research, Vellore Institute of Technology, Vellore, India

6.1 Introduction

Nowadays, the modern electronic devices demand multifunctional traits such as flexibility miniaturization, portability, and the ability to function in self-power mode. Most of the electronic gadgets are still powered using conventional electrochemical batteries that can function only over a short span of time and need to be replaced at regular intervals. In the case of biomedical devices, additional surgeries are performed to replace these batteries, which also increases the risks of infection and the associated costs. Moreover, batteries occupy more space and also increase the weight of these implantable devices which hinders the possibility of miniaturization. This has led to a widespread interest in harnessing energy from ambient environments, such as mechanical vibrations, body motion, electromagnetic radiation, heat, wind, and tidal waves, into electrical energy. Numerous energy-harvesting strategies have been proposed at meso-, micro-, and nanoscale dimensions using electromagnetic, piezoelectric, electrostatic, triboelectric, thermoelectric, and pyroelectric transduction mechanisms as self-power sources for a broad range of

Engineered Polymer Nanocomposites for Energy Harvesting Applications
DOI: https://doi.org/10.1016/B978-0-12-824155-4.00006-5

portable electronic gadgets. This rising demand to harness small-scale energy for various miniature and wearable devices has led to the development of nanogenerators (NGs). NGs are devices that effectively convert ambient mechanical/thermal energy produced due to small-scale physical change into electrical energy. The integral concept of NGs was introduced by Prof. Zhong Lin Wang of the Georgia Institute of Technology, United States [1,2]. Electrical output can be harvested from different kinds of mechanical oscillations such as low-frequency human motions, raindrops, tidal waves, automobiles, flowing water, etc. via different nanogenerator assemblies. NGs are broadly classified based on the transduction mechanism used to generate the power, such as piezoelectric nanogenerators (PENG) [3–5], triboelectric nanogenerators (TENG) [6,7], electromagnetic nanogenerators (EMG) [8], and thermoelectric nanogenerators (TEG) [9,10]. Amongst them, TENGs and PENGs have been extensively studied and are widely used due to their simplicity in design, better efficiency, flexibility, and moderate power density at low frequencies. Such NGs could be availed for self-powered smart wearables and medical implantable devices for diagnostics. In general, PENGs are more functional in the range of 60–100 Hz while TENGs are favorable for much lower frequencies (less than 10 Hz), mainly human motions [11,12]

Undoubtedly, the best candidates for designing NGs are combinations of inorganic materials and synthetic polymers, which can generate very high electrical output. But the majority of these materials are not biodegradable and are toxic and pose serious concern to the environment. Hence, there is an urgent need to replace these synthetic materials derived from fossil resources with natural and sustainable biopolymers. Polysaccharides are the major constituent of biomass and account for more than 90% of carbohydrate mass found in nature. These are polymers consisting of high-molecular-weight carbohydrate molecules with many monosaccharide units. A number of polysaccharides, like cellulose, chitin, chitosan, pullulan, alginates, starch, and their derivatives have been extensively studied for use in energy-harvesting applications. This chapter discusses the working mechanisms, configurations, and design aspects of polysaccharide-based nanocomposites with a focus on piezoelectric and TENGs. These NGs devised from biodegradable polymers have enormous potential for use as smart, wearable, self-powered, and disposable electronic sensors, and implantable energy harvesters for biomedical devices,

6.2 Piezoelectric nanogenerators

The unique mechanoelectrical conversion efficiency of piezoelectric materials has a significant role in the development of PENGs. The first PENG reported was based on inorganic piezocrystals, such as PZT (lead zirconate titanate), barium titanate ($BaTiO_3$), ZnO nanowires, etc., which display a high electromechanical coupling coefficient and high power densities [13–16]. In general, piezoceramics showcase a high electromechanical coupling factor with large dielectric and piezoelectric coefficients. But they are very brittle in nature and cannot withstand large strains, and are also toxic. On the other hand, processable electroactive polymers like poly(vinylidene fluoride) (PVDF), poly(vinylidene fluoride-co-trifluoroethylene) (PVDF-TrFE) have better flexibility but low electromechanical coupling factor. Hence, piezoelectric polymer–ceramic composites are widely explored for PENGs that combine the benefits of both these materials. The incorporation of nanomaterials, surface functionalization, and varying of the structural aspects can enhance the characteristic β-phase of the piezoelectric materials. The mechanism of transduction in the piezoelectric harvester depends on the nature of the applied energy. When a mechanical strain is applied on a piezoelectric material, electricity will be generated which is known as the direct piezoelectric effect. In the converse effect, the piezo material becomes mechanically deformed due to the input electrical energy. Fig. 6.1A shows electromechanical transduction in piezo materials. The performance of the harvester depends on the operation modes and different structural configurations for the range of applications.

6.2.1 Operation modes

Every piezoelectric material manifests an imaginary axis called the polar axis, existing along the separation of polar or opposite charges. Depending on the direction of the applied strain relative to the direction of polar axis, the operation modes are defined. This greatly influences the performance of the PENG. The two common modes are 33 and 31. The polar axis is considered as "3" and if the direction of the applied strain is at right angles "1," it forms 31 and 33, when applied strain is parallel to polar axis. The two modes are shown in Fig. 6.1B. In mode 33, the applied stress and the voltage generated are in the same direction, while for mode 31, voltage is generated at a perpendicular direction. The different structural configurations that can enhance the effect of these modes include cantilever-based structure, cymbals, and stacked structures, as shown in Fig. 6.2.

A cantilever beam can be unimorph or bimorph depending on the number of piezo layers utilized. A cantilever beam made up of piezopolymers is attached by a calibrated mass at one end for optimizing the resonance frequency below 100 Hz and the other fixed end is attached to the vibration source. An alternating voltage is produced upon vibration through the backed electrodes on the piezomaterials. In a cymbal structure, the piezoelectric material is placed between two metal cymbal-shaped endcaps. An

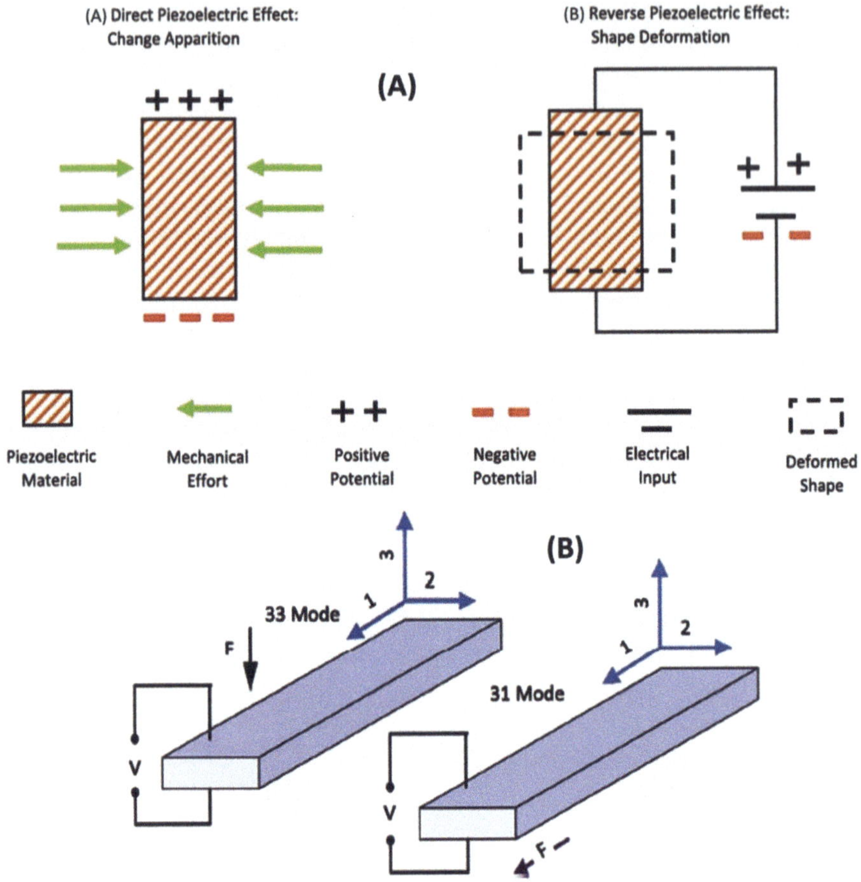

FIGURE 6.1 (A) Electromechanical conversion of piezoelectric phenomena. (B) Operation modes of piezoelectric transducer. [17] Source: *Reproduced with permission from N. Sezer, M. Koç, Nano Energy 80 (2021) 105567.*

applied axial stress is transferred to radial and tangential stress in the structure, thus amplifying the performance of the generator. Materials are stacked together in the direction of the electric field to improve the output in the case of high load applications.

6.3 Triboelectric nanogenerators

In the case of TENGs, the nanogenerator converts the applied friction or mechanical energy to electrical energy. The current is generated due to the potential difference produced upon the rubbing of two surfaces.

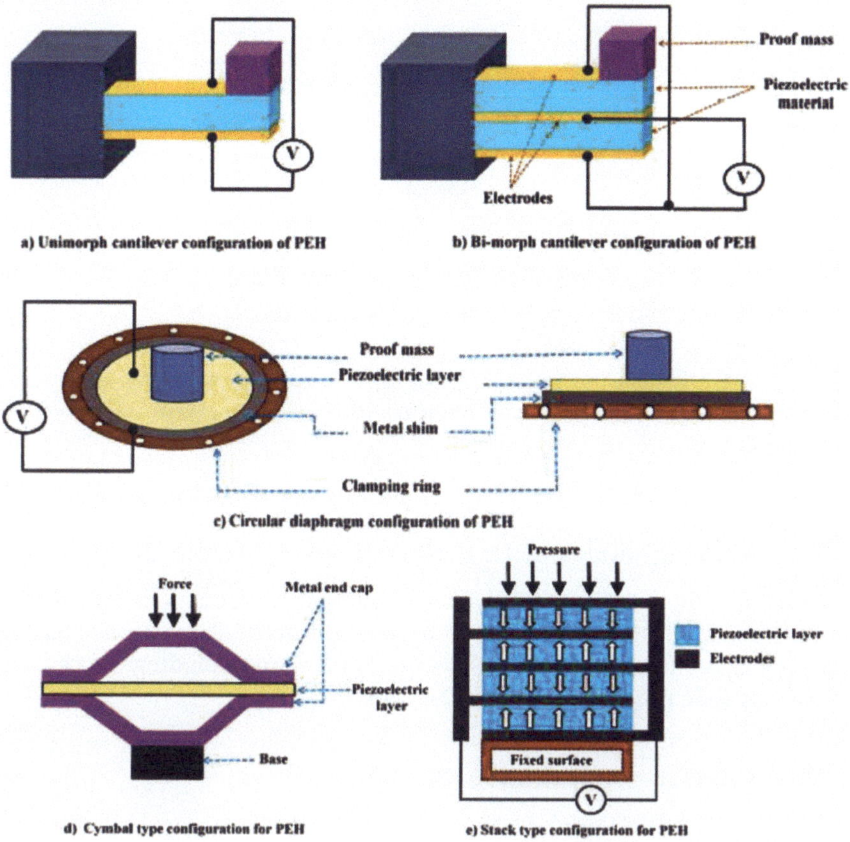

FIGURE 6.2 Different structural configurations of piezoelectric energy harvester. [18] Source: *Reproduced with permission from S. Mishra, L. Unnikrishnan, S.K. Nayak, S. Mohanty, Macromol. Mater. Eng. 304 (2019) 1800463.*

The dielectric layers in a TENG are selected depending on the triboelectric series, where a wide range of materials are categorized according to their polarity and triboelectric charge densities into tribonegative and tribopositive materials. The greater the disparity in polarities, the more friction will be produced, hence leading to better coupling and therefore good current generation. Polysaccharides have been arranged in triboelectric series according to their varying triboelectric charge densities. The addition of electron withdrawing and electron donating groups to cellulose and other polysaccharides can enhance triboelectric charge densities accordingly. For instance the addition of methyl groups can improve the electron donating ability while functionalization with nitro groups enhances the electron withdrawing ability of polysaccharides.

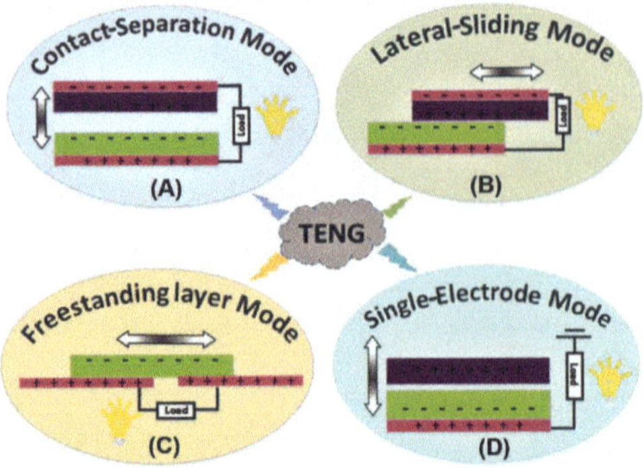

FIGURE 6.3 Fundamental modes of triboelectric nanogenerators: (A) contact–separation mode; (B) lateral sliding mode; (C) freestanding mode; (D) single-electrode mode. [19] Source: *Reproduced with permission from A. Ahmed, I. Hassan, M.F. El-Kady, A. Radhi, C.K. Jeong, P.R. Selvaganapathy, et al., Adv. Sci. 6 (2019) 1802230.*

6.3.1 Working modes of triboelectric nanogenerators

Generally, the working of TENGs is classified into four modes depending on how the applied friction leads to triboelectification. These modes include contact–separation mode, lateral sliding mode, free-standing mode, and single-electrode mode, as shown in Fig. 6.3.

6.3.1.1 Contact–separation mode

In Contact–separation mode, when the dielectric layers come into contact under application of mechanical force, there is change in polarities generated on the tribolayers. Upon the release of external force, electrostatic coupling occurs leading to potential difference in the terminal, and hence current is produced in the external circuit. This cycle then continues upon further press and release.

6.3.1.2 Lateral sliding mode

In Lateral sliding mode, there is a lateral separation of charge centers caused by rubbing of different terminals in a tangential direction. This potential difference results in current in the external circuit.

6.3.1.3 Free-standing mode

In this mode, the relative movement between the electrode and dielectric layer result in triboelectric charge followed by electrostatic induction leading to current generation. This mode is capable of high efficiency due to its unique design.

6.3.1.4 *Single-electrode mode*

In single-electrode mode, the current is produced when the generated charge overcomes the electric field. Upon separation between the electrode and dielectric layer, the charge flows to the ground. An effective polarization occurs when displaced parallel to the interfaces and this mode is highly suited for static objects. Of all, the contact–separation mode is most widely studied due to its high power output with a simple and robust design. The utilization of polysaccharides in energy harvesters is a sustainable approach to fabricate fully sustainable and green NGs for wearable and disposable electronics.

6.4 Nanocellulose-based energy-harvesting nanogenerators

Cellulose is one of the most abundant bioresources available on the planet, and it can be found in plants, algae, bacteria, animals (e.g., tunicates), and various other biomass sources (see Fig. 6.4). It is a linear homopolymer consisting of D-glucose units linked with 1,4-β-glycosidic bonds with three hydroxyl groups per glucose unit. Cellulose has a fibrous structure and is the main constituent present in the cell wall of trees/plants along with hemicellulose, lignin, and pectin, and the hierarchal structure from wood to cellulose nanofibers (CNF) is depicted in Fig. 6.5. The elementary fibrils, known as CNF, with as average diameter of 3–5 nm can be extracted from the plant cell walls by well-known techniques such as acid hydrolysis, enzymatic hydrolysis, and mechanical processes. As the fossil fuel resources are depleting and serious environmental concerns posed by the misuse of polymers and plastics, sustainable biomaterials like cellulose and its derivatives have gathered much attention as a potential replacement to synthetic materials. Due to the exemplary properties of CNFs like high aspect ratio, tensile strength, flexibility, and elastic modulus, it is widely being explored for a plethora of applications such as water purification membranes, tissue engineering scaffolds, wound healing, air filtration, and packaging. Moreover, upon suitable modifications, nanocelluloses are one of the potential biomaterials for electronic applications like sensors, flexible antennas, diodes, and electromagnetic interference (EMI) shielding [20–23]. In recent years, the focus was shifted toward the utilization of cellulose and its derivatives for energy storage applications such as batteries, supercapacitors, fuel cells, and energy-harvesting devices like PENGs and TENGs.

6.4.1 Bacterial cellulose-based triboelectric nanogenerators

Bacterial cellulose (BC), also known as microbial cellulose, is a type of nanocellulose that can be synthesized by bacteria via oxidative

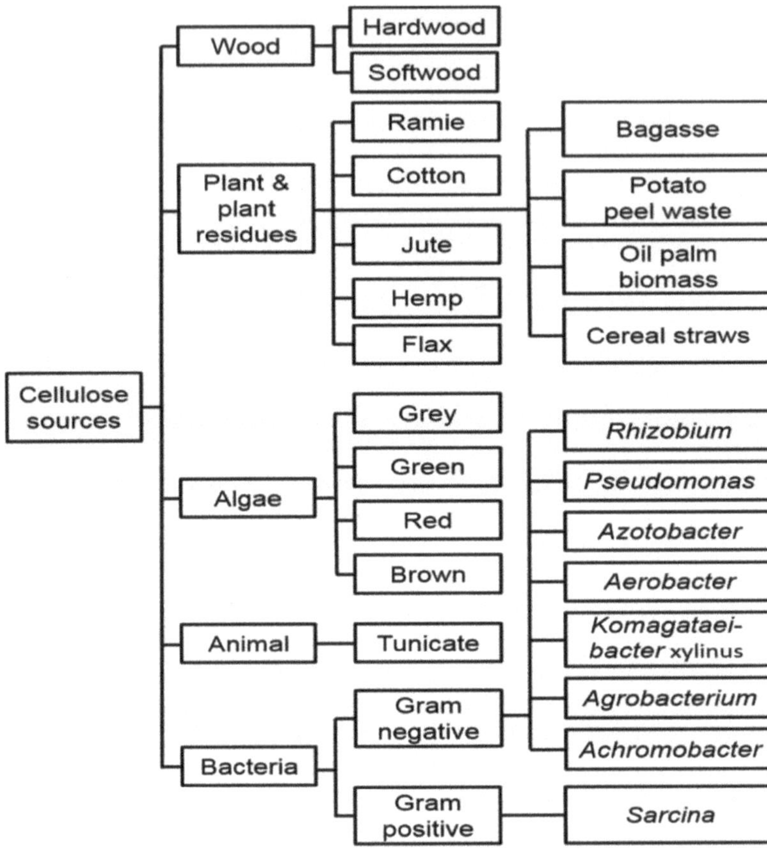

FIGURE 6.4 Classification of cellulose based on its source: wood, plant, algae, animal (tunicate), and bacteria. [24] Source: *Reproduced with permission from H. Seddiqi, E. Oliaei, H. Honarkar, J. Jin, L.C. Geonzon, R.G. Bacabac, et al., Cellulose 28 (2021) 1893–1931.*

fermentation in both synthetic and nonsynthetic media. These nanofibers have an average diameter of 10–20 nm and possess high crystallinity and a large surface area with good water retention capacity [26]. Due to its biocompatible nature, it is mostly used as scaffolds for wound-healing applications. These unique properties have also interested researchers in the use of BC for designing sustainable and green electronics materials for sensors, electrodes, EMI shielding, and energy-harvesting applications. Yan Shao et al., fabricated flexible and robust TENGS based on tribopositive BC/$BaTiO_3$ films paired against tribonegative PDMS films. The BC/$BaTiO_3$ films were fabricated by vacuum filtration technique, whereas the PDMS films were spin-coated onto glass substrates and later peeled off after the crosslinking process. The addition of $BaTiO_3$ particles in BC enhanced the

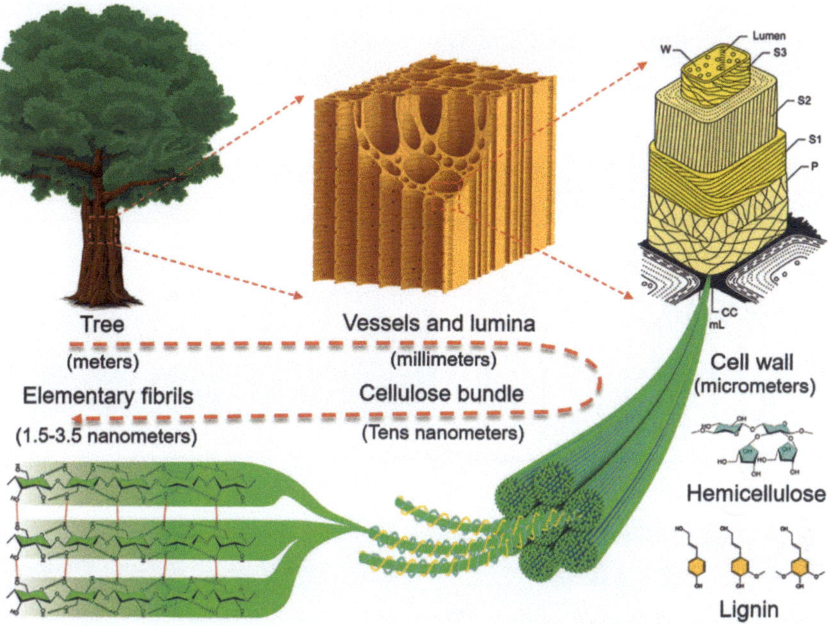

FIGURE 6.5 A schematic representing the hierarchical structure of wood and its elementary constituents, viz. CNFs, lignin and hemicellulose. [25] Source: *Reproduced with permission from C. Chen, L. Hu, Acc. Chem. Res. 51 (2018) 3154–3165.*

dielectric constant and also improved the surface roughness and surface area of BC films, which in turn boosted the performance of the TENG. With the combination modified BC and PDMS layers, a maximum open voltage of 181 V and current of 21 μA was achieved at 2 Hz with a peak force of 42 N [27]. This TENG showcased good flexibility and stability and could be easily integrated inside fabrics to harvest the electrical energy generated by human motion.

6.4.2 Bacterial cellulose-based piezoelectric nanogenerators

Ceramics are conventional piezoelectric materials which showcase a high piezoelectric performance. The main drawback of bulk ceramics is their brittle nature and poor resistance to combat irregular mechanical deformations. However for practical applications, flexible and biodegradable materials like polysaccharides and other biopolymers are being widely explored for energy-harvesting applications. A flexible piezoelectric paper based on BC/$BaTiO_3$ (BTO) nanoparticles with outstanding output voltage was fabricated by Guangjie Zhang and coworkers for nanogenerator applications. A facile vacuum filtration method was used

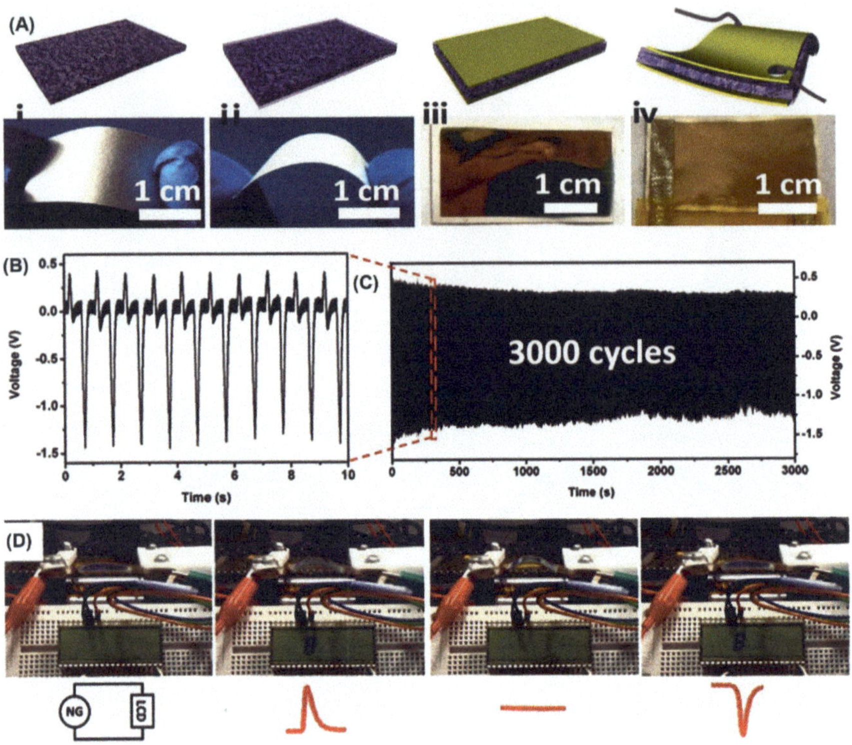

FIGURE 6.6 (A) Fabrication process of the BTO/BC piezoelectric paper based NG device. (B) Output voltage of the BTO/BC piezoelectric paper based NG under the bending frequency of 1 Hz. (C) Cyclic bending test for 3000 cycles of bending/releasing motions. (D) Four states when a commercial LCD screen is driven by the NG under bending conditions. [28] Source: *Reproduced with permission from G. Zhang, Q. Liao, Z. Zhang, Q. Liang, Y. Zhao, X. Zheng, et al., Adv. Sci. 3 (2016) 1500257.*

to fabricate the composite piezoelectric BC film and assemble it into a nanogenerator (see Fig. 6.6A). These paper-based PENG demonstrated an open-circuit voltage of 14 V and short-circuit current density of 190 nA/cm^2 [28]. It is also noteworthy that the maximum power density generated by BC/BTO paper was 10 times higher than the BTO/PDMS structure. A peak voltage of 1.5 V was achieved with 1 Hz deformation frequency. The durability of the device was tested up to 3000 bending/releasing cycles with very minor degradation in output voltage to 1.2 V, which can also trigger an LCD screen, as shown in Fig. 6.6B,C,D. Such lightweight, flexible, biobased and cost-effective, paper-based NGs display great potential as wearable energy harvesters and as self-powered electronics.

6.4.3 Nanocellulose-based hybrid piezoelectric nanogenerator-triboelectric nanogenerator

With increasing demand for clean and green energy energy, the concept of hybrid tribo/piezoelectric nanogenerator (HTPENG) has gathered much attention due to their ability to harness ambient energy from various motions, such as sound/water waves and body movements, by combining both piezoelectric and triboelectric effects. This concept was utilized by Ming Li et al. to assemble a fully cellulose-based HTPENG using $BaTiO_3$/MWCNT/BC paper as the piezoelectric layer and nitrocellulose nanofibril paper as the triboelectric layer. Both the papers were fabricated by facile vaccum filtation procces and assembled into a hyrid nanogenerator, which functions in different working modes, as shown in Fig. 6.7(A,B,C). An output of 18 V and 1.6 $\mu A/cm^2$ was achived, which can light up three LED bulbs (see Fig. 6.7(D,E)). The equivalent circuit diagram for the HTPENG for charging a capacitor is depicted in Fig. 6.7(D) and can charge a 1 μF capacitor to 2.5 V in 80 s [29]. This work provides new insights to design, assemble, and integrate all cellulose-based hybrid PENG/TENG

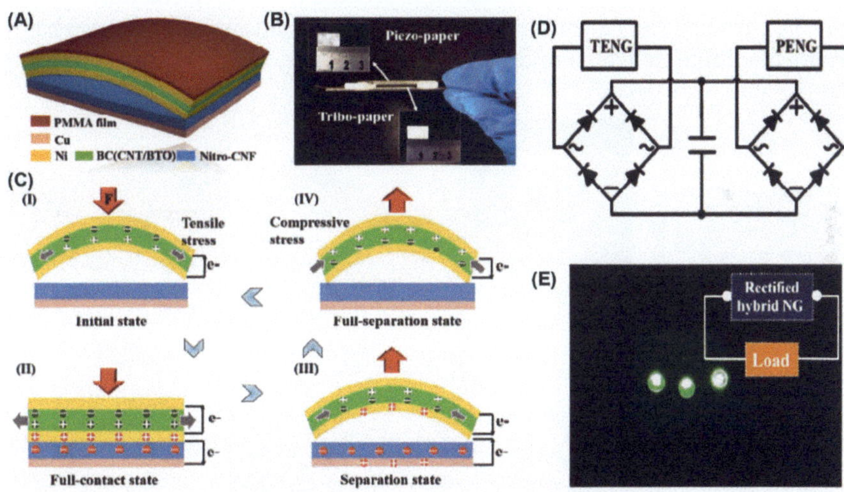

FIGURE 6.7 The structure and working mechanism of HTPENG. (A) The 3D schematic view of the arch-shaped HTPENG. (B) The photo of the HTPENG. (C) The working mechanism of the HTPENG in a contact-separation cycle. (I) Piezoelectric effect charge distribution when the external force applied. (II) Piezoelectric and triboelectric charge distribution at full-contact state. (III) Negative PENG and TENG at separating state. (IV) The maximum negative piezoelectric potential under full separation state. (D) Circuit diagram of the HTPENG for charging the capacitor. (E) LEDs lit by the rectified HTPENG, the inset shows a schematic diagram of the connection circuit. [29] Source: *Reproduced with permission from M. Li, Y. Jie, L.H. Shao, Y. Guo, X. Cao, N. Wang, et al., Nano Res. 12 (2019) 1831–1835.*

with a high degree of flexibility for smart and self-powered applications in the near future.

6.5 Chitin and chitosan-based energy-harvesting nanogenerators

Chitin is the most abundant biopolymers available on the planet next to cellulose. It is a linear polysaccharide consisting of N-acetyl-2-amido-2-deoxy-d-glucose units linked by $\beta(1\rightarrow4)$ bonds [30]. It is the main structural component present in the exoskeletons of arthropods like crab, shrimp, lobster, and insects. Apart from these sources, chitin is also present in the cell walls of mushrooms and other fungi, as depicted in Fig. 6.8. A typical shrimp shell constitutes about 20%–30% chitin, 30%–40% proteins, 30%–50% calcium carbonate, lipids, astaxanthin, and other minerals, and its hierarchical structure within the exoskeleton is shown in Fig. 6.9 [31]. Chitin exists as highly ordered microfibrils in these shells and depending on the orientation and packing of the chitin macromolecules, three polymorphs, namely α, β, and γ, are found in nature. The extraction of chitin nanofibers from waste shrimp shells involves a deproteinization process to remove proteins and a demineralization process to remove calcium carbonate. The strong inter/intramolecular hydrogen bonding in chitin makes it almost insoluble in water and dilute acids. Hence, chitin is converted into

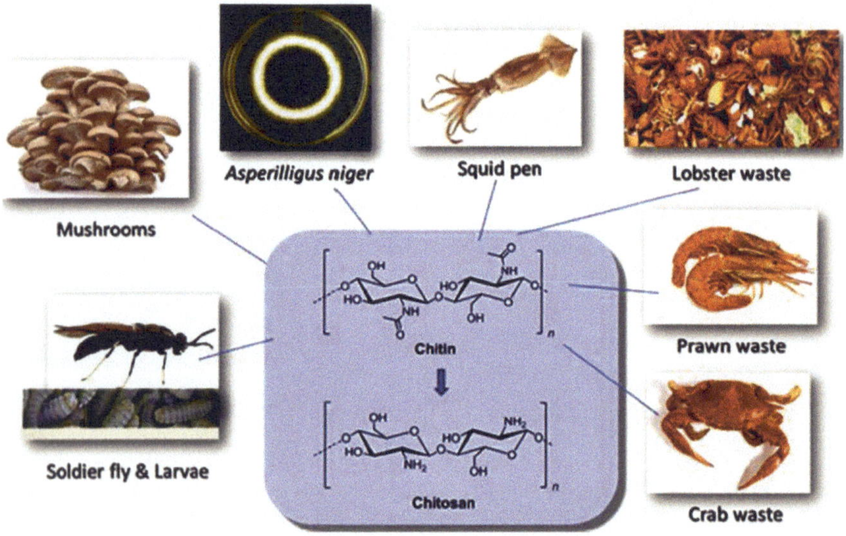

FIGURE 6.8 Various sources for extraction of chitin and its derivatives. [32] Source: *Reproduced with permission from A. Jardine, S. Sayed, Curr. Opin. Green. Sustain. Chem. 2 (2016) 34–39.*

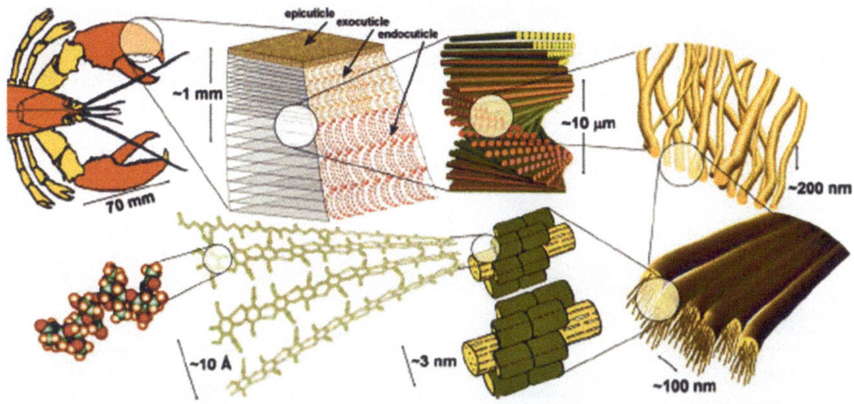

FIGURE 6.9 Exoskeleton of lobster depicting hierarchical and ordered structure of chitin nanofibers. [31] Source: *Reproduced with permission from D. Raabe, C. Sachs, P. Romano, Acta Matér. 53 (2005) 4281–4292.*

chitosan by hydrolyzing the acetamide groups to primary amine groups. This makes chitosan easily soluble in aqueous solutions of organic acids like acetic and formic acids. Owing to its biodegradability, nontoxicity, and biocompatible nature, chitin and its derivatives are being used widely for biomedical applications like scaffolds for tissue engineering, wound healing, and drug delivery. The surface functionality of chitin could be readily tuned to alter the physiochemical properties like solubility, transparency, and electrical conductivity, which makes it also suitable for high end electrical and optical applications like EMI shielding, sensors, batteries, solar cells, and in energy-harvesting NGs.

6.5.1 Chitin-based triboelectric nanogenerators

Fluoropolymers such as PVDF and its copolymers are the most promising functional organic materials which acts as tribonegative part of TENGs. Tuning the crystal orientation within PVDF-TrFE to enhance the β phase is of paramount importance for various practical applications. Kijoo Eom et al. fabricated a flexible TENG using epitaxially grown PVDF-TrFE on chitin film. The crystallographic match between PVDF-TrFE and chitin enables the orientation of polarization axis of PVDF-TrFE to be aligned perpendicular to the substrate, which was found to enhance the TENG performance and also reduce the hygroscopic nature of chitin film [33]. The working mechanism and functional modes of this TENG are shown in Fig. 6.10A. Furthermore, these modified TENGs showed a significantly higher output current and performance compared to nonepitaxial PVDF-TrFE/chitin film (see

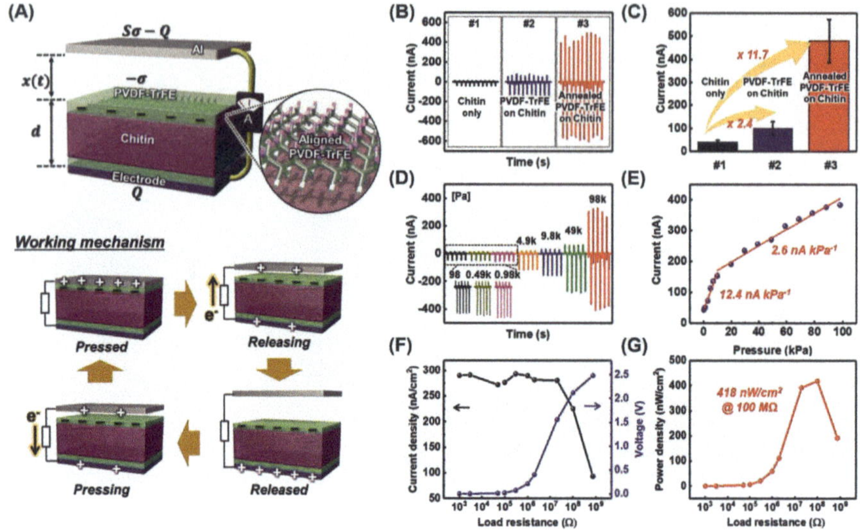

FIGURE 6.10 (A) A cartoon depicting the triboelectric charge generation and electron flow between the PVDF-TrFE/chitin hybrid film and Al electrode. (B–G) Triboelectric output current, current density, voltage, power density of different film-based TENGs as a function of time, applied pressures, and external load resistance. [33] Source: *Reproduced with permission from K. Eom, Y.E. Shin, J.K. Kim, S.H. Joo, K. Kim, S.K. Kwak, et al., Nano Lett. 20 (2020) 6651–6659.*

Fig. 6.10B–G) which implies that tailoring the crystal orientation of PVDF-TrFE using chitin is a promising approach for developing high-performance TENGs.

6.5.2 Chitin based piezoelectric nanogenerators

Chitin nanofibers (ChNF) are mainly used to trigger the nucleation of highly electroactive and piezoelectric β phase in PVDF and its copolymers. Nur Amin Hoque et al. made use of chitin nanofibers extracted from waste crab shell to design efficient, lightweight, and flexible PENGs based on pristine ChNF (CPENG) and ChNF/PVDF (PCPENG), respectively (see Fig. 6.11A,B). Pristine CNF thin-film-based CPENG showed excellent open circuit output voltage (V_{oc}) of 22 V and 0.12 μA short circuit current (I_{sc}) with high piezoelectric coefficient d_{33} = 9.49 pC/N. Moreover, with the addition of ChNF to PVDF the output performance of PCPENG (Voc = 49 V and Isc = 1.9 μA) was enhanced compared to CPENG [34]. Furthermore, the PCPENG was able to charge a 2.2 μF capacitor to 3.6 V within a short time span of 20 seconds and could light up to 22 blue LEDs when connected in series (see Fig. 6.11C–F).

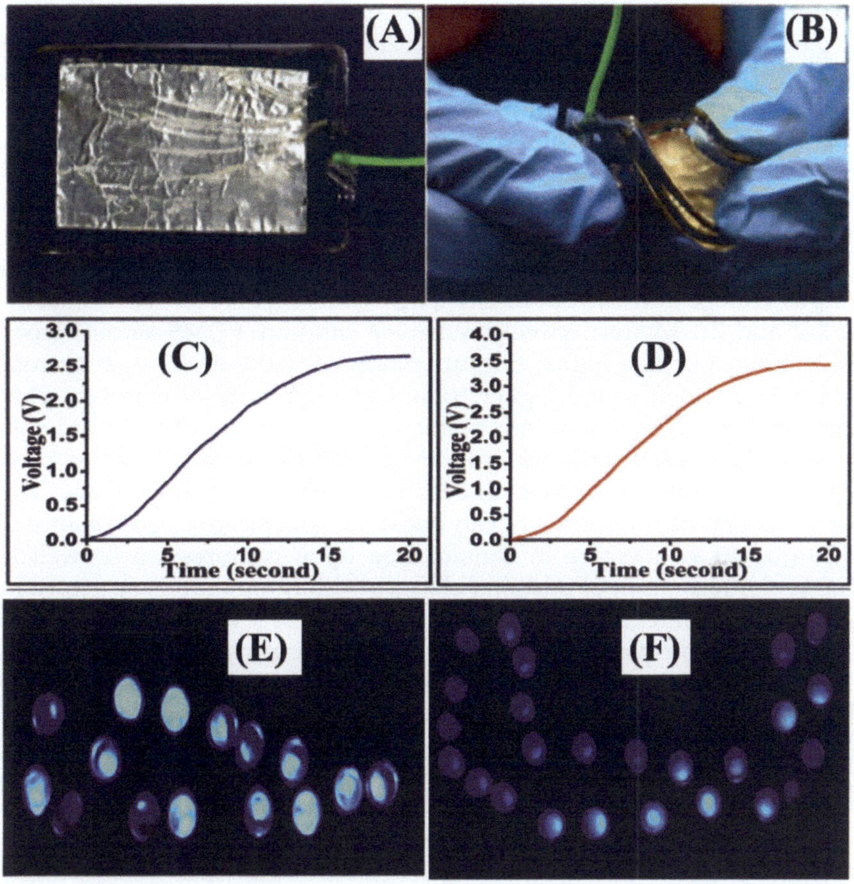

FIGURE 6.11 Digital photograph of (A,B) fabricated device PCPENG and its flexibility test. (C,D) Capacitor charging performance of CPENG and PCPENG. Photograph of LEDs light up in dark background (E) by CPENG and (F) by PCPENG. [34] Source: *Reproduced with permission from N.A. Hoque, P. Thakur, P. Biswas, M.M. Saikh, S. Roy, B. Bagchi, et al. J. Mater. Chem. A 6 (2018) 13848–13858.*

6.6 Porous nanocellulose/chitosan aerogel film-based triboelectric nanogenerators

Aerogels are a class of porous materials which possess excellent and unique physicochemical properties, such as high porosity (80%–99.8%), large specific surface area (100–1600 m^2/g) with ultralow densities (0.003–0.500 g/cm^3) [35]. In recent years, polysaccharide-based aerogels have gathered great attention due to their renewability, low cost, and biodegradability for varied technological applications, like scaffolds for tissue engineering, wound healing, water purifications, sensors, and

EMI shielding. In the case of energy-harvesting applications, the effect of porosity on the triboelectric and piezoelectric performance of aerogel-based NG needs to be explored. Qifeng Zheng and coworkers designed a novel class of high-performance aerogel-based NGs using tribopositive nanocellulose and chitosan aerogel films paired against tribonegative polyimide (PI)/PDMS aerogels. In order to discern the effect of porosity, the performance of aerogel-based NGs was compared with their dense film counterparts. The working mechanism of both aerogel and dense film NGs of CNF and chitosan is given in Fig. 6.12A–D. The output voltage and current of CNF/PDMS aerogel-based NG was about 22.3 V and 2.2 μA, respectively. Whereas chitosan/PDMS aerogel-based NGs showed even a higher output voltage of 30.0 V and 3.4 μA current, which is ascribed to the higher tribopositive polarity of chitosan when compared to CNFs [36]. Interestingly, the performance of CNF and chitosan aerogel-based NGs was found to be 8 times and 11 times more than their dense film NGs under the same mechanical stress. This clearly indicates that porosity can boost the triboelectric performance of NGs due to an increase in contact area upon compression as well as due to the formation of addition charges within the porous structure by electrostatic induction.

Another technique to improve the triboelectric performance of biobased TENGs is by functionalizing polysaccharides with suitable functional groups that can enhance the tribopositive nature. For instance, aminosilane modified

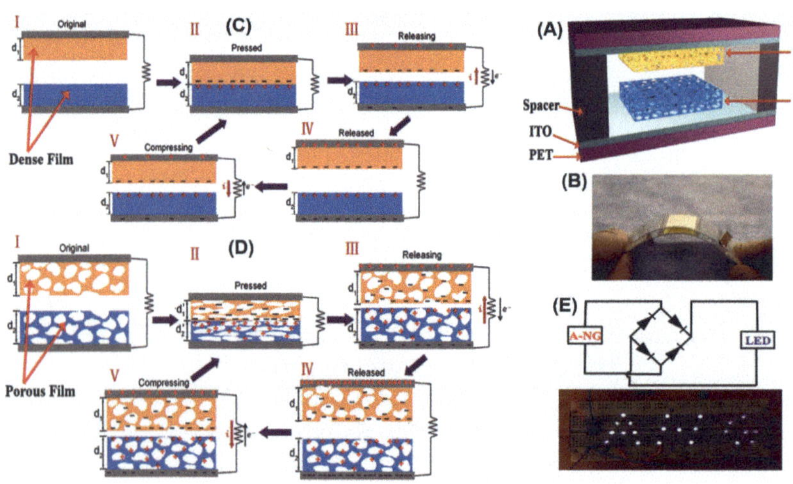

FIGURE 6.12 (A,B) A schematic diagram and photograph of a highly porous aerogel film-based TENG (A-NG). (C,D) Schematic demonstrating the operating principle of the dense film (D-NG) and aerogel film (A-NG)-based TENG. [36] Source: *Reproduced with permission from Q. Zheng, L. Fang, H. Guo, K. Yang, Z. Cai, M.A.B. Meador, et al., Adv. Funct. Mater. 28 (2018) 1706365.*

nanocellulose aerogel-based TENG showed a dramatically higher output voltage of 57.2 V and 8.6 μA output current. Among all NG combinations, the highest voltage of 60.6 V and current of 7.7 μA was obtained for porous chitosan/polyimide NG which was sufficient to power 22 blue LEDs (see Fig. 6.12E) [36]. This opens up new horizons to fine-tune the tribopositive polarity of polysaccharides for designing high-performance TENGs for wearable and self-powered applications.

6.7 Miscellaneous polysaccharides-based energy-harvesting nanogenerators

6.7.1 Pullulan-based triboelectric nanogenerator

Pullulan is a water-soluble, linear polysaccharide consisting of maltotriose units connected to each other by an α-1,6 glycosidic bond. It is produced from starch by the yeast-like fungus *Aureobasidium pullulans* [37]. Unlike most other polysaccharides, pullulan is easily water soluble, as it has a low degree of hydrogen bonding. It also possess excellent film-forming ability, good oxygen barrier properties, thermal stability, and printability, which make it one of the promising candidates for edible food packaging applications. Moreover, its biodegradability and unique properties have attracted researchers to exploit its use in disposable electronics, wearable sensors, and energy-harvesting applications. Yuanchao Lu et al., fabricated a high-performance, flexible, and recyclable pullulan-based TENG for wearable applications. The different working modes of pullulan-based TENG is shown in Fig. 6.13A. A maximum open-circuit voltage of 46 V was recorded at 10 Hz motion frequency and 10 W loading power for neat pullulan TENG [38]. These biodegradable TENGs also portrays its practical applicability by lighting up an array of LEDs and respond as signals to hand-bending movements (see Fig. 6.13B,C), thus demonstrating its feasibility in next-generation wearable devices.

6.7.2 Sodium alginate based piezoelectric nanogenerators

Sodium alginate (SA) is a linear polysaccharide derivative of alginic acid which comprises (1–4)-linked β-d-mannuronic acid and α-l-guluronic acid units. It is a component present in the cell wall of marine brown algae and consists of about 30%–60% alginic acid [39]. During its extraction process, alginic acid is converted to sodium alginate so as to make it soluble in water, which aids in extraction. Due to its nontoxic and biodegrade nature, it is most widely used in biomedical, food, and pharma sectors. Due to a large number of free hydroxyl and carboxyl groups present on the chain backbone, there is enormous possibility for chemical

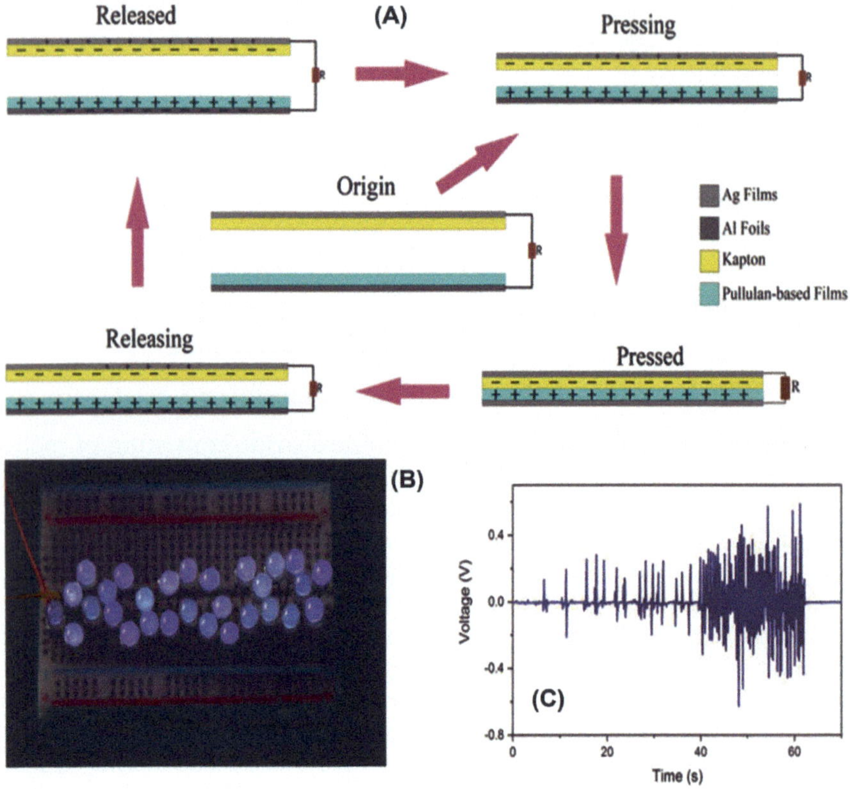

FIGURE 6.13 (A) Device structure scheme and working principle of the pullulan-based TENG. (B) Optical photo of the LED arrays lighted by the P-P TENG with 10 Hz motion frequency and 10 W loading power. (C) Electric open-circuit voltage of wearable F-P TENG with various hand-bending movements. [38] Source: *Reproduced with permission from Y. Lu, X. Li, J. Ping, J. song He, J. Wu, Adv. Mater. Technol. 5 (2020) 1900905.*

functionalization to alter its physiochemical properties for use in other applications, like sensors, EMI shielding, and energy-harvesting applications. Nagamalleswara Rao Alluri et al. designed a flexible and laterally aligned composite linear worm-based piezoelectric energy (WPNG) harvester made up of a 3D gel network of calcium alginate–$BaTiO_3$ nanoparticles transformed into a 1D worm-like structure using ionotropic gelation (IG) technique. The circuit assembly and different working modes of this WPNG device and the electrical output are shown in Fig. 6.14(A–E). A single flexible TENG was able to produce a maximum open circuit voltage (V_{OC}) of 10.90 V and a short circuit current (I_{SC}) of 1.88 mA while rapidly tapping with hand force [40]. This voltage was sufficient to light up a monochrome LCD and also 11 blue and green LEDs. For portable and wearable application, a single and double WPNG was inserted into the shoe sole to

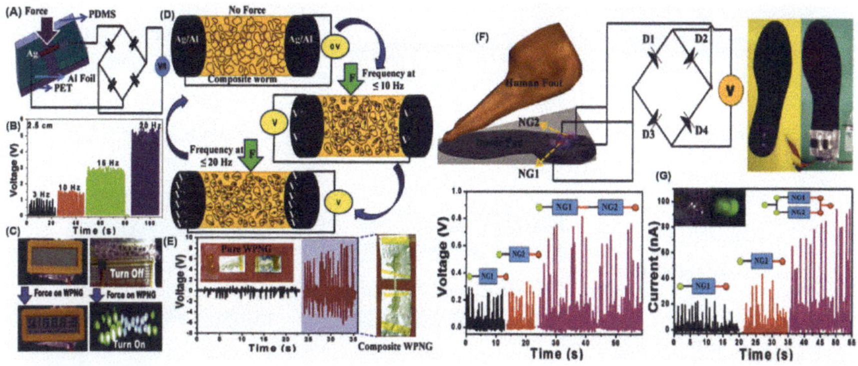

FIGURE 6.14 (A) Schematic of the WPNG device connected to full wave bridge rectifier circuit. (B) Open circuit voltage of WPNG device under different frequency (3, 10, 15, and 20 Hz). (C) The WPNG device lighting up a monochromic LCD display and 22 LEDs (11 green, 11 blue). (D) Schematic illustration of working mechanism of the assembled WPNG. (E) Voltage generated with time performance of WPENG and inset shows optical photographs of the fabricated device. (F) Cartoon demonstrating power generation by WPNG inside a shoe insole when connected to a full wave bridge rectifier along with digital photographs of single, double WPNGs placed within the shoe insole. (G) Open circuit voltage of WPNG device generated by low frequency human foot motion and the inset shows a green LED turning OFF/ON due to periodic human stress on the WPNGs. [40] Source: *Reproduced with permission from N.R. Alluri, S. Selvarajan, A. Chandrasekhar, B. Saravanakumar, G.M. Lee, J.H. Jeong, et al., Energy 118 (2017) 1146–1155.*

harness electrical output from the mechanical motion of the human foot. A maximum voltage of 0.8 V was recorded while connecting and integrating two WPNGs in parallel mode (see Fig. 6.14(F,G)). Moreover, the desired electrical performance can be altered using these devices which makes them promising candidates for self-powered wearable applications

6.7.3 Starch-based triboelectric nanogenerators

Starch is a polysaccharide which consists of two types of polymer chains, i.e., amylose and amylopectin. Amylose has a linear structure with α1–4 glycosidic linkage, whereas amylopectin has a branched structure with both α1–4 and α1–6 glycosidic linkages. Typically, starch consists of 20%–30% amylose and 70%–80% amylopectin [41]. The amylose to amylopectin ratio is a crucial parameter that affects the starch structure and crystallinity and other end properties. Robert Ccorahua et al. fabricated a series of bio-TENGs using starch electrolyte films complexed with a cationic salt $CaCl_2$ for increasing the triboelectric performance and was paired against tribonegative PTFE and silicone. The different working modes of this starch-based TENG is show in Fig. 6.15A. Interestingly, starch films loaded with 0.5% of salt showcased a higher output voltage of 1.2 V and was three-fold the value of noncomplexed starch (0.4 V)-based TENG [42]. This is due to the

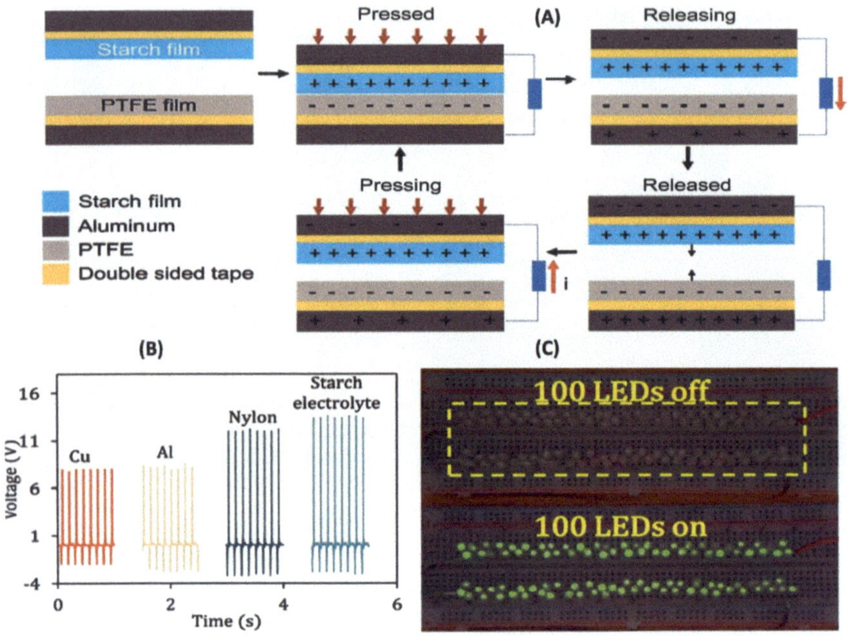

FIGURE 6.15 (A) The working modes of starch film based TENG. (B) Output voltage of various TENGs when paired against silicone (Ecoflex 00–30). (C) Practical demonstration by lighting 100 green LEDs with starch-based TENG. [42] Source: *Reproduced with permission from R. Ccorahua, J. Huaroto, C. Luyo, M. Quintana, E.A. Vela, Nano Energy 59 (2019) 610–618.*

fact that the presence of cations/anions will induce a net positive/negative charge to the polymer which will in turn boost its triboelectric property. The performance of modified starch-based TENG was found to be higher when compared to other conventional materials, such as Al, Cu, and nylon fabric (see Fig. 6.15B). In order to showcase the practical applicability of these TENGs an array of 100 green LEDs were turned on with only 2×2 cm^2 contact area and 4 N applied load, as demonstrated in Fig. 6.15C. This demonstrates the potential of starch-based TENGS for self-powered energy-harvesting applications to combat the future energy demands.

6.8 Conclusion and future outlook

In recent years, polysaccharide-based nanocomposites are showing promising results for the development of high-performance green NGs. The conventional approaches to fabricate polymer nanocomposites employ a lot of toxic chemicals and solvents. A paradigm shift towards green protocols for manufacturing sustainable nanocomposites is also quite crucital to promote circular economy [43]. Mostly synthetic polymers and inorganic materials are employed as active layers in NGs due to their high piezoelectric or triboelectric coefficients. It is well-known that the piezoelectric/triboelectric properties of polysaccharides are not comparable with those of traditional materials. But, upon suitable modifications they can deliver even higher performance than their pristine forms. For instance, the addition of piezoelectric nanofillers like zinc oxide or barium titanate can enhance the overall piezoelectric coefficient and output voltage in polysaccharides-based PENGs. In general, most of the polysaccharides exhibit a weak tribopositive nature. Upon surface functionalization with an electron withdrawing/donating group, one can enhance the triboelectric charge densities of polysaccharides to the desired range. Improving the electrical conductivity and surface roughness of polysaccharide-based materials was found to enhance the overall electrical output of these sustainable TENGs. Nevertheless, the presence of porosity also showed a positive impact toward boosting the electrical performance of polysaccharide aerogel-based TENGs. Polysaccharides have a strong tendency for moisture absorption and can adversely affect the NG performance; it is recommended to encapsulate the entire assembly with hydrophobic materials. In future, integrating both TENG and PENG to a single hybrid NG has huge prospects to improve the output voltage of fully sustainable and biodegradable energy-harvesting devices for self-powered, wireless, and wearable applications for the Internet of things, implantable medical devices like pacemakers, cochlear implants, and biosensors.

References

[1] F.R. Fan, L. Lin, G. Zhu, W. Wu, R. Zhang, Z.L. Wang, Nano Lett. 12 (2012) 3109–3114.
[2] Z.L. Wang, J. Song, Science (80–.) 312 (2006) 242–246.
[3] Y. Qin, X. Wang, Z.L. Wang, Nature 451 (2008) 809–813.
[4] X. Wang, J. Song, J. Liu, L.W. Zhong, Science (80–.) 316 (2007) 102–105.
[5] M. Lee, C.Y. Chen, S. Wang, S.N. Cha, Y.J. Park, J.M. Kim, et al., Adv. Mater. 24 (2012) 1759–1764.
[6] X. Lu, Y. Xu, G. Qiao, Q. Gao, X. Zhang, T. Cheng, et al., Nano Energy 72 (2020) 104726.
[7] Z.L. Wang, ACS Nano 7 (2013) 9533–9557.
[8] B. Yang, C. Lee, W. Xiang, J. Xie, J. Han He, R.K. Kotlanka, et al., Micromech. Microeng. 19 (2009) 035001.
[9] Y. Yang, W. Guo, K.C. Pradel, G. Zhu, Y. Zhou, Y. Zhang, et al., Nano Lett. 12 (2012) 2833–2838.
[10] J.P. Rojas, D. Conchouso, A. Arevalo, D. Singh, I.G. Foulds, M.M. Hussain, Nano Energy 31 (2017) 296–301.
[11] B.Y. Lee, D.H. Kim, J. Park, K. Il Park, K.J. Lee, C.K. Jeong, Sci. Technol. Adv. Mater. 20 (2019) 758–773.

[12] Y. Gong, Z. Yang, X. Shan, Y. Sun, T. Xie, Y. Zi, Energies 12 (2019) 2184.
[13] S.R. Anton, H.A. Sodano, Smart Mater. Struct. 16 (2007) R1.
[14] D. Vatansever, R.L. Hadimani, T. Shah, E. Siores, Smart Mater. Struct. 20 (2011) 055019.
[15] V. Corral-Flores, J.J. Pérez-Herrera, E. Torres-Moye, J. Romero-García, D. Bueno-Baqués, R.F. Ziolo, Mater. Sci. Forum (2010) 33–37.
[16] K.Y. Lee, J. Bae, S.M. Kim, J.H. Lee, G.C. Yoon, M.K. Gupta, et al., Nano Energy 8 (2014) 165–173.
[17] N. Sezer, M. Koç, Nano Energy 80 (2021) 105567.
[18] S. Mishra, L. Unnikrishnan, S.K. Nayak, S. Mohanty, Macromol. Mater. Eng. 304 (2019) 1800463.
[19] A. Ahmed, I. Hassan, M.F. El-Kady, A. Radhi, C.K. Jeong, P.R. Selvaganapathy, et al., Adv. Sci. 6 (2019) 1802230.
[20] D.A. Gopakumar, A.R. Pai, Y.B. Pottathara, D. Pasquini, L. Carlos De Morais, M. Luke, et al., ACS Appl. Mater. Interfaces 10 (2018) 20032–20043.
[21] A.R. Pai, T. Binumol, D.A. Gopakumar, D. Pasquini, B. Seantier, N. Kalarikkal, et al., Carbohydr. Polym. 246 (2020) 116663.
[22] A.R. Pai, C. Paoloni, S. Thomas, in Nanocellulose Based Compos. Electron. (2021) 237–258.
[23] D.A. Gopakumar, A.R. Pai, Y.B. Pottathara, D. Pasquini, L.C. de Morais, A. Khalil H. P.S, et al., J. Appl. Polym. Sci. 138 (2021) 50262.
[24] H. Seddiqi, E. Oliaei, H. Honarkar, J. Jin, L.C. Geonzon, R.G. Bacabac, et al., Cellulose 28 (2021) 1893–1931.
[25] C. Chen, L. Hu, Acc. Chem. Res. 51 (2018) 3154–3165.
[26] R. Narayan, Encyclopedia of Biomedical Engineering, 2019.
[27] Y. Shao, C. ping Feng, B. wen Deng, B. Yin, M. bo Yang, Nano Energy 62 (2019) 620–627.
[28] G. Zhang, Q. Liao, Z. Zhang, Q. Liang, Y. Zhao, X. Zheng, et al., Adv. Sci. 3 (2016) 1500257.
[29] M. Li, Y. Jie, L.H. Shao, Y. Guo, X. Cao, N. Wang, et al., Nano Res. 12 (2019) 1831–1835.
[30] B. Joseph, R.M. Sam, P. Balakrishnan, H.J. Maria, S. Gopi, T. Volova, et al., Polymers (Basel), 12, 2020, p. 1664.
[31] D. Raabe, C. Sachs, P. Romano, Acta Mater. 53 (2005) 4281–4292.
[32] A. Jardine, S. Sayed, Curr. Opin. Green. Sustain. Chem. 2 (2016) 34–39.
[33] K. Eom, Y.E. Shin, J.K. Kim, S.H. Joo, K. Kim, S.K. Kwak, et al., Nano Lett. 20 (2020) 6651–6659.
[34] N.A. Hoque, P. Thakur, P. Biswas, M.M. Saikh, S. Roy, B. Bagchi, et al., J. Mater. Chem. A 6 (2018) 13848–13858.
[35] L. Long, Y. Weng, Y. Wang, Polymers (Basel), 10, 2018, pp. 1–28.
[36] Q. Zheng, L. Fang, H. Guo, K. Yang, Z. Cai, M.A.B. Meador, et al., Adv. Funct. Mater. 28 (2018) 1706365.
[37] V.D. Prajapati, G.K. Jani, S.M. Khanda, Carbohydr. Polym. 95 (2013) 540–549.
[38] Y. Lu, X. Li, J. Ping, J. song He, J. Wu, Adv. Mater. Technol. 5 (2020) 1900905.
[39] L. Pereira, J. Cotas, in Alginates - Recent. Uses This Nat. Polym. (2020).
[40] N.R. Alluri, S. Selvarajan, A. Chandrasekhar, B. Saravanakumar, G.M. Lee, J.H. Jeong, et al., Energy 118 (2017) 1146–1155.
[41] H. Nawaz, R. Waheed, M. Nawaz, D. Shahwar, Chem. Prop. Starch (2020).
[42] R. Ccorahua, J. Huaroto, C. Luyo, M. Quintana, E.A. Vela, Nano Energy 59 (2019) 610–618.
[43] B. Joseph, et al., Sustainable Chemistry 2 (2) (2021) 254–270. Available from: https://doi.org/10.3390/suschem2020015. In this issue.

CHAPTER

7

Polymer-based composite materials for triboelectric energy harvesting

Ammu Anna Mathew[1], *S. Vivekanandan*[1] and *Arunkumar Chandrasekhar*[2]

[1]School of Electrical Engineering, Vellore Institute of Technology, Vellore, India, [2]Nanosensors and Nanoenergy Lab, Department of Sensors and Biomedical Technology, School of Electronics Engineering, Vellore Institute of Technology, Vellore, India

7.1 Introduction

The quality of day-to-day life is dictated by energy. With restricted fossil energy for the future, looking for new energy has been a significant challenge confronting the present society. With the expanding energy demands around the world, discovering new methods for energy-harvesting is a way to meet human prerequisites for the advancement of the current culture. The plentiful mechanical energy in the surroundings has drawn attention' because of its wide availability and minimal expense. Lately, different mechanical energy-harvesting advancements have been created, which convert energy in the surroundings into usable energy. The overall necessities for harvesting these different energies are minimal expense, high productivity, and high stability [1].

Energy-harvesting is defined as the process of transformation of one type of energy into another for a specific reason. Generally, mechanical energy in one area is collected and utilized for mechanical movement in another. There are four fundamental energy sources accessible for energy-harvesting in the surroundings: mechanical energy, radiant

Engineered Polymer Nanocomposites for Energy Harvesting Applications
DOI: https://doi.org/10.1016/B978-0-12-824155-4.00008-9

energy, thermal energy, and chemical energy. Particular materials and processes are required for the energy-harvesting of all of these sources. The surroundings regularly restrict the technique for harvesting energy that can be utilized for an application, as these encompassing sources are not accessible in each area, or during each period of the year. These four techniques are electromagnetic, electrostatic, piezoelectric, and triboelectric energy-harvesting, as shown in Fig. 7.1 with the energy sources and various applications [2].

Electromagnetic energy-harvesting depends on Faraday's law. According to this law, at the point when an electrical coil is placed in a magnetic field, a change in the magnetic field or the change in the coil movement brings about a charge generation [3]. This can be applied to numerous surrounding energy sources, the most widely recognized of which are vibration and radio waves. Utilizing vibration or actual movement to cause this interactive

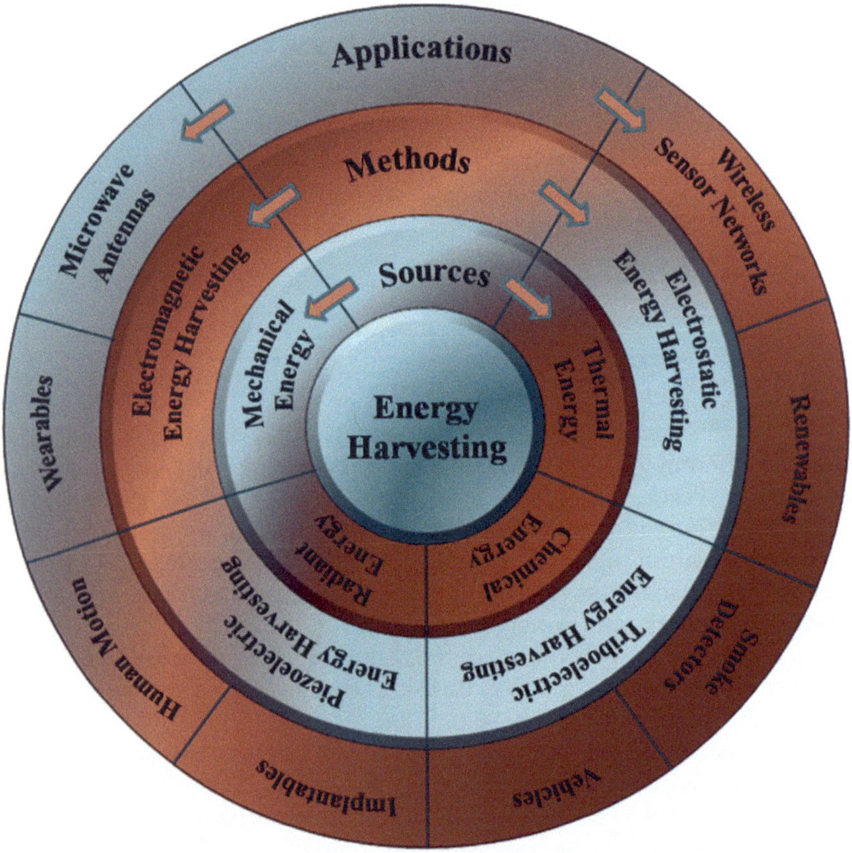

FIGURE 7.1 Energy-harvesting: sources, methods, and applications.

movement of the magnet and coil to happen numerous times per second can create a lot of electricity with a voltage value ranging from 0.5 to 2 V [4]. In electrostatic energy-harvesting, usually structures are made out of two metal capacitor plates (with equal and opposite electrical charges) that are disengaged from each other via air, a vacuum, or some other sort of insulators, frequently dielectric materials are used [5]. An electric field is created between the charged plates. The actual partition of these plates after charging produces a current. An input that would make these two plates continually move comparative with each other would produce a lot of electricity [6]. Electrostatic energy harvesters are complicated metal designs that are connected to a battery to charge the capacitor plates. Neither of these energy-harvesting techniques are suitable for packaging systems.

The method gaining most attention in energy-harvesting is piezoelectric energy-harvesting based on the piezoelectric effect where certain materials become electrically energized because of the applied mechanical strain [7]. Numerous materials show this behavior, all of which fall into five fundamental classes: single crystal; thin film; piezoceramics; polymeric materials, and forced piezoelectric materials, or foams. Triboelectric energy-harvesting based on the triboelectric effect is another promising technique for vibration energy collecting, which has aroused the curiosity of numerous scientists and experienced critical progression in examination and practice. It can be viewed as a subset of electrostatic energy-harvesting, where an electrical field is utilized for energy collection [8].

The huge advancement in triboelectric nanogenerator (TENG) has opened another part of energy-harvesting innovation in a vital challenge of scaling down, and huge advances in the fundamental standards and utilizations of TENG have been accomplished. Because of numerous appealing highlights of TENG, for example, choice of flexible materials, less complex manufacturing techniques, low-frequency operation, high yield voltage, and capacity to work under unpredictable mechanical movements, it has been utilized in self-powered gadgets. It enjoys the benefits of high effectiveness, convenience, economical, and environmental friendliness. TENG has for quite some time been exposed to mechanical effects during mechanical energy-harvesting. These materials are vulnerable to cracking and damage, which leads to a sharp reduction in their life span, security, and sustainability. Several solutions have been proposed based on these issues by researchers mainly based on material aspects [9].

The triboelectric impact which spins around electrostatics material science is experienced by everybody in their day-to-day life. It is a combination of contact electrification impact and electrostatic induction, which includes the association of two distinctive dielectric materials. Earlier triboelectric signals were considered as noise. After the

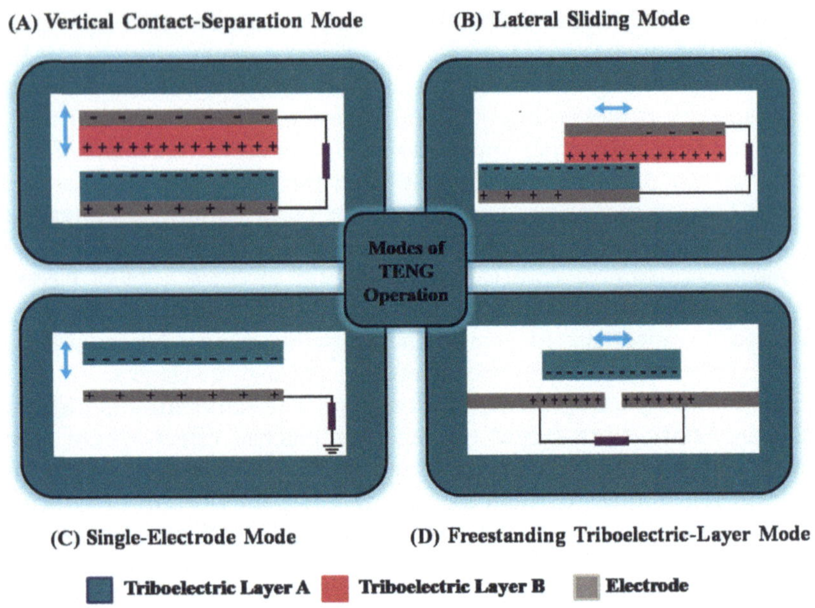

FIGURE 7.2 TENG modes.

breakthrough in electricity by a Van de Graaff generator, TENG was hugely perceived for tending to the difficulties engaged with building a practical and advanced society. The energy-harvesting is done by putting the electrodes on the positive and negative triboelectric materials. Due to the external force, these layers reach one another, which brings about the change of mechanical energy into electrical energy. It is mainly because of various electron affinities of the two thin films: natural/inorganic, which actuate the progression of electrons, accordingly adding to a potential difference that is imagined as an AC signal. TENG operates in four different modes: vertical contact–separation (CS) mode, lateral sliding (LS) mode, single-electrode (SE) mode, and freestanding triboelectric layer mode, as shown in Fig. 7.2. Each mode of operation is selected based on the application, dimension required, and several other factors [10].

7.2 Material selection

The main design concern in nanogenerator (NG) construction is the material choice. The materials opted for this purpose should have the ability to develop either triboelectric or piezoelectric nanogenerator effect. The surface charge density, efficiency, energy work function, capacitance characteristics based on dynamic charge transfer ability, charge injection depth, electron, and ion movements, etc. should be considered while choosing a material for NG construction [11].

Simplicity and cost-effectiveness are the two main factors considered in the fabrication of the piezoelectric nanogenerators (PENGs). This can be achieved by considering materials with wurtzite (gallium nitride—GaN; zinc oxide—ZnO; and cadmium sulfide—CdS) and perovskite (barium titanate—$BaTiO_3$; polyvinylidene fluoride—PVDF; lead magnesium niobate-lead titanate—PMN-PT, and nanofibers) structures. Further experimentation resulted in new versions of the wurtzite structure: p-type and n-type. The piezoelectric characteristics are displayed more in perovskite structures [12].

TENGs with triboelectric effect allow a wide range of material selection covering almost all materials. Material polarity is determined by the movement of electrons inside and outside the material's bonds, thereby determining the charge of that particular material. Material that drops electrons is assigned positive and others are negative. TENG construction is dependent on the triboelectric series, where a list of materials is displayed based on the polarity ranging from positive polarity material to negative polarity material with neutral materials in between. Neutral materials do not have the affinity to attract or drop electrons. The material position in this series is almost fixed based on the material work function and its affinity toward electrons, but in rare cases there may be a variation due to external effects, namely ecological and surface circumstances [13].

7.2.1 Triboelectric series

Materials that can acquire or lose electrons are called triboelectric materials [13]. The order of the tendency to acquire or lose electrons is known as the triboelectric series, as shown in Fig. 7.3. The arrangement of materials in the series depends on the conductivity of the individual material. The degree of charge is connected to a material's position in this series. The distance of the two materials from each other on the series builds the charge impact instead of the particular location in the series. Hence if two materials in contact are nearer on the scale, there is less charge transfer almost near to zero. When the materials are far separated, regardless of where on the scale, charge transfer happens. Maximum charge transfer occurs when the topmost material in the triboelectric series interacts with the bottommost material. Here the latter gains electrons, thus becoming negatively charged, and the former loses electrons, thereby becoming positively charged. The charge transfer is greater when the distance between the materials in the series increases,

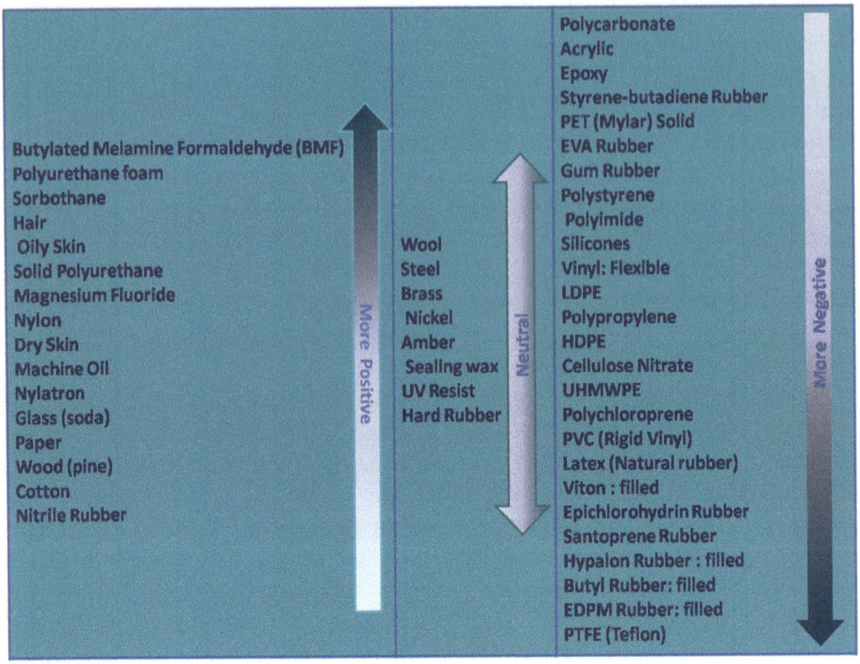

FIGURE 7.3 Summarized chart of triboelectric series.

thereby improving the TENG performance. The conductivity level after triboelectrification is independent of the electron affinity of materials [14].

The physical structure of materials should be considered. The electron transfer possibility is more when there is more surface region in contact between materials. In addition, the environmental conditions also matter. At a low relative humidity condition, there is a more rapid humidity charge transfer for hydrophilic materials than at higher relative humidity. Hydrophilic materials are usually called polar molecules [15]. Other factors which can improve the TENG performance include the proper decision in terms of material composition, surface potential, and roughness, temperature–humidity effects, motion parameters, etc. The triboelectric series may differ from one analyst to another because of the intricacies associated with the experimentation, particularly the charge control in materials [16].

7.2.2 Triboelectrification

The triboelectric effect is viewed as similar to the adhesion phenomenon in which two materials made out of various atoms tend to remain together on contact because of a chemical reaction. When two dissimilar materials come in physical contact with each other, one of the materials produces an electrical output due to contact electrification called triboelectricity. Contact electrification is the process that produces surface charges on two divergent materials at the point when they are in contact. During this contact, every material fosters a charge of inverse polarity. The process of contact electrification occurs between any combination of two materials: ionic and nonionic materials. Ionic materials include conductors (metals) and semiconductors, whereas nonionic materials are insulators [17].

In the case of ionic polymers with large covalent bound ions along with mobile offset ions at their surface, the charge transfer is done by ions. Contact electrification and hence charge transfer occurs in nonionic polymers (poly(methyl methacrylate)—PMMA; polytetrafluoroethylene—PTFE; polystyrene—PS; and polyethylene—PE), as well as where surface mobile ions are absent. Three proposed charge transfer mechanisms which involve nonionic polymers are ion-, electron-, and material/mass transfer.

Free ions are absent on the insulator surface but most of the studies suggest that there is still ion transfer in insulators during contact electrification. In any nonvacuum climate, there is water present in the environment, which collects on the outside of materials, known as "the water layer." Studies recommend that hydroxide and hydronium particles inside this water layer are responsible for the charge movement. Further studies have proved that charge can be created and move in a vacuum in the absence of any dampness, recommending that numerous charge movements may take place at the same time [18].

The electron transfer mechanism is brought about by the exchange of energetic electrons situated on the surface of these nonionic polymers. At the point when insulators are unsettled, the actual contact on the surface causes stress, considering these electrons to be delivered. When free, electrons are pulled into the positively charged surface in contact, thereby causing contact electrification. This hypothesis has been affirmed utilizing electrochemistry to show that electron transfer happens between insulators [19].

Surface capacitance analysis, charge injection depth, interface electrical efficiency analysis, surface charge density, and morphology determine the triboelectrification in materials. The improvement in the surface morphology can be realized by creating micro-/nanostructures on the surface. The structures such as hemispheres, pyramids, cylinders, squares, etc. on the surface enhance the area of contact, thereby improving the triboelectric effect. Surface functionalization is also a factor for improved triboelectrification [11].

7.2.3 Material transfer mechanism and polymer electrets

The surface molecule transfer, and thereby charge, even by insulators, to opposite materials in contact electrification can be analyzed by the surface analysis technique, where the surface of both the materials will be displayed, called X-ray photoelectron spectroscopy (XPS). By XPS, a localized material transfer on the surface is displayed, thereby forming localized positive and negative charge groups on the surface of the material, known as the surface charge mosaic. In triboelectric contact electrification, there is no mass transfer [20].

Contact electrification with certain materials ends up with a positive or negative charge development that controls the body of the material, making it permanently or quasipermanently charged. A material that may sustain charge with the visible electrical field at its surface is called electrets and is classified into two: space-charge electrets and dipolar electrets [21].

The dipolar electrets materials have both electrically positive and negative poles as they have a quasipermanent dipole or permanent dipole across the majority of the polymers. Under the influence of a strong electrical field, dipolar electrets can be fabricated from a polymer by cooling from over its glass transition temperature (Tg) to underneath it. At the point when cooled, the material holds the electrical affinities and has conversely charged poles. Dipolar electrets are utilized in numerous applications, especially for electrostatic energy-harvesting.

Electrets utilized in triboelectric energy-harvesting purposes are space-charge electrets. Space charge electrets acquire their charges uniquely in contrast to dipolar: here electric poles are absent but have a net naturally visible electrical charge. These are produced using the triboelectric charging effect of polymers, or by contact electrification. The electrets material seizes the charge upon contact electrification. Both the polymer categories can be space-charge electrets. Examples of nonionic polymers having space-charge electrets are PTFE, PS, and PE.

To be in either of the electrets group, a polymer must have high dielectric strength, i.e., it should withstand enormous electrical potential without breaking down and also should have the ability to hold that electrical charge inside its framework for quite a while. Overall the material should be with moderately low conductivity. There will be a large tendency to accept or lose electrons in materials resulting in higher contact electrification potential thus establishing the role of triboelectric series. The high electrical potential is due to the insulation property and large dielectric strength in electrets [21].

7.3 Polymer and Composite polymer materials

The cyclic covalent bond of individual functional monomers forms a large molecule called polymers. This polymer occurrence can be partitioned into two unique types: natural polymers and synthetic polymers. Synthetic polymers are developed in laboratories. Triboelectric materials incorporate an enormous amount of synthetic polymers. They show great resistance toward water, low dielectric constant, good plasticity, better thermal stability, and great mechanical strength, which are significant for use in TENG hardware with good performance. Natural polymers for TENGs as dielectrics are utilized to supplant different nonbiocompatible materials and synthetic polymers, as they are ideal for wearable gadgets due to their flexible, conformable, durable, washable, stretchable, and deformable characteristics. Likewise, biomaterials and products started to be brought into TENG application due to their potential benefits of minimal expense, biodegradability, presence of miniature and nanostructures, sustainability, abundance, and ecological friendliness.

These two types are again classified into three categories: organic polymers, inorganic polymers, and hybrid polymers. Organic polymers have carbon content and can be in either of the two categories: natural or synthetic. The polymers which are devoid of carbon are called inorganic polymers. The mixtures of both organic and inorganic polymers which are developed in the laboratory are called hybrid polymers. Based on the structural difference and fabrication they can again be classified into two: elastomers and thermoplastics [22].

A lot of materials have been utilized to manufacture TENG since its innovation, because all materials have a triboelectrification impact, from metal to steel, to wood, and polymer materials (PM). Fig. 7.4 shows the configuration and characteristics of some of the pure polymers. Albeit discretionary sets of triboelectric materials are infinite, few out of every odd pair can get high output [15]. The high output performances of TENGs are primarily based on the differences in electron affinity of the materials selected from the triboelectric series. Polymer material with fluorine [PTFE, PVDF, and fluorinated ethylene propylene (FEP)] has consistently been utilized as an electron-pulling out part of the TENGs because of the strong force of attraction of the electrons in the fluorine component. Conversely, PMs containing electron-giving groups (wool, silk, and nylon) have been always utilized as an electrophobic part of TENG.

Among the polymeric materials containing fluorine, PTFE is the most attractive due to its electronegativity characteristics and it is very useful in TENG applications. Another fluorine-containing polymer is FEP with a strong electrophilic capacity. It is also used as an electronegative material for triboelectric applications. Polyimide is another fluorine atom-containing polymeric material with electron-withdrawing ability used for several applications, such as coating, adhesive agent, packaging, and TENGs. The nonreactive category of polymers includes PVDF and its copolymers, which are mainly used for piezo- and triboelectric

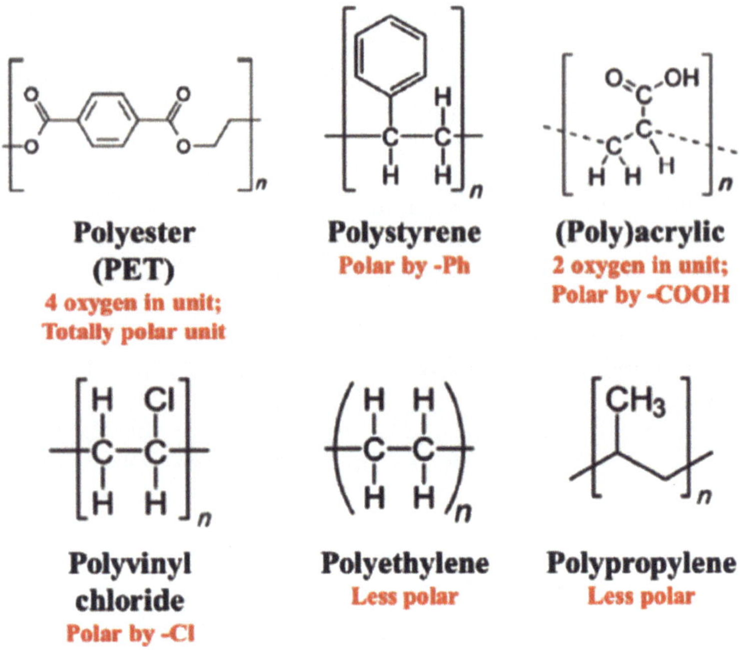

FIGURE 7.4 Pure polymer: configuration and characteristics. [11] Source: *Reproduced with permission S. Pan, Z. Zhang, Fundamental theories and basic principles of triboelectric effect: a review, Friction 7 (1) (2018) 2–17. https://doi.org/10.1007/s40544-018-0217-7. Copyright Springer Nature.*

energy-harvesting applications. This material has desirable flexibility, ample mechanical strength, better piezoelectric characteristics, and low acoustic impedance. Silicone-based polymers are another category used for TENG-based applications and polydimethylsiloxane (PDMS) is the one mostly used. PDMS has good transparency, high electronegativity, flexibility, stretchability, elasticity, biocompatibility, sufficient mechanical robustness, low cost, chemical stability, desirable processing, and electrification characteristics. This is mainly used as a friction material in TENG applications. Polyethylene terephthalate (PET) is another type of polymer material with lesser electron affinity, flexibility, mechanical strength, low cost, and good machinability. PET is used as a tribonegative material for triboelectric applications [23]. Some of the applications using PMs are shown in Fig. 7.5 [24].

Composite material can be characterized as a combination of at least two materials, with the end combination resulting in properties better than the properties of the individual constituent materials [23,25–27]. Composite materials are currently generally called reinforcement arrangements/fillers, which are implanted in a matrix, ensuring the

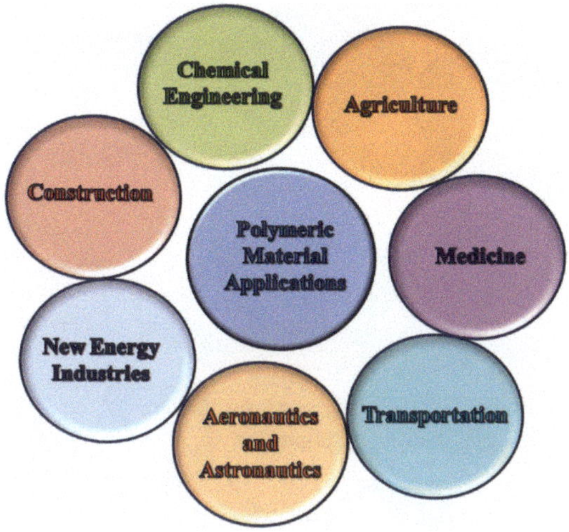

FIGURE 7.5 Applications of polymeric materials.

orientation and cohesion of the load [23,28–30]. The stress of the composite is also considered. The factors that affect the composite properties include the matrix under consideration, charge, extent and shape, the production process, and the nature of the interfacing. The lattice and the reinforcement can be plastic, metallic, or ceramic, which makes it conceivable to visualize a huge number of mixes. In composite materials, segments with correlative physical and mechanical properties are consolidated. The inclusion of reinforcements of good rigidity, of high modules in a polymer framework, makes it conceivable to improve the thermal and mechanical characteristics [31].

Concerning polymeric composite materials (PCMs), they normally comprise fillers scattered in a polymer framework to improve the mechanical properties of the polymer material. Recent researchers have centered on nanocomposites, where the filler in the nanoscale is one-dimensional, for instance, nanoparticles (NPs), 2D-sheets (nanofibers and graphene oxide), and carbon nanotubes (CNT). These nanofillers at low mass fractions have shown tremendous enhancements to material properties, basically because of the great surface area to volume proportion that nanomaterials have. The expanded interfacial region between the nanomaterial and constant polymer lattice brings about expanded polymer–filler strength [32].

The benefit of composites with a polymer network, in contrast with metals, is the assembling interaction which permits the creation of parts of complex shape together with their lower density, hence there is lower fuel utilization (for aviation and vehicle), higher transport payload and higher speed or longer reach for rockets. The composite materials are

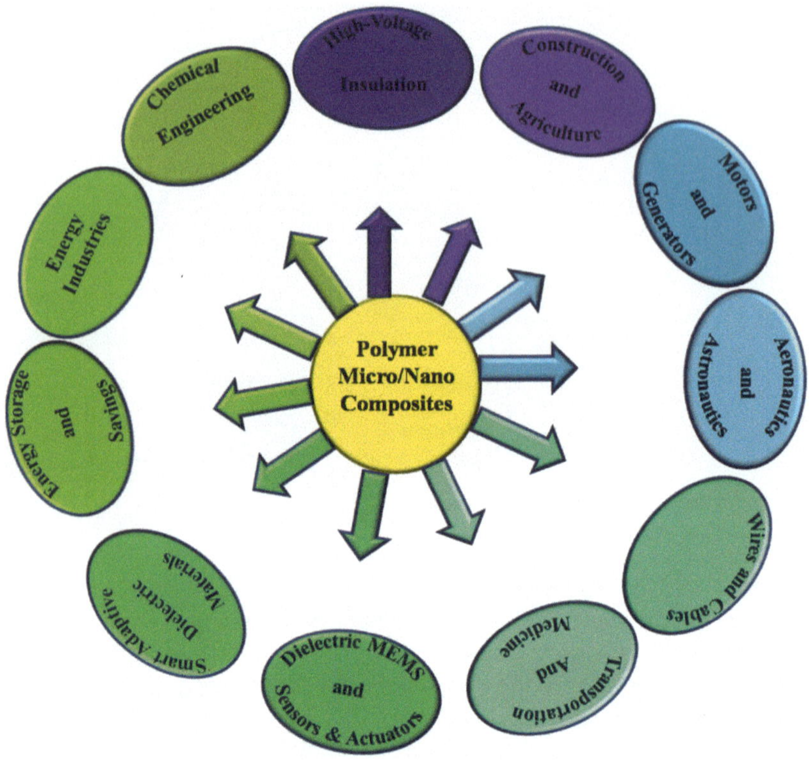

FIGURE 7.6 Applications of composite polymeric materials.

grouped by matrix types into three classes: mineral, organic, and metallic. These composite materials influence a few fields of use, in particular automotive, sports, structural design, packaging, aviation, biomedicine, aerospace, and thermomechanical components. Fig. 7.6 shows some applications of composite polymeric materials [33]. Other applications have been also proposed for nanocomposites: wastewater treatment, biomedical applications, and structural materials are a few examples [34].

When compared with noncomposite materials, polymeric nanocomposites have improved properties like expanded strength, preferred adsorption behavior, and thermal stability [35,36]. The PCM's mechanical and actual properties are controlled by their constituent properties and by the microstructural arrangement. Fibers are mostly liable for strength and solidness properties, the polymeric framework adds to stress transfer and gives microclimate assurance. Furthermore, fillers are utilized to improve the output, giving advantages such as shrinkage control, surface perfection, and resistance toward the crack. Modifier components and additives can expand the convenience of the polymeric network, improve their processability, or increase composite durability. The building up of a polymeric lattice with a high strength and modulus filaments uses the viscoelastic relocation of the network under pressure to move the load to the fiber; the outcome is a high strength and modulus composite material. The point of the blend is to deliver a two-phase material wherein the essential stage, which decides stiffness, is the fibers and is very much scattered and reinforced and ensured by an auxiliary stage, the polymeric matrix. Some of the PCMs include polyurea with modified CNT or graphene oxide, epoxy resins with silica NPs, epoxy resins with graphene oxide, phenolic polyester resin with fiberglass, cellulose acetate with NPs, and so on [37]. Some of the applications based on PCMs are explained in the section below.

7.4 Composite polymer-based triboelectric nanogenerator applications

In 2020, Cao et al. introduced a composite ferroelectric polymer interfacial layer to enhance the triboelectric output power density, as shown in Fig. 7.7 (A1–A3). The TENG performance is maximized using an electrical activation technique based on composite structures in the ferroelectric interfacial layer. The durability and triboelectricity of Nylon 11 (positive frictional surface) are improved by doping with PMMA. The negative triboelectric layer made of PVDF-TrFE was additionally upgraded by compositing with NPs of ZrO_2. Interior electric dipoles are polarized was done in complementary directions for further boosting of triboelectric effect in the positive and negative composite polymers in addition to composite structure formation with the help of fillers (PMMA or ZrO_2 NPs). Under short circuit (S.C) condition, 220 $\mu C/m^2$ frictional surface charge density and ~500 $\mu A/cm^2$ S.C current density were achieved, thereby promoting high charge outputs. An open circuit output voltage of ~500 V is also obtained at ~42 mW/cm^2 power density output for this particular design [38].

Gupta and others in 2020 demonstrated an energy harvester using composite material ZnO-PS for self-powered TENG-based medical applications. Here, a composite material formed from solution-processed ZnO nanorods (NRs) and composite ZnO-PS is used for the device fabrication. The positive frictional layer is formed by the composite ZnO-PS and the negative layer is formed by the Kapton/Cu. The fabrication process is simple and economical giving enhanced output in terms of power density. FE-SEM images are used for nanostructure morphology analysis which shows the improvement in the performance of ZnO-PS composite NRs when ZnO NRs are diffused with PS through thermal annealing at an appropriate seeding temperature (300°C).

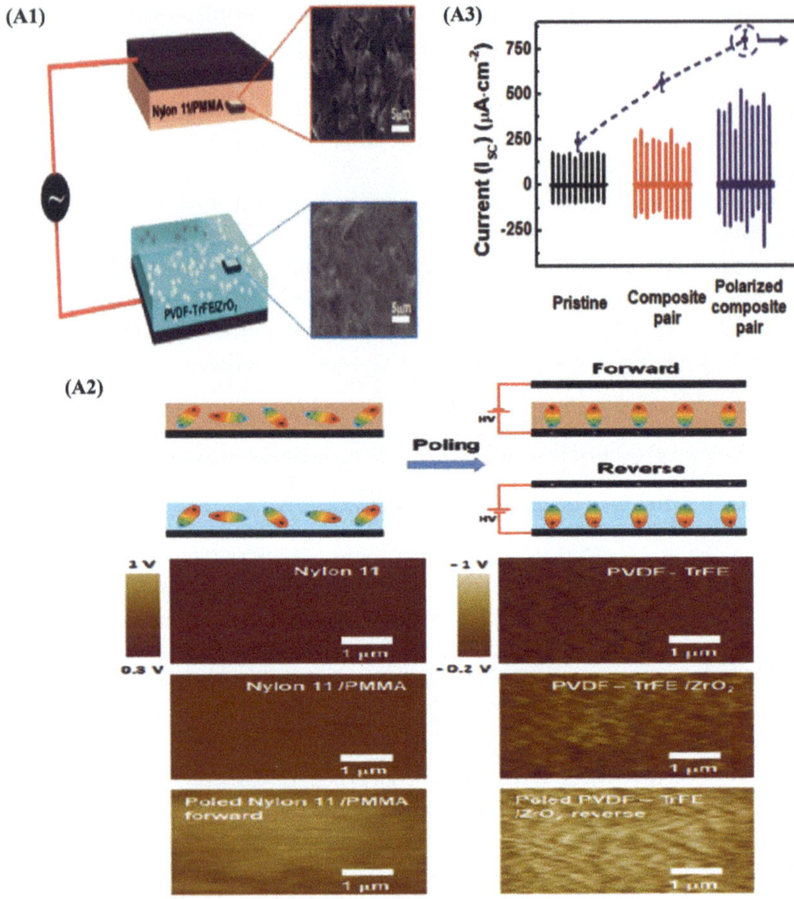

FIGURE 7.7 (A1) Graphical representation of composite layer made of Nylon 11/PMMA and PVDF-TrFE/ZrO_2 NPs along with SEM images. (A2) Surface potential measurement by KPFM. (A3) Current and charge density of TENG based on pristine PVDF-TrFE/Nylon 11 at short circuit condition. [38] Source: *Reproduced with permission V.A. Cao, S. Lee, M. Kim, M.M. Alam, P. Park, J. Nah, Output power density enhancement of triboelectric nanogenerators via ferroelectric polymer composite interfacial layers, Nano Energy 67 (2019) 104300, https://doi.org/10.1016/j.nanoen.2019.104300. Copyright Elsevier.*

The oxygen vacancy on the surface and composite nanostructure morphology is filled during the diffusion process. Other characterization techniques such as Raman spectroscopy and UV–visible absorption also confirm the same. The nonflexible substrate prohibits piezoelectric analysis. The surface improvement using the composite has resulted in a power density of $2.30 \times 10^{-4}\ W/m^2$ for the TENG whereas ZnO NR gave only $1.65 \times 10^{-4}\ W/m^2$ showing a clear improvement of 71% in the power density. The composite ZnO-PS and ZnO delivered an output voltage of ~7 and ~5.8 V, respectively, showing a higher triboelectric power production by the composite. A higher output current of $\sim 9.9 \times 10^{-8}$ A is obtained at short circuit condition from the ZnO-PS composite TENG. The surface charge density has also been measured. The proposed composite material ZnO-PS-based TENG finds its application in health monitoring where it is used as a power source [39].

In 2018, Hwang and coworkers introduced an economical, large-scale mesoporous composite polymer made by combining mesoporous aluminum oxide (Al_2O_3), obtained through one-step metal microparticles (MPs) oxidation, as shown in Fig. 7.8 (A1–A3), with a polymer PDMS which finds application in sensors for safety cushions. This composite polymer is highly deformable and TENG exhibits higher output performance (6.67 times) with improved capacitance (1.8 times) compared with polymers. The TENG structure is a gapless structure fabricated through a simple scalable process. Evaporation is performed at 80°C to the mixture of PDMS with aluminum (Al) MPs and deionized (DI) water, resulting in an increasing concentration of Al MPs with a range of 0–15 wt.% whose surface is oxidized and has nanoflake structures, thereby creating a mesoporous PDMS incorporated in pores with Al_2O_3 MPs (33×27 cm^2). The porosity of mesoporous composite films reaches 71.35% as the Al MPs concentration reaches 15%, resulting in higher output voltage and current of 259.58 V and 7.16 μA, respectively, at 4 Hz/33.3 kPa. The output of mesoporous PDMS is higher (6.67 times) compared with nonporous PDMS and is lightweight. For pressure ranging from 5.5–33.3 kPa, the resultant pressure sensitivity is 0.18 A/kPa and 6.71 V/kPa. When the mesoporous Al_2O_3-PDMS film is applied to a gapless TENG structure, 0.02 μA/kPa and 1.00 V/kPa are the respective linear pressure sensitivity. This invention has other self-powered applications like blankets, matrices, and road sensors [40].

In 2015, Chun et al. introduced a vertical CS mode TENG made of composite polymer to enhance the output performance, as shown in Fig. 7.8 (B1–B3). Here dielectric is made of a simple and versatile blend of mesoporous films infused with gold (Au) NPs. The composite polymer was prepared by mixing the PDMS solution with DI water containing Au NPs. This mixture is evaporated in a very slow manner at room temperature resulting in Au NPs accumulation in the pores on its bottom side. The AMTENG (Au NPs-embedded mesoporous TENG) fabricated due to Au NPs spatial distribution gives an output power of 13 mW which is five times more than the power produced by the flat-film-based TENG under similar conditions and mechanical force. The output power density is 160 μW/cm^2. The contact inside the pores between PDMS and Au NPs creates a charge which in turn produces aligned dipoles. These dipoles produced have an impact on the mesoporous film in terms of surface potential energy as well as flexibility.

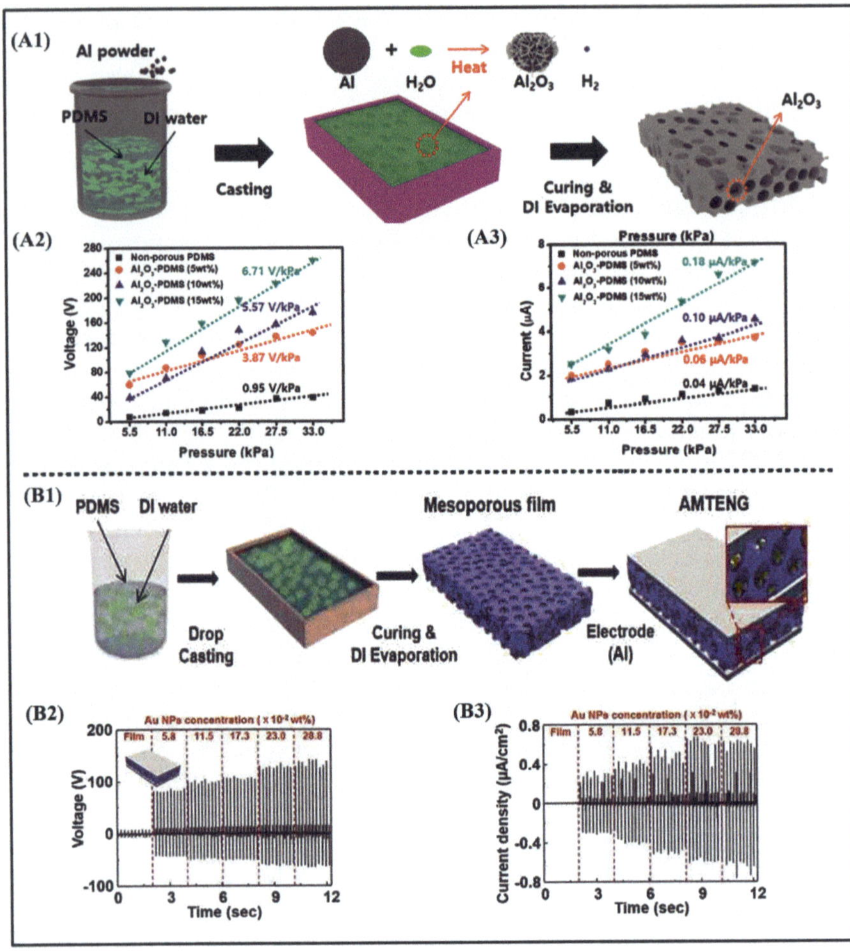

FIGURE 7.8 (A1) Schematic representation of fabrication: Mesoporous Al_2O_3-PDMS. (A2) Output voltage of mesoporous Al_2O_3-PDMS at different proportions with nonporous PDMS with increasing pressure. (A3) Output current of mesoporous Al_2O_3-PDMS at different proportion with nonporous PDMS with increasing pressure. (B1) Graphical representation of AMTENG fabrication process (B2) AMTENG output voltage for different Au NPs weight concentration (%) ranging from 5.8 to 28.8×10^{-2} (B3) AMTENG output current for different Au NPs weight concentration (%) ranging from 5.8 to 28.8×10^{-2}. Source: *Reproduced with permission H.J. Hwang et al., Mesoporous highly-deformable composite polymer for a gapless triboelectric nanogenerator via a one-step metal oxidation process, Micromachines 9 (12) (2018) 1–11, https://doi.org/10.3390/mi9120656. Adapted from MDPI: Open Access: CC BY 4.0. Reproduced with permission J. Chun, J.W. Kim, W.-S. Jung, C.-Y. Kang, S.-W. Kim, Z.L. Wang, et al., Mesoporous pores impregnated with au nanoparticles as effective dielectrics for enhancing triboelectric nanogenerator performance under harsh environment, Energy Environ. Sci. 3 (2015) 10715–10722, https://doi.org/10.1039/b000000x. Copyright Royal Society of Chemistry.*

The fabrication process is effortless and scalable, capable of wide-scale and self-powered gadgets production. The electrical performance of the AMTENG was estimated under several mechanical burdens and mesoporous structures with different porosities at various Au concentrations. The porosity of mesoporous films increases up to roughly 59% as the measure of water increases to 50%. With a particularly improved output power and distinctive device design, there are different self-powered applications in different fields [41].

Niu et al. in 2018 introduced a polymer nanocomposite with electronic self-healing capability, as shown in Fig. 7.9 (A1−A3). The proposed CS mode flexible, reproducible and soft self-healing TENG (SH-TENG) exhibits high performance with 2.5 W/m^2 peak power and also with a charge density of 100/μCm^2. The nanocomposite with high conductivity is formed with semiembedded silver nanowires (Ag NWs) in the polymer and is used as the frictional layer of TENG giving enhanced output. Both the top layer and bottom layer are formed by partially embedding the Ag NWs into the PDMS film and TPU (thermoplastic polyurethane) film, respectively. When this SH-TENG is cut off or injured, PDMS and TPU with silver NWs show the capacity of self-healing essentially by connecting the broken ends under NIR (near-infrared) light and recuperating the electrical properties. The healing cycle of PDMS is due to the hydrogen bonds available in the polymer and the healing power of TPU is attributed to hot melting. The Ag NWs are inserted in the polymer for preparing the electrode, due to which the NWs stay intact during the bending and touching process. Electrodes maintain good stability, conductivity, and durability. Even after five different cutting−healing cycles, the healed device produces an output voltage and current which reaches up to about 99% of its original output. The proposed device is stable and flexible which presents a huge step toward the creation of dependable TENGs with recoverability and low support costs [42].

Kim and others in 2017 demonstrated a composite polymer-based wearable TENG with nanostructures, as shown in Fig. 7.9 (B1−B3). The proposed flexible and stable TENG (20 × 30 mm) is made of plasma-etched PDMS-CNT. The chemical modification is due to the plasma-etching based on fluorocarbon (tetrafluoromethane CF_4), and thereby enhances the surface charge density. The plasma-etching also physically affects the surface morphology of PDMS-CNT, thereby enhancing the TENG output. The mechanical properties of the composite are due to the addition of CNT to the polymer resulting in a durable thin film. Various CNT concentrations were doped in the PDMS by varying the fluorocarbon etching time to confirm the effect and it was finalized that 4 wt.% at 60 seconds is the appropriate proportion which gives better electrical performance (77.8 V/ 1.98 mW/3.29 W/m^2). When compared

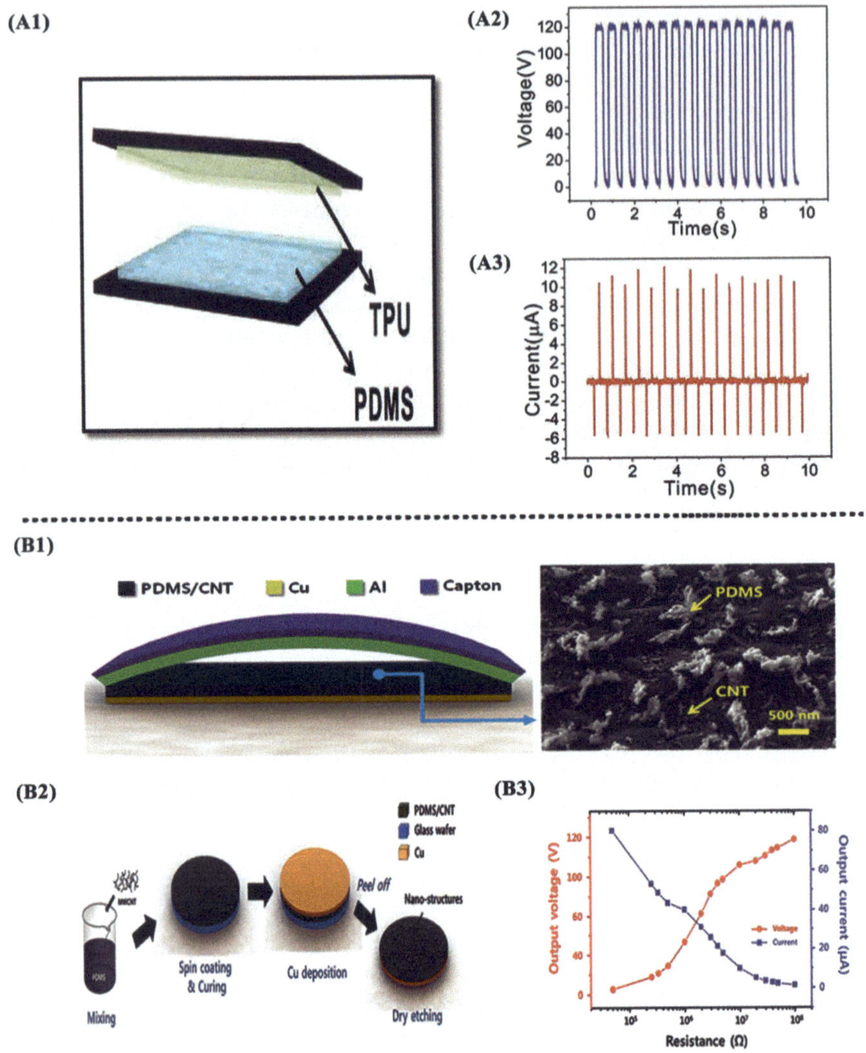

FIGURE 7.9 (A1) Experimental setup for mechanical energy-harvesting using self-healing TENG: pictorial diagram. (A2) Open-circuit voltage of SH-TENG. (A3) Short-circuit current of SH-TENG. (B1) Schematic diagram of PDMS–CNT nanostructure-based TENG with SEM image (B2) Schematic representation of the dry etching fabrication technique using PDMS–CNT nanostructures (B3) Electrical characteristics for various load resistances using the flexible TENG. Source: *Reproduced with permission H. Niu et al., Polymer nanocomposite-enabled high-performance triboelectric nanogenerator with self-healing capability, RSC Adv. 8 (54) (2015) 30661–30668, https://doi.org/10.1039/C8RA05305G.* Copyright Royal Society of Chemistry. *Reproduced with permission M.K. Kim, M.S. Kim, H.B. Kwon, S.E. Jo, Y.J. Kim, Wearable triboelectric nanogenerator using a plasma-etched PDMS-CNT composite for a physical activity sensor, RSC Adv. 7 (76) (2017) 48368–48373, https://doi.org/10.1039/c7ra07623a. Copyright Royal Society of Chemistry.*

with pure PDMS there is an increase of 248.7% in output voltage and 106.5% in the case of unetched PDMS–CNT. The proposed biocompatible, stretchable physical activity sensor finds its application in human activity measurement through capacitance charging [43].

7.5 Conclusion

As people are in quest of intelligent life and a smart city, gadgets with high usefulness becomes their priority. The invention of TENG moves forward in the field of converting mechanical energy into electrical energy to satisfy our energy needs. It is a vital innovation on account of its high yield power, high effectiveness, lightweight, minimal expense, simple manufacture, and small size. TENG can adequately utilize mechanical energies in practically any structure and any scale and give a consistent direct current source to charge different commercial hardware under the four basic modes of operation of TENG. The incorporation of PMs in the TENG design and progress is a promising methodology for the synergy of dependability and durability of TENG devices. Indeed, there are significant endeavors to improve TENG' output through a progression of techniques dependent on the modification of polymeric materials, predominantly including tribosurface morphology design, bulk composition alteration, and surface molecular functionalization. Since the TENGs result relies intensely upon the produced triboelectric charges during the friction cycle, it is overwhelmed by the idea of triboelectric materials, regardless of tribosurface morphology or physicochemical properties. PMs have shown incredible potential in TENG because of their exceptional properties, for example, humidity-resistant, biocompatible, pressure-resistant, shape memory, biodegradable, and self-healing. Greater emphasis should be given to the huge amount of PMs used for this application. Recent accomplishments in the field of synthetic polymer science give a wide choice of PMs with a huge cache of synthetic procedures for their planning, adaptability, and a worthy value/quality proportion, which provides a methodological stage for getting promising PMs for TENGs.

Moreover, a significant task is to discover a relationship between' the composition, spatial association, and properties of PMs, which in numerous regards is confused with an experimental approach to deal with organizing these frameworks and anticipating their application in TENGs. The composite polymers in proper proportion and composition also have yielded better results in different applications. The functional group mechanism in these materials is itself a huge research topic. The choice of specific combinations of PMs to obtain reversible reactions to improve the charge transfer effect, material modification, and material synthesis are other trending research fields.

It ought to be stressed that the future trend will be the advancement of multifunctional PMs. In spite of the enormous capability of polymeric materials, their commercialization, like TENGs as a rule, is still a long way from the real world. Indeed, there are significant issues that should be defeated to begin the viable mechanical advancement of these frameworks. Specifically, polymeric materials should be improved to achieve better mechanical properties and durability. This methodology is progressive; however, so far the further directions of its advancement principally rely upon difficulties and issues that have not yet been completely settled.

Acknowledgment

The authors would like to thank the Department of Science and Technology (DST), Government of India for the financial support under the Women Scientist Scheme -B (WOS -B) vide Reference No. DST/WOS-B/EIT-4/2021 to carry out this work.

References

[1] Z.L. Wang, J. Chen, L. Lin, Progress in triboelectric nanogenerators as a new energy technology and self-powered sensors, Energy Environ. Sci. 8 (8) (2015) 2250–2282. Available from: https://doi.org/10.1039/c5ee01532d.

[2] K.A. Cook-Chennault, N. Thambi, A.M. Sastry, Powering MEMS portable devices - A review of non-regenerative and regenerative power supply systems with special emphasis on piezoelectric energy harvesting systems, Smart Mater. Struct. 17 (4) (2008). Available from: https://doi.org/10.1088/0964-1726/17/4/043001.

[3] S.P. Beeby, M.J. Tudor, N.M. White, Energy harvesting vibration sources for microsystems applications, Meas. Sci. Technol. 17 (12) (2006). Available from: https://doi.org/10.1088/0957-0233/17/12/R01.

[4] S. Boisseau, G. Despesse, and B. Ahmed, Electrostatic conversion for vibration energy harvesting, in: Small-Scale Energy Harvesting, 2012, pp: 91–134.

[5] S. Roundy, P.K. Wright, J. Rabaey, A study of low level vibrations as a power source for wireless sensor nodes, Comput. Commun. 26 (11) (2003) 1131–1144. Available from: https://doi.org/10.1016/S0140-3664(02)00248-7.

[6] T. Sterken, K. Baert, R. Puers, G. Borghs, Power extraction from ambient vibration, in: Proceedings 3rd Workshop on Semiconductor Sensor and Actuator—SeSens, Leuven Belgium, 29 November, 2002, pp: 680–683.

[7] P. Glynne-Jones, S.P. Beeby, E.P. James, N.M. White, The modelling of a piezoelectric vibration powered generator for microsystems, Transducers '01 Eurosensors XV (2001) 46–49. Available from: https://doi.org/10.1007/978-3-642-59497-7_10.

[8] Z.L. Wang, Triboelectric nanogenerators as new energy technology and self-powered sensors - Principles, problems and perspectives, Faraday Discuss. 176 (Xx) (2014) 447–458. Available from: https://doi.org/10.1039/c4fd00159a.

[9] D.A. Barkas, C.S. Psomopoulos, P. Papageorgas, K. Kalkanis, D. Piromalis, A. Mouratidis, Sustainable energy harvesting through triboelectric nano – generators: a review of current status and applications, Energy Procedia 157 (2019) 999–1010. Available from: https://doi.org/10.1016/j.egypro.2018.11.267.

[10] G. Khandelwal, N.P. Maria Joseph Raj, S.J. Kim, Triboelectric nanogenerator for healthcare and biomedical applications, Nano Today 33 (2020) 100882. Available from: https://doi.org/10.1016/j.nantod.2020.100882. 1–29.
[11] S. Pan, Z. Zhang, Fundamental theories and basic principles of triboelectric effect: a review, Friction 7 (1) (2018) 2–17 [Online]. Available. Available from: https://doi.org/10.1007/s40544-018-0217-7.
[12] C. Te Huang, et al., GaN nanowire arrays for high-output nanogenerators, J. Am. Chem. Soc. 132 (13) (2010) 4766–4771. Available from: https://doi.org/10.1021/ja909863a.
[13] H. Zou, et al., Quantifying the triboelectric series, Nat. Commun. 10 (1) (2019) 1–9. Available from: https://doi.org/10.1038/s41467-019-09461-x.
[14] A.A. Mathew, A. Chandrasekhar, S. Vivekanandan, A review on real-time implantable and wearable health monitoring sensors based on triboelectric nanogenerator approach, Nano Energy 80 (2020) 105566. Available from: https://doi.org/10.1016/j.nanoen.2020.105566. no. November 2020.
[15] W. Chesworth, A. Perez-Alberti, E. Arnaud, Ice erosion, Encycl. Earth Sci. Ser. (2008) 333–338. Available from: https://doi.org/10.1007/978-1-4020-3995-9_282.
[16] M.W. Williams, Triboelectric charging of insulating polymers-some new perspectives, AIP Adv. 2 (1) (2012). Available from: https://doi.org/10.1063/1.3687233.
[17] A.F. Diaz, R.M. Felix-Navarro, A semi-quantitative tribo-electric series for polymeric materials: the influence of chemical structure and properties, J. Electrostat. 62 (4) (2004) 277–290. Available from: https://doi.org/10.1016/j.elstat.2004.05.005.
[18] F. Galembeck, T.A.L. Burgo, L.B.S. Balestrin, R.F. Gouveia, C.A. Silva, A. Galembeck, Friction, tribochemistry and triboelectricity: recent progress and perpectives, RSC Adv. 4 (109) (2014) 64280–64298. Available from: https://doi.org/10.1039/b000000x.
[19] C. Liu, A.J. Bard, Electrostatic electrochemistry at insulators, Nat. Mater. 7 (6) (2008) 505–509. Available from: https://doi.org/10.1038/nmat2160.
[20] W.R. Salaneck, A. Paton, D.T. Clark, Double mass transfer during polymer-polymer contacts, J. Appl. Phys. 47 (1) (1976) 144–147. Available from: https://doi.org/10.1063/1.322306.
[21] L.S. McCarty, G.M. Whitesides, Electrostatic charging due to separation of ions at interfaces: contact electrification of ionic electrets, Angew. Chem. Int. (Ed.) 47 (12) (2008) 2188–2207. Available from: https://doi.org/10.1002/anie.200701812.
[22] S. Chandra, Wearable flexible sensors: fabrication and characterization, in: Wearable Sensors: Applications, Design and Implementation, January, 2017, 1–27.
[23] R. Hsissou, R. Seghiri, Z. Benzekri, M. Hilali, M. Rafik, A. Elharfi, Polymer composite materials: a comprehensive review, Compos. Struct. 262 (2021) 0–3. Available from: https://doi.org/10.1016/j.compstruct.2021.113640. no. January.
[24] A. Chen, C. Zhang, G. Zhu, Z.L. Wang, Polymer materials for high-performance triboelectric nanogenerators, Adv. Sci. 7 (14) (2020) 1–25. Available from: https://doi.org/10.1002/advs.202000186.
[25] H.P.S.A. Khalil, M.A. Tehrani, Y. Davoudpour, A.H. Bhat, M. Jawaid, A. Hassan, Natural fiber reinforced poly(vinyl chloride) composites: a review, J. Reinf. Plast. Compos. 32 (5) (2013) 330–356. Available from: https://doi.org/10.1177/0731684412458553.
[26] R. Hsissou, et al., Evaluation of corrosion inhibition performance of phosphorus polymer for carbon steel in [1 M] HCl: computational studies (DFT, MC and MD simulations), J. Mater. Res. Technol. 9 (3) (2020) 2691–2703. Available from: https://doi.org/10.1016/j.jmrt.2020.01.002.
[27] O. Dagdag, et al., Epoxy pre-polymers as new and effective materials for corrosion inhibition of carbon steel in acidic medium: computational and experimental studies, Sci. Rep. 9 (1) (2019) 1–14. Available from: https://doi.org/10.1038/s41598-019-48284-0.
[28] M. Asim, N. Saba, M. Jawaid, M. Nasir, M. Pervaiz, O.Y. Alothman, A review on phenolic resin and its composites, Curr. Anal. Chem. 13 (2018) 2017. Available from: https://doi.org/10.2174/1573411013666171003154410. no. June.

[29] N. van de Werken, H. Tekinalp, P. Khanbolouki, S. Ozcan, A. Williams, M. Tehrani, Additively manufactured carbon fiber-reinforced composites: state of the art and perspective, Addit. Manuf. 31 (2020). Available from: https://doi.org/10.1016/j.addma.2019.100962.
[30] R. Hsissou, et al., Experimental, DFT and molecular dynamics simulation on the inhibition performance of the DGDCBA epoxy polymer against the corrosion of the E24 carbon steel in 1.0 M HCl solution, J. Mol. Struct. 1182 (2019) 340–351. Available from: https://doi.org/10.1016/j.molstruc.2018.12.030. no. November.
[31] G.Z. Li, et al., High-performance transparent and flexible triboelectric nanogenerators based on PDMS-PTFE composite films, Adv. Electron. Mater. 5 (4) (2019) 1–8. Available from: https://doi.org/10.1002/aelm.201800846.
[32] R. Petre, N. Petrea, G. Epure, T. Zecheru, Polymer composite materials and applications for chemical protection equipments, Int. Conf. Knowl.-based Organ. 21 (3) (2015) 873–877. Available from: https://doi.org/10.1515/kbo-2015-0148.
[33] I. Pleşa, P.V. Noţingher, S. Schlögl, C. Sumereder, M. Muhr, Properties of polymer composites used in high-voltage applications, Polymers (Basel). 8 (5) (2016). Available from: https://doi.org/10.3390/polym8050173.
[34] T. Rodrigues-marinho, N. Castro, V. Correia, P. Costa, S. Lanceros-méndez, Triboelectric energy harvesting response of different polymer-based materials, Materials (Basel). 13 (21) (2020) 1–12. Available from: https://doi.org/10.3390/ma13214980.
[35] S. Fu, Z. Sun, P. Huang, Y. Li, N. Hu, Some basic aspects of polymer nanocomposites: a critical review, Nano Mater. Sci. 1 (1) (2019) 2–30. Available from: https://doi.org/10.1016/j.nanoms.2019.02.006.
[36] S. Ponyrko, R.K. Donato, L. Matějka, Tailored high performance shape memory epoxy-silica nanocomposites. Structure design, Polym. Chem. 7 (3) (2016) 560–572. Available from: https://doi.org/10.1039/c5py01450f.
[37] P. Manchi, S.A. Graham, B. Dudem, H. Patnam, J.S. Yu, Improved performance of nanogenerator via synergetic piezo/triboelectric effects of lithium niobate microparticles embedded composite films, Compos. Sci. Technol. 201 (2021) 108540. Available from: https://doi.org/10.1016/j.compscitech.2020.108540. no. October 2020.
[38] V.A. Cao, S. Lee, M. Kim, M.M. Alam, P. Park, J. Nah, Output power density enhancement of triboelectric nanogenerators via ferroelectric polymer composite interfacial layers, Nano Energy 67 (2020) 104300. Available from: https://doi.org/10.1016/j.nanoen.2019.104300. no. November 2019.
[39] A.K. Gupta, C.-H. Hsu, S.-N. Lai, C.-S. Lai, ZnO-polystyrene composite as efficient energy harvest for self-powered triboelectric nanogenerator, ECS J. Solid State Sci. Technol. 9 (11) (2020) 115019. Available from: https://doi.org/10.1149/2162-8777/aba7fa.
[40] H.J. Hwang, et al., Mesoporous highly-deformable composite polymer for a gapless triboelectric nanogenerator via a one-step metal oxidation process, Micromachines 9 (12) (2018) 1–11. Available from: https://doi.org/10.3390/mi9120656.
[41] J. Chun, J.W. Kim, W.-S. Jung, C.-Y. Kang, S.-W. Kim, Z.L. Wang, et al., Mesoporous pores impregnated with Au nanoparticles as effective dielectrics for enhancing triboelectric nanogenerator performance under harsh environment, Energy Environ. Sci. 3 (2015) 10715–10722. Available from: https://doi.org/10.1039/b000000x.
[42] H. Niu, et al., Polymer nanocomposite-enabled high-performance triboelectric nanogenerator with self-healing capability, RSC Adv. 8 (54) (2018) 30661–30668. Available from: https://doi.org/10.1039/C8RA05305G.
[43] M.K. Kim, M.S. Kim, H.B. Kwon, S.E. Jo, Y.J. Kim, Wearable triboelectric nanogenerator using a plasma-etched PDMS-CNT composite for a physical activity sensor, RSC Adv. 7 (76) (2017) 48368–48373. Available from: https://doi.org/10.1039/c7ra07623a.

CHAPTER

8

Magnetoelectric polymer nanocomposites for energy harvesting

Sobi K. Chacko, M.T. Rahul and Raneesh Balakrishnan

Department of Physics, Catholicate College, Pathanamthitta, Kerala, India

8.1 Introduction

In this era of modern technology, energy is one of the most critical factors to lead a sustainable life. But the nonrenewable resources such as fossil fuels and natural gases are too costly to harvest, and their supplies are also limited. Because of their lack of renewability and its contributions to environmental pollution, these nonrenewable resources are widely criticized. So the research community has made effort to find alternative renewable energy resources, which has resulted in solar cells, batteries, etc. [1]. The rapidly increasing "Internet of Things" (IoT), through which the various technologies and devices are interconnected and transfer data, demands uninterrupted continuous self-powered devices [2–4]. At this time, a significant challenge is to develop continuous energy harvesting with environment-friendly, maximum efficiency, and without wastage of energy. A self-powered energy-harvesting system will reduce the dependence on external power sources. The advancements in nanotechnology, now capable of fabricating different self-powered energy harvesters, can convert the magnetic, mechanical, and thermal forms of energy into electrical energy, thereby avoiding the external electric power sources.

There exist different types of energy harvesting, such as pyroelectric energy harvesting, piezoelectric energy harvesting, photovoltaic

Engineered Polymer Nanocomposites for Energy Harvesting Applications
DOI: https://doi.org/10.1016/B978-0-12-824155-4.00004-1

energy harvesting, thermoelectroenergy harvesting, magnetostatic energy harvesting, etc. Pyroelectricity means the ability of certain materials to generate voltage depending on the temperature; hence, in these materials, there is a coupling between climate and polarization. In the case of piezoelectric energy harvesting, the voltage is generated concerning the applied stress/strain. Photovoltaic energy harvesters use solar light to create energy, and thermoelectric energy harvesters convert heat into the form of energy. In the case of multiferroics, combining the piezoelectric energy harvesting and magnetostrictive energy harvesting, these multifunctionalities are accommodated in a single-phase, resulting in a drastic enhancement in energy-harvesting property. Therefore these multiferroic materials have gained significant attention in various fields due to their multifunctionality and technological applications [5].

8.2 Magnetoelectric materials

Magnetoelectrics are a unique category of materials, and their natural occurrence in nature is extremely rare. The criteria to show a ferroelectric or magnetic nature are mutually exclusive in a material; electron spin contributes to the magnetic properties, while charges of electrons and ions determine the electrical properties [6–10]. Based on the presence of ferroic orders and the interaction between them, magnetoelectric materials are classified into different categories [11]. Coexistence and coupling between the ferroelectric and magnetic orders and potential applications in various fields have made these ME materials a hot research area [10,12].

By the constituent materials, ME materials can be classified into single-phase and composite [8,13–15]. In simple words, single-phase multiferroic materials are those which show two or more ferroic orders simultaneously, and the ME coupling arises directly between the different ferroic orders [15]. Most of the single-phase multiferroic materials have a magnetic transition temperature below room temperature. $YMnO_3$, $YbMnO_3$, $BiFeO_3$, $ErMnO_3$, $GaFeO_3$, etc., are some of the single-phase multiferroics materials [16–19]. Among the single-phase multiferroics, bismuth ferrite ($BiFeO_3$) [20] is the only one that shows the room temperature multiferroic property. The large leakage current, low permittivity and permeability, weak magnetic moments, and insufficient magnetoelectric coupling in single-phase multiferroic materials hinder practical applications [21].

In 1972 Van Suchtelen proposed the idea of indirect spin-mediated magnetoelectrics composites [22]. To overcome the cons of single-phase ME materials, the researchers extended their work into composite ME materials. The composite multiferroics incorporate different materials.By combining the different materials which supports either magnetic or electric ferroic orders. multiferroicity in composites are achieved. In the case of composite multiferroic materials, the coupling takes place indirectly [23]. The origin of the magnetoelectric effect on the blended materials is the synergy of the different phases present in the composite. This ME effect in composite materials is also known as the product tensor property [22]. There must be a minimum of two constituents in ME composites, which contribute to either a magnetic or electroactive phase. These ME composites can be either ceramic or polymer-based. Ceramic ME composites still have some cons, such as being expensive, brittle, dense, and heavy in the device configuration. To avoid these shortcomings, polymer-based magnetoelectric composites come to the scenario, and these give good mechanical properties, large magnetoelectric coupling, low leakage current, lightweight, easy fabrication, etc. [24–28].

In the magnetoelectric effect, the applied magnetic field will make mechanical deformation or strain via the magnetostriction effect. This strain is transferred to the electroactive material. The ME coupling is often referred to in terms of magnetoelectric coupling coefficient (MECC).

$$\text{Magnetoelectric effect; } ME_H = \frac{\text{magnetic}}{\text{mechanical}} \times \frac{\text{mechanical}}{\text{electric}} \tag{8.1}$$

$$\text{Magnetoelectric effect; } ME_E = \frac{\text{electric}}{\text{mechanical}} \times \frac{\text{mechanical}}{\text{magnetic}} \tag{8.2}$$

where E and H represent the electric and magnetic fields, respectively. In the ME_H effect, the sample undergoes magnetostriction when it is subjected to the external magnetic field. This will induce a strain on the electroactive part in the composite. MECC α can be measured using the lock-in amplifier method.

$$\alpha = \frac{dE}{dH} = \left(\frac{1}{t}\right)\left(\frac{dV}{dH}\right) = \frac{Vout}{ht} \tag{8.3}$$

where V_{out} represents the ME voltage, t denotes the thickness of the sample, and h denotes the amplitude of the applied magnetic field. A detailed description of the magnetoelectric coupling measurement setup can be found elsewhere [16,29].

In this chapter, we discuss the polymer ceramic magnetoelectric nanocomposites solely, where the nanomaterials contribute the magnetic part, and the electroactive polymer provides the ferroelectric part.

8.3 Materials

8.3.1 Magnetic/magnetostrictive materials

Ferromagnetic materials exhibit a net magnetization/spontaneous magnetization even in the absence of an external magnetic field. Fig. 8.1 represents the domain alignment of ferromagnetic materials. Among the wide variety of magnetic metal oxides, spinel ferrites AB_2O_4 type structure nanoparticles are widely studied due to their excellent magnetic and magnetostrictive properties. Among the spinel family, AFe_2O_4 (where A = Ni, Co, Zn, Mg, Mn, and Cu) are broadly studied. Rare earth iron garnets, such as $RE_3Fe_5O_{12}$, where RE = Y, Er, Sm, Dy, etc., are also exhibited good magnetic properties [30–32]. Some of the ABO_3 family members, also exhibit better magnetic properties; rare earth perovskites compounds belong to this category [33]. Most of the works in polymer-based magnetoelectrics are based on the fact that the magnetostriction effect in ferromagnetic materials is the origin of the magnetoelectric effect. Magnetostriction, which is also known as the magnetomechanical effect, is the dimensionality change that occurs to a magnetic material when it is magnetized [34]. The magnetostriction effect is due to the rotations/changes in the alignment of the domains. The domains, which are previously randomly oriented, try to orient in the direction of the applied field. Magnetostriction is an intrinsic property of magnetic materials [35]. It is defined as the fractional change in length, $\lambda = \Delta l/l$. The magnetostriction effect has a wide range of application in various fields such as actuators, sensors, etc. Its ability to modify magnetic energy to physical motion makes it possible. The high magnetostrictive properties of cubic spinel ferrites help achieve a high magnetoelectric coefficient; hence these cubic phased spinel ferrites are used as the inorganic magnetic counterparts in the magnetoelectric polymer composite formation [36–39].

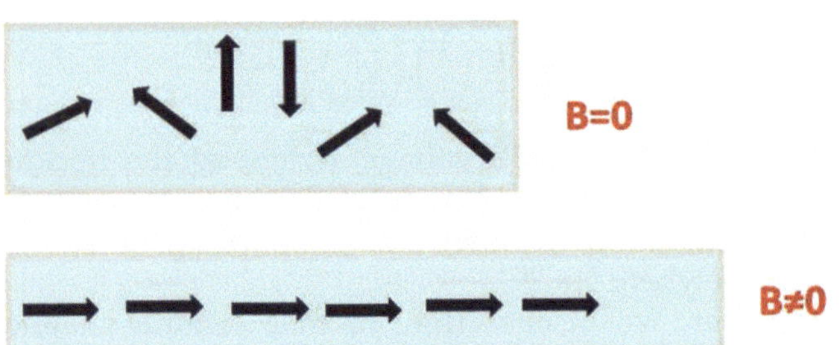

FIGURE 8.1 Domain alignment of ferromagnetic materials at magnetic field B = 0 and B≠0.

8.3.2 Ferroelectric materials

Many materials exhibit a ferroelectric nature, varying from crystals, ceramics, amorphous and semicrystalline polymers, composites using ceramic, polymers, and ceramic and polymer composites [40–42]. The Landolt–Bornstein database classifies the ferroelectric materials into three major groups based on their chemical composition and crystal structure. The three groups are (1) inorganic oxides, (2) inorganic nonoxides, and (3) organic crystals, liquid crystals, and polymers [41,43].

A ferroelectric material must have spontaneous polarization and it can be reversed by applying an external electric field. Ferroelectrics are the electrical analogs of ferromagnetics in magnetism. Ferroelectric materials are a subdivision of pyroelectric, and pyroelectrics are a subclass of piezoelectrics. Hence the ferroelectric materials exhibit both piezo- and pyroelectric properties [42,44]. The ferroelectric property was first observed by Joseph Vasalek in Rochelle salt in 1921 [45]. The next ferroelectric substance discovered was potassium dihydrogen phosphate in 1935 [46]. But a drastic development in the field of ferroelectric materials was begun only after discovering a new class of ferroelectric perovskites materials such as barium titanate ($BaTiO_3$)and lead zirconate titanate(PZT) in the early 1940s [47]. Barium titanate is the first polycrystalline ceramic material found which exhibits ferroelectricity. Even this decade, barium titanate remains one of the most widely studied ceramic piezoelectric materials because of its outstanding piezoelectric and dielectric properties, and its low cost [48–50]. Barium titanate and lead-based perovskite titanate were the two most studied ferroelectric materials [40,51,52]. Still, the environmental concerns led to the search for lead-free ferroelectric ceramics which ended up in the rediscovery of $KNaNbO_3$, $NaNbO_3$, $KNbO_3$, $LiNbO_3$, sodium bismuth titanate, etc. [53–57]. The most extensively studied ferroelectric materials belong to the oxide-based ABO_3 type perovskite family. Another kind of ferroelectric material that is not considerably checked is Aurivillius compounds (e.g., $Bi_4Ti_3O_{12}$), bismuth titanate-based materials, with a general formula of $(Bi_2O_2)^{2+}$ $[A_{m-1}B_mO_{3m+1}]^{2-}$ [58]. They can be used in ferroelectric RAM devices, sensors, etc. The brittle and dense nature and high-temperature preparation of ceramics complicate the fabrication and decrease the stability; these restrict the application of ceramic ferroelectric materials [59]. To overcome the drawbacks of ceramic ferroelectric materials, ferroelectric polymers have come to the fore.

8.3.3 Ferroelectric polymers

Ferroelectric polymers are sensational for their excellent electroactive properties, good dielectric properties and outstanding flexibility. The ferroelectric polymers, also known as smart materials, provide excellent piezo- and pyroelectric properties. Their fields of application have been extended to piezo-/pyroenergy harvesters, memory devices, solar cells, sensors, actuators, biomedical applications, etc. Polyvinylidene fluoride (PVDF) is the second largest produced fluoropolymer and the most extensively studied ferroelectric polymer [60]. Ferroelectricity in PVDF was first discovered in the 1970s. After that, some odd nylons (nylons with an odd number of carbon atoms in the monomer unit), PLLA, and polylactic glycolic acid (PLGA) were also recognized as ferroelectric polymers [44,61–63]. The ferroelectric polymer materials are soft, flexible, lead-free, biocompatible, etc. Therefore the polymer-based materials have high demand in flexible device fabrication [64–66]. Among the polymer ferroelectrics, PVDF and its copolymers, such as P(VDF-TrFE), poly(vinylidene fluoride-*co*-hexafluoropropylene) (PVDF-HFP), or poly (vinylidene fluoride-*co*-chlorotrifluoroethylene) (PVDF-CTFE), are considered as the best candidates due to the high dielectric constant, piezo-, pyro-, and ferroelectric nature [60,67,68]. Along with PVDF, ferroelectric polymers like polyamides, polyureas, polythioureas, polypeptides, cyanoethyl, and cellulose have also been investigated to a reasonable level in previous years [42,69–73].

Apart from this, naturally occurring biopolymers also exhibit piezoelectricity. Cellulose is the most abundant natural polymer on Earth and it comes under the category of electroactive polymers [74,75]. It is eco-friendly, cost-effective, and renewable also. Hence it can be efficiently utilized in various applications. Cellulose-based magnetoelectric laminate composites reported a MECC of 1.41 V/cm Oe [75]. The fano-like resonance observed in the cellulose-based composite allows sharp tuning of MECC from peak value to zero. The discovery of piezoelectricity in cellulose paves a path to seek piezoelectricity in other biomaterials [75].

8.3.3.1 *Polyvinylidene fluoride and its copolymers*

The PVDF is a semicrystalline thermoplastic electroactive polymer with a chemical formula of (-CH_2-CF_2-) which shows five distinct crystalline phases, named α, β, γ, δ, and ε, depending on the processing conditions of the polymer [76]. It is regarded as the "material of merit" due to its outstanding properties and significant role in practical applications of magnetoelectric materials. The different polymorph classifications are based on the chain conformations [67,77–79]. The β phase is the most investigated phase among the five crystalline phases because of its high ferroelectric nature and it is the most thermodynamically stable compared to other polymorphs. PVDF is nonsoluble in water, and it is chemically inert in various solvents and oils, and shows lesser permeability to gases and liquids [68]. To induce and enhance the β phase in PVDF, different processing methods have been used [28,80–83]. The β phase is achieved from PVDF by different fabrication methods. Electrical poling, uniaxial stretching, solution crystallization, the addition of suitable nanofillers, and high-temperature annealing are used to introduce and improve the ferroelectric nature of PVDF. A detailed review on the different processing conditions, characterizations, and applications of electroactive phases of PVDF was given by Martins et al. [67].

To make changes in PVDF properties, such as crystallinity, electroactive nature, melting temperature, stability, and flexibility, copolymers based on PVDF are formed. Co-polymers of PVDF, such as PVDF-HFP, PVDF-TrFE, PVDF-CTFE (with chemical formula $[(CH2CF2)_x(CF2CFCl)_y]_n)$, and poly (vinylidene fluoride-trifluoroethylene-chlorotrifluoroethylene) (PVDF-TRFE CTFE) possess their own peculiarities [84–86].

8.4 Types of polymer-based magnetoelectric composites

Interfacial bonding and connectivity have significant roles in the properties of magnetoelectrics. Magnetoelectric composites have been prepared with either 0–3, 1–3, and 2–2 connectivity. Fig. 8.2 shows the graphical representation of various types of connectivity in polymer-based magnetoelectric materials. The number indicates the connectivity of magnetic and piezoelectric phases. Most of the works are based on the first two connectivities, that is, 0–3 and 2–2. In the 0–3 connectivity, the magnetic particles are embedded in the electroactive matrix. Thin-film polymer–ceramic multiferroics systems represent the 0–3 connectivity. The 2–2 phase represents the laminate composite consisting of sandwiched magnetic and piezoelectric layers in an alternating manner. In the 1–3 connectivity cylinder matrix composite, they are formed by embedding magnetic rods, fiber needles, etc., in the piezoelectric matrix. Compared to other systems, the 2–2 connectivity systems possess high MECCs [14].

Thin films of $(1-x)$PVDF–x(($Bi_{0.5}$ $Ba_{0.25}$ $Sr_{0.25}$)($Fe_{0.5}$ $Ti_{0.5}$)O_3 with varying compositions of x ($x = 0,0.5,0.1,0.15$) composites having 0–3 connectivity

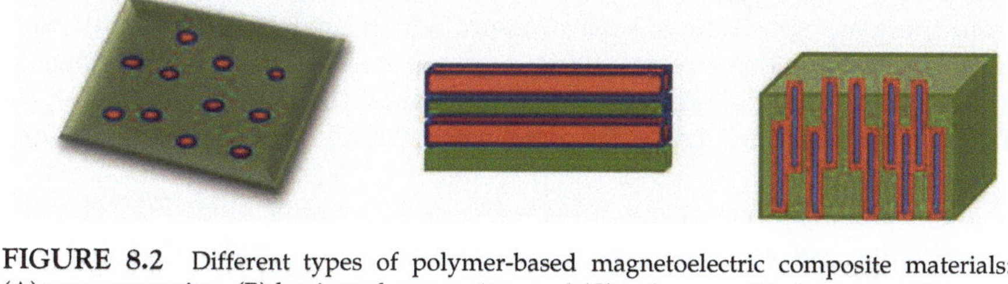

FIGURE 8.2 Different types of polymer-based magnetoelectric composite materials: (A) nanocomposites, (B) laminated composites, and (C) polymer as binder composites.

were fabricated using a simple solution casting method. *N*-methyl-2-pyrrolidone (NMP) was used as the solvent for the composite solution preparation. The solution was stirred at 60°C–70°C, and the prepared solution was coated in a glass substrate with a micrometer controlled tape caster. For the evaporation of solvents, it was kept in the air for a few days. After complete drying, thin films were peeled off from the substrate. Electrical and magnetoelectrical properties were enhanced and a MECC value of 18. mV/cm Oe was obtained for the maximum filler concentration, which is, $x = 0.15$ [87].

Particulate two-phase composite films of Terfenol-D (TD)/PVDF-TrFE have been prepared by Periera et al. and a MECC of 38 mV/cm Oe was obtained [88].

These different connectivities help to tune the magnetic and ferroelectric properties independently. Composite structures offer more significant magnetoelectric coupling than single-phase multiferroics.

8.5 Fabrication methods of polymer-based multiferroic composites

Depending on the target application and the property requirements, different fabrication methods are used for composite preparation. Increasing the flexibility along with an enhancement in the magnetoelectric coupling is the essential parameters. A higher degree of crystallinity is necessary to get a high ferroelectric β phase and piezoelectric phase, which is directly influenced by the processing conditions.

The mechanical properties and electroactive nature of the ferroelectric polymer are greatly affected by the fabrication methods. Solvent casting, spin coating, electrospinning, etc., are the easiest and commonly used fabrication methods for preparing polymer-based magnetoelectric composite. The spray printing technique was introduced to prepare composites in recent years due to the transformative industrial approach. Ink for the spray painting technique was prepared by 20 wt.% of ferrite added to the desired amount of DMF and subjected to ultrasonication for 8 hours, and to which the 2 g of PVDF-TrFE were added. Flexible films of 50 μm were prepared by spray printing the composite ink using an airbrush at 10 cm from the glass substrate at 3 psi pressure and heated at 210°C for 10 minutes. The final composite results in a magnetoelectric coupling voltage of 21.2 mV/cm Oe [89]. Rehana et al. fabricated ($BiFeO_3$-$NaNbO_3$) PVDF-TrFE composite films using the spin-coating method. Room temperature multiferroic measurements of the prepared films showed an MECC value of 2.4 V/cm Oe, which is much higher than the ceramic alone [90]. Here in this chapter, we are focusing on the solvent casting and electrospinning methods.

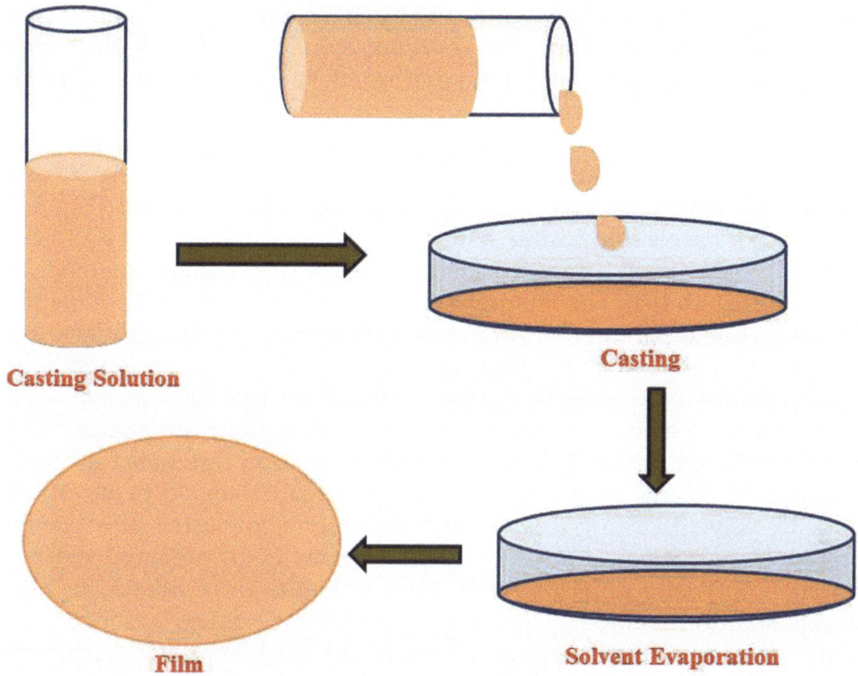

FIGURE 8.3 Graphical representation of solvent casting.

8.5.1 Solvent casting

Solvent casting/solution casting is one of the easiest and cost-effective methods for the fabrication of polymer-based multiferroic composites in film form. Graphical representation of the solvent casting method is shown in Fig. 8.3. For solution casting, the required polymer is dissolved in the appropriate solvent and transferred to a petri dish or coated on a mold. Mostly, acetone, dimethylformamide (DMF), *N*-methyl pyrrolidone (NMP), dimethyl sulfoxide (DMSO), or dimethylacetamide (DMAc) are used as the solvents. After the complete evaporation of the solvent by drying it at room temperature or low heating temperatures, the resultant films peel off from the substrate as free-standing films. To produce the magnetoelectric polymer composite, the magnetic nanofillers are introduced into the polymer solution with proper stirring. Solvent casting can make films with a thickness in the range of micrometer range. When we deal with ferroelectric polymers, such as PVDF, solvent casting methods can improve the flexibility and electroactive nature of the polymer [76].

PVDF/$SmFeO_3$ (SFO) films were fabricated in such a way that the desired amount of SFO nanoparticles was added to the PVDF solution in which the dimethylsulfoxide was used as the solvent, and it was ultrasonicated for 2 hours. The resultant solution was cast on a glass substrate and annealed at a temperature of 90°C. Thin films of 10 μm thickness were obtained. The maximum filler loading produced a magnetoelectric coupling voltage of 5.2 mV/cm Oe for 1 kHz [33]. Behra et al., reported an MECC of 6.12 mv/cm Oe for 2 kOe in PVDF/$MnFe_2O_4$ composite films, using the solution casting method [91]. The PVDF-TrFE-based composites with varying particle size of Fe_3O_4 nanoparticles have been prepared using the solvent casting method by Goncalves et al. [92]. For the preparation of flexible composite films, the prepared Fe_3O_4 nanoparticles have been dispersed in DMF under ultrasonication; then, the PVDF-TrFE has been added to the prepared solution. To dissolve well, a Teflon stirrer is used along with the ultrasonication bath. Then the final solution was spread on a glass plate, and it was kept under 210 degrees in an oven. The resultant film has a thickness of 50 μm. Jayakumar et al. prepared silica-coated Fe_3O_4/PVDF films by solvent casting and studied the magnetodielectric effect of the prepared multiferroic composite films. Martins et al. prepared the particulate composite films of PVDF with $CoFe_2O_4$ and $NiFe_2O_4$ using a simple solution method [93]. The prepared films possess a thickness of 30 μm. The well-dispersed ferrites nucleate the piezoelectric phase of the polymer at different concentrations. Martins et al. studied the magnetoelectric effect using three different ferrite nanoparticles in the PVDF-TrFE matrix [94]. The $Zn_{0.2}Mn_{0.8}Fe_2O_4$ (ZMFO), $CoFe_2O_4$ (CFO), and Fe_3O_4 (FO) nanoparticles have been prepared using the hydrothermal method for the preparation of multiferroic composite; the required amount of ferrite nanoparticles was dissolved in DMF and subjected to ultrasound sonication for 8 hours. PVDF-TrFE was added to the resultant solution and stirred for 2 hours in an ultrasound bath. After that, the solution was coated on a glass substrate and heated at 210°C for 10 minutes. The prepared films had a thickness of $\approx$50 μm and showed interesting variations in the coupling constant depending on the type of filler added. For CFO/PVDF-TrFe composite, a magnetoelectric coupling voltage of 6.5 mV/cm Oe was obtained. Anshida et al. prepared multiferroic films with a thickness in the range of 180–200 μm of $BaTiO_3$ and dopamine functionalized $BaTiO_3$ incorporated PVDF-TrFE. For all the samples, they kept the amount of PVDF-TrFE as 15 wt.%. Instead of DMF, they used methyl ethyl ketone as the solvent. The solvent casting and solvent evaporation were done at room temperature and the peeled off films were heated at 138°C, which were done to ensure the β phase of PVDF-TrFE was obtained. They measured the AC and DC values of MECC, and they were reported as 18.2 and 22.2 mv/cm Oe, respectively, for the maximum filler loading, which is dopamine functionalized. It also reports the role of the functionalization of $BaTiO_3$

in the various properties of the composites [95]. PVDF/CFO nanoparticles and PVDF/CFO nanorods composites were prepared using the solvent casting method. The poling and stretching process plays a good role in enhancing the beta phase up to 78.5%, where the presence of magnetic fillers helps to stabilize it in beta form [96]. Multiferroic composite films of PVDF-Fe_3O_4 with reduced graphene oxide and ZnO have been fabricated. Where the inclusion of Fe_3O_4 and ZnO resulted in a reduction of the electroactive phase of the polymer compared to the PVDF-RGO combination. Fe_3O_4 acts as the magnetostrictive phase in the composite [97].

8.5.2 Electrospinning

Electrospinning is considered as the most efficient and popular method for fabricating fibers in the nanometer range. Through electrospinning the ferroelectric phase can be achieved directly from the solution. Nanofibers exhibit a high surface to volume ratio and flexibility [98]. Fig. 8.4 shows a schematic representation of the electrospinning process.

During electrospinning, a sufficiently high electric field is used to create a jet from the liquid drop, followed by stretching and elongation to

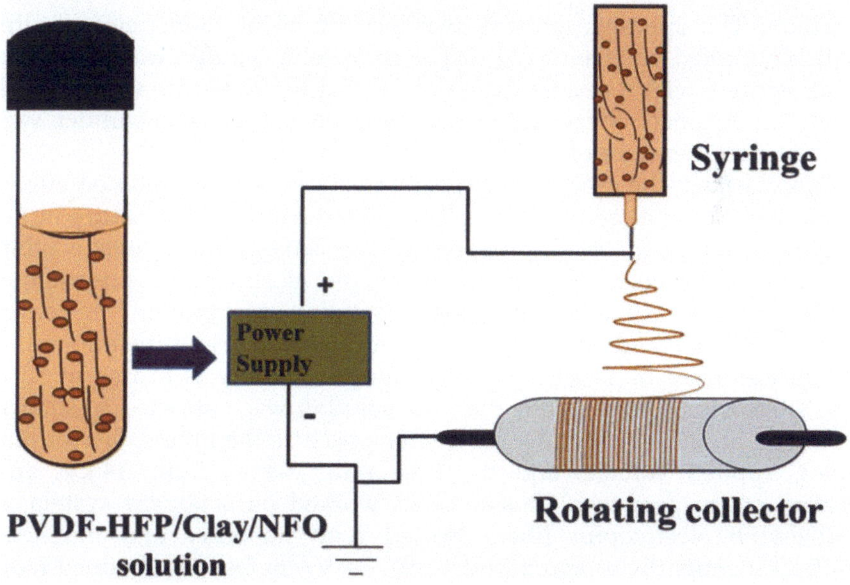

FIGURE 8.4 Schematical representation of electrospinning process. Source: *Reproduced with permission from S.K. Chacko, M.T. Rahul, B. Raneesh, N. Kalarikkal, Enhanced magnetoelectric coupling and dielectric constant in flexible ternary composite electrospun fibers of PVDF-HFP loaded with nanoclay and $NiFe_2O_4$ nanoparticles. N. J. Chem. 44 (26) (2020) 11356–11364.*

produce fibers. The main components for the electrospinning setup consist of a high-voltage power supply, a syringe pump and needle, and a conductive collector; it can be either rotating or stationary. The positive terminal of the power supply is connected to the needle, and the collector is grounded. The electrospinning process can be divided into four steps: the formation of the Taylor cone from the liquid drop, elongation of the formed jet on a straight line, thinning of the jet due to the effect of high electric field, and the collection jet as a solid fiber on the collector. To prepare polymer–ceramic composite electrospun fibers, the nanofillers are dispersed in the prepared polymer solution and are subjected to electrospinning. By controlling the different parameters, such as applied voltage, the distance between the collector and needle, gauge diameter of the needle, flow rate viscosity of the polymer solution, nature of filler, and fiber characteristics can be controlled. Andrew et al. studied the effect of spinning conditions in the formation of the β phase of PVDF fibers [99].

There are many works reported in the area of electrospinning of PVDF [100,101]. Multiferroic nanofibers of $Nd_{0.05}Bi_{0.95}Fe_{0.95}Co_{0.05}O_3$/PVDF-TrFE with an average diameter of 450 nm were fabricated using electrospinning [102]. Initially, PVDF-TrFE (80/20) was dissolved in *N*,*N*-dimethylacetamide (DMAc) and the desired amount of ceramic particle was added to the system. The composite solution was sonicated for 15 minutes to achieve a good dispersion of nanoparticles in the polymer solution. The spinning process was performed at room temperature, with an applied voltage of 12 kV. The composite solution was taken in a 5 mL syringe with a needle gauge of 21, and the flow rate was fixed as 1 mL/h. Obtained fibers were dried to remove the solvent under vacuum at a temperature of 40°C.

Moreover, in the case of electroactive polymers, the combined effects of electrical poling and stretching help to induce the polar nature in the polymer. Hemalatha et al. prepared PVDF/$ZnFe_2O_4$ fiber mats exhibiting 92% of ferroelectric β phase using the electrospinning technique [103]. For the polymeric composite solution preparation, an equal amount of DMF and acetone were taken as solvents, and the amount of PVDF was fixed at 14 wt.%. The spinning solution was heated at 50°C and stirred for 5 hours; after that, the solution was subjected to sonication for 1 hour. The spinning parameters, such as the tip to collector distance, applied voltage, and the flow rates, were 15 cm, 14 kV, and 0.40 mL/h, respectively. Chacko et al. worked on a ternary system of PVDF-HFP electrospun fibers loaded with nanoclay and magnetic $NiFe_2O_4$ nanoparticles and a MECC of 17 mV/cm Oe was obtained [28].

In recent years, more complex fiber structures are also gaining attention. Coaxial/core–shell spinning, triaxial spinning, etc., are some of them. In one of the recent studies using co-axial spinning Deepalekshmi et al. reported a piezoelectric nanogenerator with a output voltage of

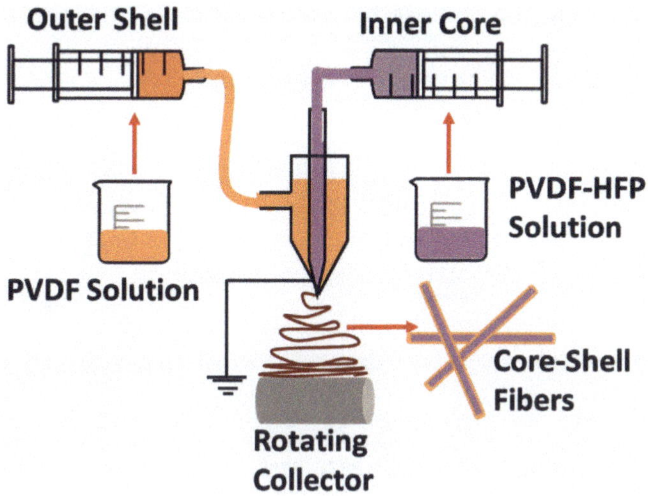

FIGURE 8.5 Graphical representation of coaxial spinning. Source: *Reproduced with permission from D. Ponnamma, M.M. Chamakh, A.M. Alahzm, N. Salim, N. Hameed, M.A.A. AlMaadeed, Core-shell nanofibers of polyvinylidene fluoride-based nanocomposites as piezoelectric nanogenerators. Polymers 12 (10) (2020) 2344.*

14 V, using the coaxial spinning method to fabricate nanofibers of PVDF/PVDF-HFP in a core–shell manner. Metal oxide nanoparticles of TiO_2 and ZnO were used as the fillers [104] (Fig. 8.5).

8.6 Energy harvesting aspects of magnetoelectric material

Energy harvesting from various sources, such as light, wind, vibrations, magnetic fields, etc., is necessary to satisfy the increasing demand of current for next-generation electronic devices [36,105–110]. Many researchers have worked on the methods to obtain self-powered, continuous energy harvesting. Polymer-based magnetoelectric materials grab the attraction, mainly due to their highly flexible nature, and they can realize energy generation through the piezoelectric effect [111–114]. Triboelectric nanogenerators are also an emerging field in energy harvesters [115–118]. Energy-harvesting applications include the conversion of energy from mechanical, magnetic, electrical, and thermal forms. MECCs obtained for various ferropolymer-based materials are sufficiently high and meet the practical requirements [38,119–122]. Based on the magnetomechanoelectric mechanism, electricity can be harvested from magnetic noises. When a magnetoelectric system is placed in an external magnetic field, the magnetostrictive component undergoes dimensional changes, which creates a strain in the piezoelectric

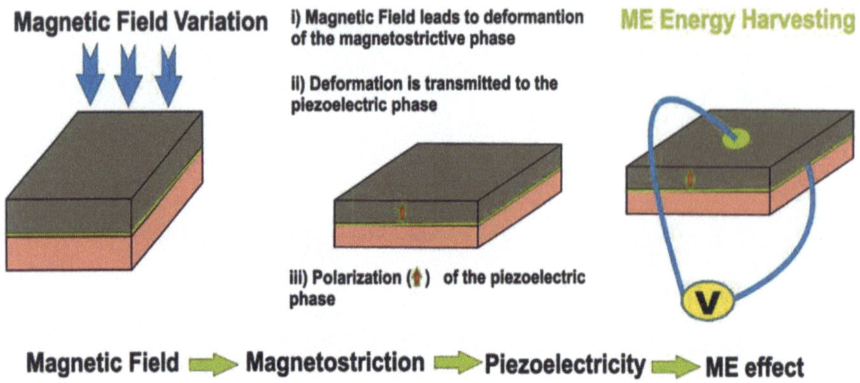

FIGURE 8.6 Schematical representation of magnetoelectric energy harvesting. Source: *Reproduced with permission from P. Martins, S. Lanceros-Méndez, Polymer-based magnetoelectric materials. Adv. Funct. Mater. 23 (27) (2013) 3371–3385.*

component which generates an output voltage. Therefore the selection of magnetostrictive and piezoelectric material is a crucial process in order to obtain excellent values for magnetoelectric coupling. Fig. 8.6 shows a schematical representation of magnetoelectric energy harvesting. Terfenol-D with the chemical formula $Tb_{0.3}Dy_{0.7}Fe_2$ and Metglass are widely used as magnetostrictive parts in magnetoelectric composites [120,123] due to their high magnetostrictive coefficient and high permeability.

Ghosh et al. demonstrated an IoT sensor based on a magnetomechanoelectric nanogenerator, which was fabricated using $NiFe_2O_4$ nanoparticles and PVDF-TrFE in 0–3 configuration. It was capable of producing MECC of 11.43 mV/cm Oe. And it generated a peak-to peak open circuit voltage of 1.4 V [124]. In 2021 Hemalatha et al. prepared $NiFe_2O_4$ fiber-loaded PVDF films using the solvent casting method. The nanogenerator fabricated using the prepared nanocomposite was capable of producing a peak-to-peak voltage of 10 V. It was also used in the fabrication of a magnetomechanoelectric generator that produced an output voltage of 4.8 V at an AC magnetic field of 10 Oe [110]. Reis et al. discussed the performance of a Metglas/PVDF/Metglas composite, which generated the power and power density of 12 μW and 0.9 mW/cm^3 respectively. They discussed five distinct interface circuits to optimize the better energy harvesting process. Maximum power and power density were obtained for the Dickson voltage multiplier for a load resistance of 180 kΩ at 7 Oe DC magnetic field [125]. The magnetoelectric coupling measurement setup for the ternary system has been shown in Fig. 8.7. An energy-harvesting device using magnetostrictive $Fe_{64}Co_{17}Si7B_{12}$ ribbons and PVDF was fabricated by Lasheras et al. The role of the length of the magnetoelectric laminate on the output power was studied. It was reported that the length of the magnetoelectric laminate shows a

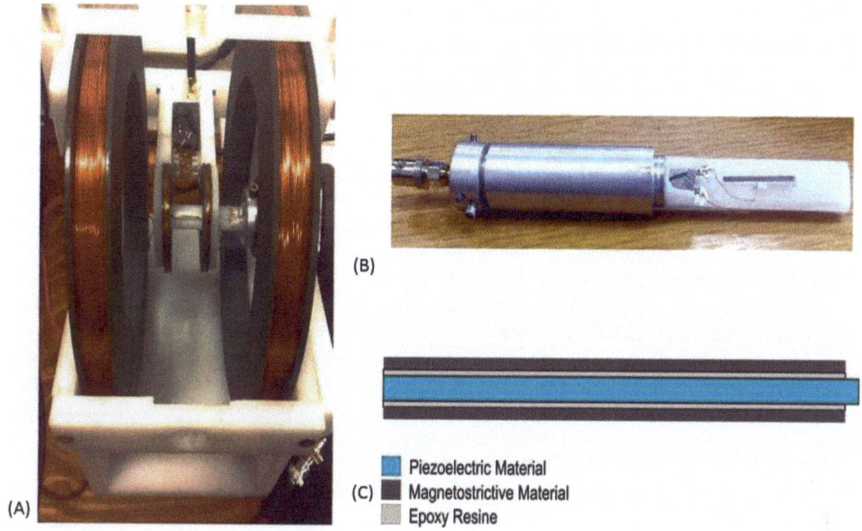

FIGURE 8.7 (A) ME measurement setup; (B) ME sample holder; (C) schematic representation of the trilayer ME composite. Source: *Reproduced with permission from S. Reis, M.P. Silva, N. Castro, V. Correia, J.G. Rocha, P. Martins, et al. Electronic optimization for an energy harvesting system based on magnetoelectric metglas/poly (vinylidene fluoride)/metglas composites. Smart Mater. Struct. 25 (8) (2016) 085028.*

connection with the output power, and it decays in proportion with the former. In the 0.5 cm, length it gave a power of 6.4 μW and a power density of 1.5 mW/cm^3 at 337 kHz [120].

8.7 Conclusion

In recent years much attention has been gained by multiferroic materials due to their exclusive properties. Energy-harvesting technologies play a crucial role in the modern era of technology. Multiferroic materials offer a sustainable, self-powered way to harvest energy from various sources. Based on the magnetomechanical effect, multiferroic materials can be successfully utilized in energy-harvesting applications. The higher ME response obtained for the polymer-based magnetoelectric composites, along with their flexibility, ease in processing, and fabrication make them preferable candidates in various technological applications.

References

[1] F. Narita, M. Fox, A review on piezoelectric, magnetostrictive, and magnetoelectric materials and device technologies for energy harvesting applications, Adv. Eng. Mater. 20 (5) (2018) 1700743.

[2] A.R. Al-Ali, Internet of things role in the renewable energy resources, Energy Procedia 100 (2016) 34–38.
[3] S.C. Mukhopadhyay, N.K. Suryadevara, Internet of things: challenges and opportunities, Internet of Things, Springer, Cham, 2014, pp. 1–17.
[4] N. Hossein Motlagh, M. Mohammadrezaei, J. Hunt, B. Zakeri, Internet of Things (IoT) and the energy sector, Energies 13 (2) (2020) 494.
[5] V. Annapureddy, H. Palneedi, G.T. Hwang, M. Peddigari, D.Y. Jeong, W.H. Yoon, et al., Magnetic energy harvesting with magnetoelectrics: an emerging technology for self-powered autonomous systems, Sustain. Energy Fuels 1 (10) (2017) 2039–2052.
[6] M. Fiebig, T. Lottermoser, D. Meier, M. Trassin, The evolution of multiferroics, Nat. Rev. Mater. 1 (8) (2016) 1–14.
[7] W. Eerenstein, N.D. Mathur, J.F. Scott, Multiferroic and magnetoelectric materials, Nature 442 (7104) (2006) 759–765.
[8] C.W. Nan, M.I. Bichurin, S. Dong, D. Viehland, G. Srinivasan, Multiferroic magnetoelectric composites: historical perspective, status, and future directions, J. Appl. Phys. 103 (3) (2008) 1.
[9] G. Lawes, G. Srinivasan, Introduction to magnetoelectric coupling and multiferroic films, J. Phys. D Appl. Phys. 44 (24) (2011) 243001.
[10] N.A. Spaldin, R. Ramesh, Advances in magnetoelectric multiferroics, Nat. Mater. 18 (3) (2019) 203–212.
[11] D. Khomskii, Trend: classifying multiferroics: mechanisms and effects, Physics 2 (2009) 20.
[12] J. Ma, J. Hu, Z. Li, C.W. Nan, Recent progress in multiferroic magnetoelectric composites: from bulk to thin films, Adv. Mater. 23 (9) (2011) 1062–1087.
[13] Y. Wang, J. Hu, Y. Lin, C.W. Nan, Multiferroic magnetoelectric composite nanostructures, NPG Asia Mater. 2 (2) (2010) 61–68.
[14] P. Martins, S. Lanceros-Méndez, Polymer-based magnetoelectric materials, Adv. Funct. Mater. 23 (27) (2013) 3371–3385.
[15] W. Prellier, M.P. Singh, P. Murugavel, The single-phase multiferroic oxides: from bulk to thin film, J. Phys. Condens. Matter 17 (30) (2005) R803.
[16] B. Raneesh, A. Saha, D. Das, N. Kalarikkal, Structural and magnetic properties of geometrically frustrated multiferroic $ErMnO_3$ nanoparticles, J. Alloy. Compd. 551 (2013) 654–659.
[17] A.K. Pradhan, K. Zhang, D. Hunter, J.B. Dadson, G.B. Loiutts, P. Bhattacharya, et al., Magnetic and electrical properties of single-phase multiferroic $BiFeO_3$, J. Appl. Phys. 97 (9) (2005) 093903.
[18] B. Raneesh, A. Saha, N. Kalarikkal, Effect of gamma radiation on the structural, dielectric and magnetoelectric properties of nanostructured hexagonal $YMnO_3$, Radiat. Phys. Chem. 89 (2013) 28–32.
[19] S.G. Bahoosh, J.M. Wesselinowa, Origin of the different multiferroism in $BiFeO_3$ and $GaFeO_3$, J. Appl. Phys. 113 (6) (2013) 063905.
[20] I.S. Banu, S.D. Lakshmi, Simultaneous enhancement of room temperature multiferroic properties of $BiFeO_3$ by Nd doping at Bi site and Co doping at Fe site, J. Mater. Sci. Mater. Electron. 28 (21) (2017) 16044–16052.
[21] J.F. Scott, Room-temperature multiferroic magnetoelectrics, NPG Asia Mater. 5 (11) (2013) e72.
[22] J. Van Suchtelen, Product properties: a new application of composite materials, Phillips Res. Rep. 27 (1972) 28–37.
[23] D.K. Pradhan, S. Kumari, P.D. Rack, Magnetoelectric composites: applications, coupling mechanisms, and future directions, Nanomaterials 10 (10) (2020) 2072.
[24] A. Mayeen, M.S. Kala, M.S. Jayalakshmy, S. Thomas, J. Philip, D. Rouxel, et al., Flexible and self-standing nickel ferrite–PVDF-TrFE cast films: promising candidates for high-end magnetoelectric applications, Dalton Trans. 48 (45) (2019) 16961–16973.

[25] F.A. Pearsall, J. Lombardi, N. Farahmand, B. van Tassel, E.S. Leland, L. Huang, et al., Polymer–nanocrystal nanocomposites: device concepts in capacitors and multiferroics, IEEE Trans. Nanotechnol. 19 (2019) 255–268.
[26] Y. Pu, J. Li, X. Wang, Y. Shi, R. Shi, M. Yang, et al., Strong non-volatile voltage control of magnetization and the magnetodielectric properties in polymer-based sandwich-structured composites, Compos. Sci. Technol. 186 (2020) 107931.
[27] S. Koner, P. Deshmukh, A.A. Khan, A. Ahlawat, A.K. Karnal, S. Satapathy, Multiferroic properties of $La_{0.7}Ba_{0.3}MnO_3$/P (VDF-TrFE)(0–3) nano-composite films, Mater. Lett. 261 (2020) 127161.
[28] S.K. Chacko, M.T. Rahul, B. Raneesh, N. Kalarikkal, Enhanced magnetoelectric coupling and dielectric constant in flexible ternary composite electrospun fibers of PVDF-HFP loaded with nanoclay and $NiFe_2O_4$ nanoparticles, N. J. Chem. 44 (26) (2020) 11356–11364.
[29] B. Raneesh, H. Soumya, J. Philip, S. Thomas, K. Nandakumar, Magnetoelectric properties of multiferroic composites $(1 - x)$ $ErMnO_{3-x}Y_3Fe_5O_{12}$ at room temperature, J. Alloy. Compd. 611 (2014) 381–385.
[30] M.N. Akhtar, A.B. Sulong, M.A. Khan, M. Ahmad, G. Murtaza, M.R. Raza, et al., Structural and magnetic properties of yttrium iron garnet (YIG) and yttrium aluminum iron garnet (YAIG) nanoferrites prepared by microemulsion method, J. Magnetism Magnetic Mater. 401 (2016) 425–431.
[31] A. Maignan, K. Singh, C. Simon, O.I. Lebedev, C. Martin, H. Tan, et al., Magnetic and magnetodielectric properties of erbium iron garnet ceramic, J. Appl. Phys. 113 (3) (2013) 033905.
[32] O. Opuchovic, A. Kareiva, K. Mazeika, D. Baltrunas, Magnetic nanosized rare earth iron garnets $R_3Fe_5O_{12}$: sol–gel fabrication, characterization and reinspection, J. Magnetism Magnetic Mater. 422 (2017) 425–433.
[33] A. Ahlawat, S. Satapathy, R.J. Choudhary, M.M. Shirolkar, M.K. Singh, P.K. Gupta, Tunable room temperature magnetoelectric response of $SmFeO_3$/poly (vinylidene fluoride) nanocomposite films, RSC Adv. 6 (50) (2016) 44843–44850.
[34] E.W. Lee, Magnetostriction and magnetomechanical effects, Rep. Prog. Phys. 18 (1) (1955) 184.
[35] P.N. Anantharamaiah, B.P. Rao, H.M. Shashanka, J.A. Chelvane, V. Khopkar, B. Sahoo, Role of Mg^{2+} and In^{3+} substitution on magnetic, magnetostrictive and dielectric properties of $NiFe_2O_4$ ceramics derived from nanopowders, Phys. Chem. Chem. Phys. 23 (2) (2021) 1694–1705.
[36] D. Ponnamma, O. Aljarod, H. Parangusan, M.A.A. Al-Maadeed, Electrospun nanofibers of PVDF-HFP composites containing magnetic nickel ferrite for energy harvesting application, Mater. Chem. Phys. 239 (2020) 122257.
[37] A. Chaurasiya, P. Pal, J.V. Vas, D. Kumar, S.N. Piramanayagam, A.K. Singh, et al., Nickel ferrite embedded polyvinylidene fluoride composite based flexible magnetoelectric systems, Ceram. Int. 46 (16) (2020) 25873–25880.
[38] S. Prathipkumar, J. Hemalatha, Magnetoelectric response and tunneling magnetoresistance behavior of flexible P (VDF-H FP)/cobalt ferrite nanofiber composite films, Ceram. Int. 46 (1) (2020) 258–269.
[39] S. Divya, K. Jeyadheepan, J. Hemalatha, Magnetoelectric P (VDF-HFP)-$CoFe_2O_4$ films and their giant magnetoresistance properties, J. Magnetism Magnetic Mater. 492 (2019) 165689.
[40] G.H. Haertling, Ferroelectric ceramics: history and technology, J. Am. Ceram. Soc. 82 (4) (1999) 797–818.
[41] J. Varghese, R.W. Whatmore, J.D. Holmes, Ferroelectric nanoparticles, wires and tubes: synthesis, characterisation and applications, J. Mater. Chem. C. 1 (15) (2013) 2618–2638.
[42] P. Costa, J. Nunes-Pereira, N. Pereira, N. Castro, S. Gonçalves, S. Lanceros-Mendez, Recent progress on piezoelectric, pyroelectric, and magnetoelectric polymer-based energy-harvesting devices, Energy Technol. 7 (7) (2019) 1800852.

[43] W. Martienssen, H. Warlimont (Eds.), Springer Handbook of Condensed Matter and Materials Data, Springer Science & Business Media, 2006.
[44] A.J. Lovinger, Ferroelectric polymers, Science 220 (4602) (1983) 1115–1121.
[45] J. Valasek, Piezo-electric and allied phenomena in Rochelle salt, Phys. Rev. 17 (4) (1921) 475.
[46] G. Busch, P. Scherrer, Eine neue seignette-elektrische Substanz, Naturwissenschaften 23 (43) (1935) 737.
[47] A. Von Hippel, R.G. Breckenridge, F.G. Chesley, L. Tisza, High dielectric constant ceramics, Ind. Eng. Chem. 38 (11) (1946) 1097–1109.
[48] M. Tyunina, J. Miksovsky, T. Kocourek, A. Dejneka, Hysteresis-free piezoresponse in thermally strained ferroelectric barium titanate films, Electron. Mater. 2 (1) (2021) 17–23.
[49] X. Dong, X. Li, H. Chen, C. Sun, J. Shi, F. Pang, et al., Effective strategy to realise excellent energy storage performances in lead-free barium titanate-based relaxor ferroelectric, Ceram. Int. 47 (5) (2021) 6077–6083.
[50] A. Mayeen, M.S. Kala, S. Sunija, D. Rouxel, R.N. Bhowmik, S. Thomas, et al., Flexible dopamine-functionalized $BaTiO_3/BaTiZrO_3/BaZrO_3$-PVDF ferroelectric nanofibers for electrical energy storage, J. Alloy. Compd. 837 (2020) 155492.
[51] E. Brunengo, M. Castellano, L. Conzatti, G. Canu, V. Buscaglia, P. Stagnaro, PVDF-based composites containing PZT particles: how processing affects the final properties, J. Appl. Polym. Sci. 137 (20) (2020) 48871.
[52] K.C. Kao, Dielectric Phenomena in Solids, Elsevier, 2004.
[53] P. Li, J. Zhai, B. Shen, S. Zhang, X. Li, F. Zhu, et al., Ultrahigh piezoelectric properties in textured (K, Na) NbO_3-based lead-free ceramics, Adv. Mater. 30 (8) (2018) 1705171.
[54] L. Xu, F. Chen, F. Jin, H. Huang, L. Qu, K. Zhang, et al., Robust ferroelectric properties in (K, Na) NbO_3-based lead-free films via a self-assembled nanocomposite approach, ACS Appl. Mater. Interfaces 12 (4) (2020) 4616–4624.
[55] K.E. Johnston, C.C. Tang, J.E. Parker, K.S. Knight, P. Lightfoot, S.E. Ashbrook, The polar phase of $NaNbO_3$: a combined study by powder diffraction, solid-state NMR, and first-principles calculations, J. Am. Chem. Soc. 132 (25) (2010) 8732–8746.
[56] T. Wada, K. Tsuji, T. Saito, Y. Matsuo, Ferroelectric $NaNbO_3$ ceramics fabricated by spark plasma sintering, Jpn. J. Appl. Phys. 42 (9S) (2003) 6110.
[57] S.C. Abrahams, J.M. Reddy, J.L. Bernstein, Ferroelectric lithium niobate. 3. Single crystal x-ray diffraction study at 24C, J. Phys. Chem. Solids 27 (6–7) (1966) 997–1012.
[58] T. Jardiel, A.C. Caballero, M. Villegas, Aurivillius ceramics: $Bi_4Ti_3O_{12}$-based piezoelectrics, J. Ceram. Soc. Jpn. 116 (1352) (2008) 511–518.
[59] R.M. Thankachan, R. Balakrishnan, Synthesis strategies of single-phase and composite multiferroic nanostructures, Synthesis of Inorganic Nanomaterials, Woodhead Publishing, 2018, pp. 185–211.
[60] V.S. Voet, G. ten Brinke, K. Loos, Well-defined copolymers based on poly (vinylidene fluoride): from preparation and phase separation to application, J. Polym. Sci. Part A Polym. Chem. 52 (20) (2014) 2861–2877.
[61] M. Poulsen, S. Ducharme, Why ferroelectric polyvinylidene fluoride is special, IEEE Trans. Dielectr. Electr. Insul. 17 (4) (2010) 1028–1035.
[62] J.W. Lee, Y. Takase, B.A. Newman, J.I. Scheinbeim, Effect of annealing on the ferroelectric behavior of nylon-11 and nylon-7, J. Polym. Sci. Part B Polym. Phys. 29 (3) (1991) 279–286.
[63] V. Sencadas, C. Ribeiro, A. Heredia, I.K. Bdikin, A.L. Kholkin, S. Lanceros-Méndez, Local piezoelectric activity of single poly (L-lactic acid)(PLLA) microfibers, Appl. Phys. A 109 (1) (2012) 51–55.
[64] X. Chen, X. Han, Q.D. Shen, PVDF-based ferroelectric polymers in modern flexible electronics, Adv. Electron. Mater. 3 (5) (2017) 1600460.

[65] S. Badatya, A. Kumar, C. Sharma, A.K. Srivastava, J.P. Chaurasia, M.K. Gupta, Transparent flexible graphene quantum dot-(PVDF-HFP) piezoelectric nanogenerator, Mater. Lett. 290 (2021) 129493.

[66] Z. Luo, J. Chen, Z. Zhu, L. Li, Y. Su, W. Tang, et al., High-resolution and high-sensitivity flexible capacitive pressure sensors enhanced by a transferable electrode array and a micropillar–PVDF film, ACS Appl. Mater. Interfaces 13 (6) (2021) 7635–7649.

[67] P. Martins, A.C. Lopes, S. Lanceros-Mendez, Electroactive phases of poly (vinylidene fluoride): determination, processing and applications, Prog. Polym. Sci. 39 (4) (2014) 683–706.

[68] B. Ameduri, From vinylidene fluoride (VDF) to the applications of VDF-containing polymers and copolymers: recent developments and future trends, Chem. Rev. 109 (12) (2009) 6632–6686.

[69] S. Liu, Z. Cui, P. Fu, M. Liu, L. Zhang, Z. Li, et al., Ferroelectric behavior and polarization mechanism in odd-odd polyamide 11,11, J. Polym. Sci. Part B Polym. Phys. 52 (16) (2014) 1094–1099.

[70] P. von Tiedemann, S. Anwar, U. Kemmer-Jonas, K. Asadi, H. Frey, Synthesis and solution processing of nylon-5 ferroelectric thin films: the renaissance of odd-nylons? Macromol. Chem. Phys. 221 (5) (2020) 1900468.

[71] S. Tasaka, T. Shouko, K.I. Asami, N. Inagaki, Ferroelectric behavior in aliphatic polyurethanes, Japanese J. Appl. Phys. 33 (3R) (1994) 1376.

[72] Q. Li, Q. Wang, Ferroelectric polymers and their energy-related applications, Macromol. Chem. Phys. 217 (11) (2016) 1228–1244.

[73] H. Oh, S.S. Kwak, B. Kim, E. Han, G.H. Lim, S.W. Kim, et al., Highly conductive ferroelectric cellulose composite papers for efficient triboelectric nanogenerators, Adv. Funct. Mater. 29 (37) (2019) 1904066.

[74] J. Kim, S. Yun, Z. Ounaies, Discovery of cellulose as a smart material, Macromolecules 39 (12) (2006) 4202–4206.

[75] Y. Zong, T. Zheng, P. Martins, S. Lanceros-Mendez, Z. Yue, M.J. Higgins, Cellulose-based magnetoelectric composites, Nat. Commun. 8 (1) (2017) 1–8.

[76] L. Ruan, X. Yao, Y. Chang, L. Zhou, G. Qin, X. Zhang, Properties and Applications of the β phase poly (vinylidene fluoride), Polymers 10 (3) (2018) 228.

[77] M. Benz, W.B. Euler, Determination of the crystalline phases of poly (vinylidene fluoride) under different preparation conditions using differential scanning calorimetry and infrared spectroscopy, J. Appl. Polym. Sci. 89 (4) (2003) 1093–1100.

[78] B. Bera, M.D. Sarkar, Piezoelectricity in PVDF and PVDF based piezoelectric nanogenerator: a concept, IOSR J. Appl. Phys. (IOSRJAP) 9 (3) (2017) 95–99.

[79] Z. Cui, N.T. Hassankiadeh, Y. Zhuang, E. Drioli, Y.M. Lee, Crystalline polymorphism in poly (vinylidenefluoride) membranes, Prog. Polym. Sci. 51 (2015) 94–126.

[80] P.K. Szewczyk, A. Gradys, S.K. Kim, L. Persano, M. Marzec, A. Kryshtal, et al., Enhanced piezoelectricity of electrospun polyvinylidene fluoride fibers for energy harvesting, ACS Appl. Mater. Interfaces 12 (11) (2020) 13575–13583.

[81] F.A. He, K. Lin, D.L. Shi, H.J. Wu, H.K. Huang, J.J. Chen, et al., Preparation of organosilicate/PVDF composites with enhanced piezoelectricity and pyroelectricity by stretching, Compos. Sci. Technol. 137 (2016) 138–147.

[82] G.S. Ekbote, M. Khalifa, A. Mahendran, S. Anandhan, Cationic surfactant assisted enhancement of dielectric and piezoelectric properties of PVDF nanofibers for energy harvesting application, Soft Matter 17 (8) (2021) 2215–2222.

[83] Z. He, F. Rault, M. Lewandowski, E. Mohsenzadeh, F. Salaün, Electrospun PVDF nanofibers for piezoelectric applications: a review of the influence of electrospinning parameters on the β phase and crystallinity enhancement, Polymers 13 (2) (2021) 174.

[84] F. Oliveira, Y. Leterrier, J.A. Månson, O. Sereda, A. Neels, A. Dommann, et al., Process influences on the structure, piezoelectric, and gas-barrier properties of PVDF-TrFE copolymer, J. Polym. Sci. Part B Polym. Phys. 52 (7) (2014) 496–506.

[85] Z. Wang, Z. Zhang, T.M. Chung, High dielectric VDF/TrFE/CTFE terpolymers prepared by hydrogenation of VDF/CTFE copolymers: synthesis and characterization, Macromolecules 39 (13) (2006) 4268–4271.

[86] M. Poulsen, A.V. Sorokin, S. Adenwalla, S. Ducharme, V.M. Fridkin, Effects of an external electric field on the ferroelectric-paraelectric phase transition in polyvinylidene fluoride-trifluoroethylene copolymer Langmuir–Blodgett films, J. Appl. Phys. 103 (3) (2008) 034116.

[87] C. Behera, R.N.P. Choudhary, P.R. Das, Development of multiferroic polymer nanocomposite from PVDF and ($Bi_{0.5}$ $Ba_{0.25}$ $Sr_{0.25}$) ($Fe_{0.5}$ $Ti_{0.5}$) O_3, J. Mater. Sci. Mater. Electron. 28 (3) (2017) 2586–2597.

[88] R. Brito-Pereira, C. Ribeiro, S. Lanceros-Mendez, P. Martins, Magnetoelectric response on Terfenol-D/P (VDF-TrFE) two-phase composites, Compos. Part. B Eng. 120 (2017) 97–102.

[89] P. Martins, J.S. Nunes, J. Oliveira, N. Peřinka, S. Lanceros-Mendez, Spray-printed magnetoelectric multifunctional composites, Compos. Part. B Eng. 187 (2020) 107829.

[90] R.P. Ummer, B. Raneesh, C. Thevenot, D. Rouxel, S. Thomas, N. Kalarikkal, Electric, magnetic, piezoelectric and magnetoelectric studies of phase pure ($BiFeO_3$–$NaNbO_3$)–(P (VDF-TrFE)) nanocomposite films prepared by spin coating, RSC Adv. 6 (33) (2016) 28069–28080.

[91] C. Behera, R.N.P. Choudhary, Electrical and multiferroic characteristics of PVDF-$MnFe_2O_4$ nanocomposites, J. Alloy. Compd. 727 (2017) 851–862.

[92] R. Gonçalves, A. Larrea, M.S. Sebastian, V. Sebastian, P. Martins, S. Lanceros-Mendez, Synthesis and size dependent magnetostrictive response of ferrite nanoparticles and their application in magnetoelectric polymer-based multiferroic sensors, J. Mater. Chem. C. 4 (45) (2016) 10701–10706.

[93] P. Martins, C.M. Costa, S. Lanceros-Mendez, Nucleation of electroactive β-phase poly (vinilidene fluoride) with $CoFe_2O_4$ and $NiFe_2O_4$ nanofillers: a new method for the preparation of multiferroic nanocomposites, Appl. Phys. A 103 (1) (2011) 233–237.

[94] P. Martins, Y.V. Kolen'ko, J. Rivas, S. Lanceros-Mendez, Tailored magnetic and magnetoelectric responses of polymer-based composites, ACS Appl. Mater. Interfaces 7 (27) (2015) 15017–15022.

[95] A. Mayeen, M.S. Kala, M.S. Jayalakshmy, S. Thomas, D. Rouxel, J. Philip, et al., Dopamine functionalization of $BaTiO_3$: an effective strategy for the enhancement of electrical, magnetoelectric and thermal properties of $BaTiO_3$-PVDF-TrFE nanocomposites, Dalton Trans. 47 (6) (2018) 2039–2051.

[96] M.H. Choi, S.C. Yang, $CoFe_2O_4$ nanofiller effect on β-phase formation of PVDF matrix for polymer-based magnetoelectric composites, Mater. Lett. 223 (2018) 73–77.

[97] E.H. Abdelhamid, O.D. Jayakumar, V. Kotari, B.P. Mandal, R. Rao, V.M. Naik, et al., Multiferroic PVDF–Fe_3O_4 hybrid films with reduced graphene oxide and ZnO nanofillers, RSC Adv. 6 (24) (2016) 20089–20094.

[98] J. Xue, T. Wu, Y. Dai, Y. Xia, Electrospinning and electrospun nanofibers: methods, materials, and applications, Chem. Rev. 119 (8) (2019) 5298–5415.

[99] J.S. Andrew, D.R. Clarke, Effect of electrospinning on the ferroelectric phase content of polyvinylidene difluoride fibers, Langmuir 24 (3) (2008) 670–672.

[100] K. Castkova, J. Kastyl, D. Sobola, J. Petrus, E. Stastna, D. Riha, et al., Structure–properties relationship of electrospun PVDF fibers, Nanomaterials 10 (6) (2020) 1221.

[101] P.D. Prasad, J. Hemalatha, Multifunctional films of poly (vinylidene fluoride)/$ZnFe_2O_4$ nanofibers for nanogenerator applications, J. Alloy. Compd. 854 (2021) 157189.

[102] N. Hernández, V.A. González-González, I.B. Dzul-Bautista, N. Ornelas-Soto, J.M. Barandiarán, J. Gutierrez, Electrospun poly (vinylidene fluoride-trifluoroethylene) based flexible magnetoelectric nanofibers, Eur. Polym. J. 109 (2018) 336–340.

[103] P.D. Prasad, J. Hemalatha, Dielectric and energy storage density studies in electrospun fiber mats of polyvinylidene fluoride (PVDF)/zinc ferrite ($ZnFe_2O_4$) multiferroic composite, Phys. B Condens. Matter 573 (2019) 1−6.
[104] D. Ponnamma, M.M. Chamakh, A.M. Alahzm, N. Salim, N. Hameed, M.A.A. AlMaadeed, Core-shell nanofibers of polyvinylidene fluoride-based nanocomposites as piezoelectric nanogenerators, Polymers 12 (10) (2020) 2344.
[105] H. Liu, H. Fu, L. Sun, C. Lee, E.M. Yeatman, Hybrid energy harvesting technology: from materials, structural design, system integration to applications, Renew. Sustain. Energy Rev. (2020) 110473.
[106] H.X. Zou, M. Li, L.C. Zhao, Q.H. Gao, K.X. Wei, L. Zuo, et al., A magnetically coupled bistable piezoelectric harvester for underwater energy harvesting, Energy 217 (2021) 119429.
[107] V. Slabov, S. Kopyl, M.P.S. dos Santos, A.L. Kholkin, Natural and eco-friendly materials for triboelectric energy harvesting, Nano-Micro Lett. 12 (1) (2020) 1−18.
[108] C. Wu, A.C. Wang, W. Ding, H. Guo, Z.L. Wang, Triboelectric nanogenerator: a foundation of the energy for the new era, Adv. Energy Mater. 9 (1) (2019) 1802906.
[109] Z. Fan, Y. Zhang, L. Pan, J. Ouyang, Q. Zhang, Recent developments in flexible thermoelectrics: from materials to devices, Renew. Sustain. Energy Rev. (2020) 110448.
[110] P.D. Prasad, J. Hemalatha, Energy harvesting performance of magnetoelectric poly (vinylidene fluoride)/$NiFe_2O_4$ nanofiber films, J. Magnetism Magnetic Mater. 532 (2021) 167986.
[111] B. Mahanty, S.K. Ghosh, S. Jana, K. Roy, S. Sarkar, D. Mandal, All-fiber acousto-electric energy harvester from magnesium salt-modulated PVDF nanofiber, Sustain. Energy Fuels 5 (4) (2021) 1003−1013.
[112] S. Sukumaran, S. Chatbouri, D. Rouxel, E. Tisserand, F. Thiebaud, T. Ben Zineb, Recent advances in flexible PVDF based piezoelectric polymer devices for energy harvesting applications, J. Intell. Mater. Syst. Struct. 32 (7) (2021) 746−780.
[113] I. Chinya, A. Sasmal, S. Sen, Conducting polyaniline decorated in-situ poled Ferrite nanorod-PVDF based nanocomposite as piezoelectric energy harvester, J. Alloy. Compd. 815 (2020) 152312.
[114] S. Bairagi, S.W. Ali, Flexible lead-free PVDF/SM-KNN electrospun nanocomposite based piezoelectric materials: significant enhancement of energy harvesting efficiency of the nanogenerator, Energy 198 (2020) 117385.
[115] L. Shi, H. Jin, S. Dong, S. Huang, H. Kuang, H. Xu, et al., High-performance triboelectric nanogenerator based on electrospun PVDF-graphene nanosheet composite nanofibers for energy harvesting, Nano Energy 80 (2021) 105599.
[116] M. Wang, W. Liu, X. Shi, J. Pan, B. Zhou, J. Wang, et al., Highly efficient and continuous triboelectric power harvesting based on a porous β-phase poly (vinylidene fluoride) aerogel, N. J. Chem. 45 (4) (2021) 1893−1898.
[117] X. Guan, B. Xu, M. Wu, T. Jing, Y. Yang, Y. Gao, Breathable, washable and wearable woven-structured triboelectric nanogenerators utilizing electrospun nanofibers for biomechanical energy harvesting and self-powered sensing, Nano Energy 80 (2021) 105549.
[118] L. Zhao, J. Duan, L. Liu, J. Wang, Y. Duan, L. Vaillant-Roca, et al., Boosting power conversion efficiency by hybrid triboelectric nanogenerator/silicon tandem solar cell toward rain energy harvesting, Nano Energy 82 (2021) 105773.
[119] B. Martin, E. Dantras, A. Lonjon, C. Lacabanne, L. Laffont, Magnetoelectric coupling in PVDF/Nano $NaNbO_3$/NanoNi composites: influence of Ni particles aspect ratio, IEEE Trans. Dielectr. Electr. Insulation 27 (5) (2020) 1381−1386.
[120] A. Lasheras, J. Gutiérrez, S. Reis, D. Sousa, M. Silva, P. Martins, et al., Energy harvesting device based on a metallic glass/PVDF magnetoelectric laminated composite, Smart Mater. Struct. 24 (6) (2015) 065024.

[121] K. Mori, M. Wuttig, Magnetoelectric coupling in Terfenol-D/polyvinylidenedifluoride composites, Appl. Phys. Lett. 81 (1) (2002) 100–101.
[122] W.Q. Jing, F. Fang, A flexible multiferroic composite with high self-biased magnetoelectric coupling, Compos. Sci. Technol. 153 (2017) 145–150.
[123] J. Gutiérrez, A. Lasheras, P. Martins, N. Pereira, J.M. Barandiarán, S. Lanceros-Mendez, Metallic glass/PVDF magnetoelectric laminates for resonant sensors and actuators: a review, Sensors 17 (6) (2017) 1251.
[124] S.K. Ghosh, K. Roy, H.K. Mishra, M.R. Sahoo, B. Mahanty, P.N. Vishwakarma, et al., Rollable magnetoelectric energy harvester as a wireless IoT sensor, ACS Sustain. Chem. Eng. 8 (2) (2019) 864–873.
[125] S. Reis, M.P. Silva, N. Castro, V. Correia, J.G. Rocha, P. Martins, et al., Electronic optimization for an energy harvesting system based on magnetoelectric metglas/poly (vinylidene fluoride)/metglas composites, Smart Mater. Struct. 25 (8) (2016) 085028.

CHAPTER

9

Hybrid composites with shape memory alloys and piezoelectric thin layers

Sunija Sukumaran[1,2], *Samir Chatbouri*[2], *Guven Muslum*[1], *Didier Rouxel*[2] and *Tarak Ben Zineb*[1]

[1]Université de Lorraine, CNRS, Arts et Métiers ParisTech, LEM3, Nancy, France, [2]Université de Lorraine, CNRS, Institut Jean Lamour, Nancy, France

9.1 Introduction

With the recent progress in material science, engineering, and micro/nanotechnologies, new class of materials are emerging with improved activities, new functions, and applications. Multifunctional, smart, or intelligent materials are particularly suited for many applications such as sensors, actuators, solar cells, microelectronic devices, and energy-harvesting devices [1,2]. In this regard, functional or "smart" materials like piezoelectric [3], pyroelectric [4], magnetoelectric, magnetostrictive [5], and shape memory alloys (SMAs) [6,7] have gained immense research interest. The functional materials can vary in terms of physical or chemical properties like volume, electrical conductivity, shape, Poisson's ratio, and Young's modulus by responding to a certain stimulus such as light, temperature, chemical, or by applying an external mechanical, magnetic, or electric field [8,9]. These properties make them suitable candidates for the design of advanced micro-/nanoelectromechanical system (MEMS/NEMS) [9,10]. MEMS systems have

Engineered Polymer Nanocomposites for Energy Harvesting Applications
DOI: https://doi.org/10.1016/B978-0-12-824155-4.00007-7

found wide application in the fields of energy-harvesting devices due to the Internet of Things [11,12]. Piezoelectric materials that have the ability to convert mechanical energy into electric energy are the most extensively used functional materials for small-scale energy harvesting [13,14]. However, other kinds of materials like SMAs are also used in different applications including energy harvesting [15,16].

SMAs are a unique class of materials that can recover their geometry upon the application of an external stimulus. That means after deformation they have the ability to regain their actual shape at the desired characteristic temperature [17,18]. Also, they can sustain a huge reversible deformation ranging from 5% to 10% [9,19]. These specific characteristics of SMAs are known as the shape memory effect and superelasticity (pseudoelasticity). There are different compositions of alloys, such as magnetic (Ni-Mn-Ga) [20], iron-based (Fe-Mn-Si) [21], and copper-based (Cu-Zn-Al, Cu-Al-Ni) [22,23], which are used depending on the specific requirements and applications. Among them, NiTi-based SMAs are extensively used in a wide variety of application as they exhibit strong SME, superelasticity, corrosion resistance, low energy consumption, and biocompatible nature [17,24]. Furthermore, NiTi SMAs thin films can be prepared by using standard deposition techniques like magnetron sputtering or by arc deposition. This has helped their vast applicability in MEMS/NEMS in recent years [25,26]. In this regard, SMAs has found more research attraction as smart (or intelligent) and functional materials [17,27].

In the 1980s the concept of smart hybrid composites emerged and gained more research interest [28]. SMA composite materials are developed by combining SMA materials in the form of wires, ribbons, thin films, or particles, and embedding them into matrix materials such as polymers, fiber-reinforced polymers, metals, or ceramics [29–31]. The physical properties of the composite matrix can be improved such as damping property, actuation, stiffness, changes in vibrational frequency, and internal strain energy of the material [15,32,33]. Several comprehensive analytical and numerical (finite element) studies were reported on the thermomechanical behavior of SMAs [34,35]. However, SMAs were not broadly explored for energy-harvesting applications. Lately, Chen et al. reported the combined characteristics of both shape memory alloys and piezoelectric ceramic composite and they called them "SMARTIES" [36]. They proposed improved actuation and sensing properties. In this context, the hybrid composite materials with different properties such as piezoelectric, pyroelectric, and shape memory effects may allow to improve the final performance of the device for many applications. Several numerical and experimental studies were carried out based on SMA/piezoelectric composites, most of the studies so far reported in the literature are based on the SMA/piezoceramics composite. However, the usage of ceramics for the fabrication of such composite devices is not an ideal choice. The ceramics possess some drawbacks which include rigidity, high density, high-temperature processing needed for thin-film fabrication, higher cost for manufacturing, and lesser design flexibility. All these limits slow down their application in many device applications, especially energy-harvesting applications [37,38]. Furthermore, they can't endure large deformation due to their brittleness. In this context, the choice of the material plays an important role as it needs to provide the design flexibility (bending, twisting, fold), and also it should maintain the original electronic and structural properties. Hence, piezoelectric polymers found more attention because they can satisfy most of the abovementioned characteristics. Also they exhibit good mechanical flexibility, biocompatibility, ease of processing and fabrication, and also are more resistant to mechanical shock [39,40]. Several piezoelectric polymers have been reported and studied in the literature, including PVDF and its copolymers, cellulose and its derivatives, polyurethanes, polyimides, and polylactic acid [3,41,42]. PVDF and its copolymers are well established and the most extensively studied material among all the other polymers due to their high piezoelectric coefficient, higher elastic compliances, electromechanical coupling, structural flexibility, good chemical resistance, and biocompatibility. In 2016 Gusarov et al. [43] made a composite using SMA ribbon and PVDF polymer. PVDF polymers are more suitable for coupling with SMA and also they can withstand high strain and are more flexible compared to ceramic-based materials. Among the PVDF-based copolymer family P (VDF-TrFE) shows a high piezoelectric coefficient and is widely used for its ferroelectric properties, sometimes with nanofillers [44,45]. Similarly, the other characteristics, such as low density, lightweight, low refractive index, and low cost, associated with them make them promising candidates for different energy-harvesting systems [37,46]. The aim of this chapter is to present the state-of-the-art and an example of how to go further in this topic to demonstrate both numerically and experimentally the feasibility of energy harvesting with hybrid composites. The final devices composed of SMAs and piezoelectric thin layer are able to find applications in many fields, e.g., as actuators, sensors, and energy harvesters.

9.2 Multiphysics behavior modeling and characterization

The prediction of the global response of hybrid composites needs a connected double level of modeling. The first level concerns the prediction of the multiphysical local behavior of each layer. In the present case, it concerns, on one hand, the thermomechanical nonlinear behavior of SMA layers induced by the forward and reverse martensitic transformation and its corresponding inelastic strain. Secondly, it concerns the piezoelectric and optionally the ferroelectric behaviors of piezoelectric layers induced by polarization variation and optionally by switching of ferroelectric domains. The second level of modeling allows to derive the effective nonlinear and multiphysical behavior of the hybrid composite. To this end, it is possible to apply various scale transition techniques as numerical integration through the thickness of the local behavior or homogenization with self-consistent or Mori–Tanaka or other scale transition techniques. The following subsections give an overview of the main adopted two-level behavior modeling of SMA and piezoelectric materials, respectively.

9.2.1 Modeling of the shape memory alloys thermomechanical behavior

The first emerging thermomechanical constitutive models are empirical and one-dimensional [47,48]. Two-dimensional and three-dimensional extensions are proposed based on more sophisticated formulations enhancing the accurate and the complete description of the various specificities of the SMA thermomechanical behavior. Their formulations derive from micromechanical approaches, combined or not with scale transition techniques, statistical physics approaches, elastic–plastic model-like, and thermodynamic approaches assuming a given thermodynamic potential and dissipation expressions [49–51].

Thermomechanical behavior models based on thermodynamic approaches at the microscopic scale generally consider the Ginzburg–Landau or the molecular dynamic theories [52,53]. They allow describing the effect of mechanisms related to the martensitic transformation occurring at the crystalline lattice scale. It corresponds to the martensite nucleation and its propagation and martensite variant reorientation. Models based on Ginzburg–Landau theory assume a polynomial expression of the free energy as the function of observable variables as temperature and strain. Initially proposed by Falk in 1980 [54,55], they were improved by many researchers in the coming years [56–58]. These models assume that the occurring instability of the thermodynamic equilibrium induces a martensitic phase transformation. The formulation of the thermodynamic equilibrium determines the structure and the motion of interfaces between martensite variants and between austenite and martensite phases. Researchers such as Dhote et al., Levitas and Preston, and Wang and Melnik [59–62] proposed extended and enhanced versions of these models. Models based on the molecular dynamic theory derive from a theoretical physics approach that emerged in the 1960s by Alder and Wainwright, and Rahman [63,64]. They solve Newton's equations describing the particle motion within a network of a finite number of particles. The size and the shape of the particle network evolve in a dynamic manner in order to ensure, at each time, the balance between internal and applied stresses [65]. The main difference between available models based on such an approach comes from the choice of the potential energy expression related to the interaction between particles. The main considered expressions are those of Lennard–Jones [66,67] and the embedded atom method initially proposed by Daw et al. [68]. Micro–macro models derive from the description of the martensitic transformation at the crystalline lattice scale. In the case of a polycrystalline microstructure, the introduction of a scale transition technique allows to describe the effective behavior of a given Representative Volume Element (RVE) [51,69–73]. Two kinds of micro–macro models are the most used. The first ones adopt a formulation inspired from the crystalline plasticity, whereas the second ones derive from microplanes and microspheres approaches. Regarding models inspired from crystalline plasticity, there are different possibilities of detwinning of the crystalline lattice, each characterized with a set of vectors (martensite variant) defining the habit plane normal and the detwinning direction. An intrinsic transformation strain and volume fraction characterize each variant and define its contribution on the resultant transformation strain. The assumption of free energy and dissipation expressions and the application of the first principle of thermodynamics and the Clausius–Duhem condition allow to manage the activation of transformation process for each variant and the determination the evolution of each variant volume fraction [51,72,74–77]. For polycrystalline SMAs, the introduction of a scale transition technique allows to derive the effective behavior starting from the local behavior of each grain (single crystal) [51,78–80]. The theory of microplanes, initially proposed by Tailor in 1938, allows to describe the SMA thermomechanical behavior as a superposition of a succession of an uniaxial behaviors in planes with different normal directions labeled "microplanes" [81,82]. Additional kinematic or static conditions need to be ensured in order to derive the constitutive equations.

The following is an example of constitutive equations corresponding to the description of thermomechanical behavior of an SMA single crystal based on a crystalline-plasticity-like approach [51,83,84]. The Gibbs free energy of this model presents the following expression:

$$\Delta G(\sigma_{ij}, T, f^n) = -\Delta G_{ch} - W_{\acute{e}las.} - W_{\text{interf.}} + \sigma_{ij}\varepsilon_{ij}$$

$$= \begin{pmatrix} -B(T - T_{ref})f^n + \frac{1}{2}f^n \sum_m H^{nm} f^m + \\ \alpha\delta_{ij}(T - T_0)\sigma_{ij} + \frac{1}{2}\sigma_{ij}S_{ijkl}\sigma_{kl} - g\sigma_{ij}R^n_{ij}f^n \end{pmatrix}$$

where ΔG_{ch}, $W_{\acute{e}las.}$, $W_{\text{interf.}}$, σ, ε, T and f^n are, respectively, the variation of chemical energy, elastic energy, interface energy, the stress and strain tensors, the material temperature, and the volume fraction of a given martensite variant (n). The material parameters characterizing the single crystal are $S, \alpha, B, g, T_0, T_{ref}, R^n, H^{nm}$ representing, respectively, elastic and thermal expansion tensors, chemical factor related to phase transformation, distortion amplitude, room and reference temperatures, Shmid tensor characterizing the crystallographic properties of each martensite variant, and the interaction parameters. The condition of the Clausius–Duhem allows to derive the following driving force governing the evolution of the volume fraction of each martensite variant:

$$\frac{\partial \Delta G}{\partial f^n} = -B(T - T_0) + \sum_m \left(H^{nm} f^m\right) - g\sigma_{ij} R^n_{ij} = F^n_y$$

Considering the consistency conditions for each active variant, in combination with the thermoelastic behavior, leads to the following constitutive equations describing the behavior of a single crystal SMA:

$$\dot{\sigma}_{ij} = \left(\left(C_{ijkl} - \sum_{n,m}\left(C_{ijrs}\tilde{\varepsilon}^n_{rs}\hat{H}^{-1}_{nm}\tilde{\varepsilon}^m_{pq}C_{pqkl}\right)\right)\right.$$

$$\left.\dot{\varepsilon}_{kl} - \left(\alpha C_{ijkl}\delta_{kl} - \sum_{n,m}\left(C_{ijrs}\tilde{\varepsilon}^n_{rs}\hat{H}^{-1}_{nm}\left(Be^m + \alpha\tilde{\varepsilon}^m_{pq}C_{pqkl}\delta_{kl}\right)\right)\right)\dot{T}\right)$$

with C being the compliance elastic tensor and $\hat{H}_{nm} = H^{nm} + \tilde{\varepsilon}^n_{pq}C_{pqrs}\tilde{\varepsilon}^m_{rs}$, where $\tilde{\varepsilon}^n_{ij} = gR^n_{ij}$.

The comparison between the predicted thermomechanical behavior of a given single crystal SMA and the corresponding observed experimental response shows that this approach allows to well describe the SMA thermomechanical behavior under tension and combined bending–torsion loadings (Fig. 9.1).

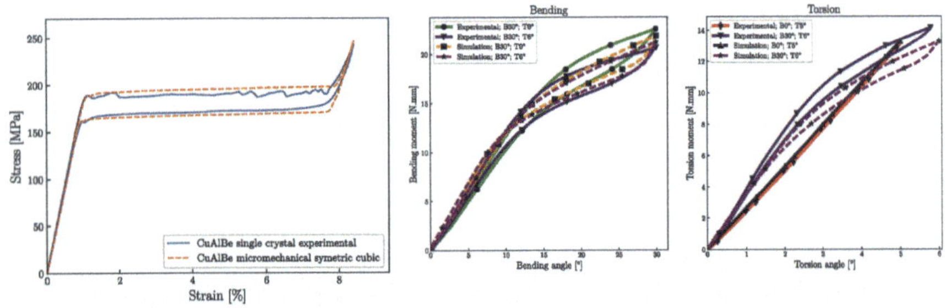

FIGURE 9.1 Comparison between experimental characterization and modeling results for tension and combined bending–torsion loadings [165]. Source: *Reproduced with permission from P. Xolin, C. Collard, M. Engels-Deutsch, et al. Finite element and experimental structural analysis of endodontic rotary file made of Cu-based single crystal SMA considering a micromechanical behavior model. Int. J. Solids Struct. (2021) S0020768321000238. Copyright 2021 Elsevier publications.*

Macroscopic models describe, in a global and effective way, the thermomechanical behavior of an SMA polycrystalline RVE. Most of the available models are gathered in three families: (1) phenomenological models, (2) models derived from an averaged thermodynamic approach, and (3) models based on the statistical physics theory at the macroscopic scale. Tanaka and Nagaki [85] were the first to propose a macroscopic phenomenological thermomechanical model with internal variables describing the effect of martensitic transformation under uniaxial loading. Liang and Rogers [48] and shortly thereafter Brinson [47] carried out its extension to tridimensional loading and proposed to distinguish the volume fraction of oriented martensite (mechanical martensite) from the self-accommodated martensite (thermal martensite). Thermomechanical mechanical models inspired from plasticity theory consider the effective transformation strain as a plastic strain but reversible. These models assume that the average effect of interactions between martensitic variants operates as a complex plastic hardening. One or many loading surfaces derive from such a formulation. Their association with the normality rule and the consistency condition allow determining of the transformation strain rate and therefore the constitutive equations governing the SMA thermomechanical behavior [86–91]. These models could consider loading surfaces with a more complex shape in order to take into account the anisotropy of the martensitic transformation evolution. They also consider additional loading surfaces describing martensite reorientation and a complex evolution law related to effects of interactions between variants inspired from combined isotropic and kinematic hardenings [92–97]. Regarding models derived from an averaged thermodynamic approach, they assume expressions of the free energy and dissipation describing in an averaged way effects of elasticity, martensitic phase transformation, and martensite reorientation. Some of these available expressions come from a simplification of micromechanical formulations [98–102]. These expressions depend on control variables (stress or strain and temperature), on internal variables associated with different mechanisms (volume fraction of martensite, volume fraction of oriented and self-accommodated martensites, mean transformation strain, etc.) and material parameters associated with elasticity and phase transformation. The application of the first principle of thermodynamics and Clausius–Duhem conditions allows to determine driving forces associated with internal variables. The comparison of these driving forces to their corresponding yield forces allows deriving the constitutive equations of the SMA thermomechanical behavior. Additional proposed developments lead to the description of transformation saturation, inner loops related to partial transformation, and the effect of cyclic loading [95,103–108]. This kind of model presents an extension in the framework of large strains and geometrical nonlinearity [108–117].

The following gives an example of constitutive equations corresponding to the description of the thermomechanical behavior of a polycrystalline SMA derived from an averaged thermodynamic approach [99,102,104]. The Gibbs free energy has the following expression:

$$\Delta G = \left| \begin{array}{l} -\frac{1}{2}\sigma_{ij}S_{ijkl}\sigma_{kl} - \sigma_{ij}\alpha_{ij}^{therm}\Delta T - \Delta T S^{A} - \sigma_{ij}f\bar{\varepsilon}_{ij}^{tr} - \sigma_{ij}f^{FA}\bar{\varepsilon}_{ij}^{twin} + B(T-T_0)f \\ + \frac{1}{2}H_f f^2 + \frac{1}{2}fH_\varepsilon\bar{\varepsilon}_{ij}^{tr}\bar{\varepsilon}_{ij}^{tr} + \frac{1}{2}f^{FA}H_{twin}\bar{\varepsilon}_{ij}^{twin}\bar{\varepsilon}_{ij}^{twin} + c_v\left((T-T_0) - T\log\frac{T}{T_0}\right) \end{array} \right.$$

In this expression, σ and T are, respectively, the stress and the temperature that control the thermomechanical loading of the RVE. This Gibbs energy expression presents four internal variables that control the state of the RVE. Two internal variables f and $\bar{\varepsilon}^{tr}$, corresponding, respectively, to the volume fraction of martensite and the mean transformation orientation (strain), describe the martensitic transformation, whereas f^{FA} and $\bar{\varepsilon}^{twin}$ are the two other internal variables describing the twinning of martensite and correspond to the volume fraction of twinned martensite and the associated mean twinning orientation (strain). S, a^{therm}, S^A, B, H_f, H_e, H_{twin}, c_v, and T_0 are material parameters corresponding, respectively, to the elastic and thermal expansion tensors, entropy of austenite, chemical parameter, and three interaction parameters related to martensitic transformation and twinning, latent heat coefficient, and the reference temperature. It is assumed that the volume fraction of twinned martensite depends on the global volume fraction and on the mean orientation strain as follows:

$$\begin{cases} \dot{f} > 0 \Rightarrow \dot{f}^{FA} = \left(\dfrac{\varepsilon_{sat}^{tr} - \varepsilon_{eq}^{tr}}{\varepsilon_{sat}^{tr}}\right)\dot{f} \\ \dot{f} \le 0 \Rightarrow f^{FA} = kf \end{cases}$$

where ε_{sat}^{tr}, ε_{equ}^{tr} and k are, respectively, the maximum mean transformation when the martensite is fully oriented, the von-Mises equivalent mean transformation strain, and a material constant. The number of internal variables is reduced to three. The application of the Clausius–Duhem inequality allows determining driving forces associated with each internal variable as follows:

$$F_m^{\sigma} = \varepsilon_{ij} - S_{ijkl}\sigma_{kl} - \alpha_{ij}^{therm}\Delta T - f\bar{\varepsilon}_{ij}^{tr} = 0; \quad F_m^{T} = S - S^A + Bf - \alpha_{ij}^{therm}\sigma_{ij} + c_v\log\frac{T}{T_0} = 0$$

$$F_m^{f} = \sigma_{ij}\bar{\varepsilon}_{ij}^{tr} - B(T-T_0) - H_f f - \frac{1}{2}H_\varepsilon\bar{\varepsilon}_{ij}^{tr}\bar{\varepsilon}_{ij}^{tr} = F_y^{f}; \quad F_m^{\varepsilon^{tr}} = f\left(\sigma_{ij}^{d} - H_\varepsilon\bar{\varepsilon}_{ij}^{tr}\right) = fF_y^{\varepsilon}$$

$$F_m^{\varepsilon^{twin}} = \sigma_{ij}^{d} - H_{twin}\bar{\varepsilon}_{ij}^{twin} = F_y^{\varepsilon^{twin}}$$

where σ^d is the deviatoric stress tensor. The comparison of the driving force of each internal variable to its corresponding yield force and taking into account consistency conditions leads to identifying the activated mechanisms and the system of equations to solve numerically in order to derive the thermomechanical behavior of the SMA. More details about this developed model and its specificities as the management of the inner loop related to partial transformation and/or twinning, saturation of transformation and twinning, tension compression asymmetry, etc. can be found in Chemisky et al. and Duval et al. [99,104]. Fig. 9.2 shows the capacity of SMA behavior prediction for this model under tension–compression and shear tension complex loadings.

The last family of models considers the statistical physics theory [118] based on the local equilibrium of an SMA single crystal by combining the model detailed in 1996 by Boyd and Lagoudas with the statistical physics theory to describe the reverse transformation recovery strain induced by heating. A homogenization technique with the assumption of homogeneous temperature through the RVE allowed deriving the effective behavior starting from the local behavior of each grain. Subsequently, Fischlschweiger and Oberaigner [119] developed a model based on the statistical physics theory by analogy to phase transformation in magnetic materials and considering an order parameter. They assumed that the thermodynamic equilibrium in the RVE level is reached for a very short time under constant stress and temperature. The entropy of the RVE is assumed to reach an extremum ensuring the canonic constraints, as proposed by Le Bellac et al. [120]. Additional models based on the statistical physics were developed by Müller and

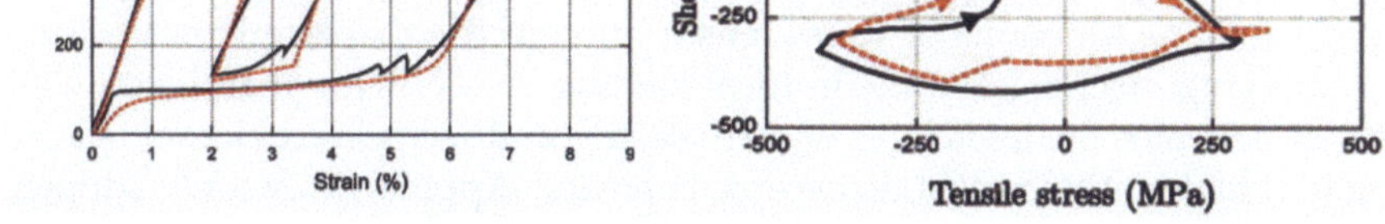

FIGURE 9.2 Comparison between the predicted SMA behavior and obtained experimental results for (A) tension loading–unloading and (B) tension–shear complex loading [104]. Source: *Reproduced with permission from A. Duval, M. Haboussi, T. Ben Zineb, Modelling of localization and propagation of phase transformation in superelastic SMA by a gradient nonlocal approach. Int. J. Solids Struct. 48 (2011) 1879–1893. Copyright 2011 Elsevier publications.*

Seelecke and Müller [121,122] and allowed simulation of the SMA pseudo-elastic behavior.

Some of these models are well adapted to implementation in a finite element code in order to carry out numerical structural analysis of the response of SMA-based applications subjected to thermomechanical loading. The implementation needs the determination, analytically or numerically, of tangent operators allowing to derive the stiffness matrices of finite elements. As an example, many SMA thermomechanical models are implemented in the finite element code Abaqus via the subroutine UMAT.

9.2.2 Modeling of the ferroelectric and ferroelastic behaviors of piezoelectric materials

The modeling of piezoelectric material behavior concerns the linear and the nonlinear parts of this behavior. The linear part corresponds to the reversible electromechanical behavior induced by the small and reversible variation of polarization under mechanical or electrical loading. Whereas the nonlinear part derives from the remnant ferroelectric and ferroelectric domain switching under high electromechanical loading. Voigt [123] was the first to propose, in 1910, a behavior model that accurately describes the linear part, namely piezoelectric behavior. This model has been widely used until now. It describes a relationship giving stress and electric displacement (polarization) as functions of strain and electric field as follows:

$$\begin{cases} \sigma_{ij} = C_{ijkl}\varepsilon_{kl} - e_{kij}E_k \\ D_i = e_{ikl}\varepsilon_{kl} + \kappa_{ij}E_j \end{cases}$$

where s, e, E, and D are, respectively, the stress and strain second order tensors, electric field, and electric displacement vectors. C, e, and k are, respectively, the fourth order elastic tensor, the third order piezoelectric coupling tensor, and the second order permittivity tensor. These tensors are intrinsic to the piezoelectric material properties and, in most cases, they present anisotropy with a different behavior between the polarization direction and the transverse plane. Other possible combinations are available, such as for example, expression of the strain tensor and the electric field vector as functions of stress tensor and electric displacement vector.

Applying mechanical and electrical loading levels higher than given yield values activates the switching of ferroelectric and ferroelastic domains leading to a highly nonlinear macroscopic behavior. Approaches widely adopted for describing elastic–plastic materials and SMA behaviors are also applied in order to derive the nonlinear part of piezoelectric material behavior at different scales. They can predict specificities of such multiphysical behavior as anisotropy, large hysteresis, and saturation of electric polarization and strain [124]. The first phenomenological model was proposed by Chen [125,126]. It considers the remnant strain and polarization as internal variables taking into account the electromechanical loading history. However, this model is one-dimensional and is limited to the case of electrical loading. A more sophisticated phenomenological model, consistent with thermodynamic laws, was proposed by Bassiouny et al. [127,128]. It is based on the concept of thermoelastoplasticity and is inspired from phenomenological descriptions adopted in ferromagnetism. It considers and additive decomposition of strain tensor and polarization vector into reversible and remnant parts. Internal variables are remnant strain and polarization. The adopted formulation applies the Clausius–Duhem inequality in order to derive the corresponding constitutive equations. Many models were proposed to enhance it [129–135]. The following equations correspond to an example of a nonlinear phenomenological electromechanical behavior model:

$$\begin{cases} \varepsilon_{ij} = \varepsilon_{ij}^{el} + \varepsilon_{ij}^{R};\ \ \varepsilon_{ij}^{R} = \varepsilon_{ij}^{P} + \varepsilon_{ij}^{F};\ \ D_i = P_i^{el} + P_i^{R} \\ \sigma_{ij} = C_{ijkl}\varepsilon_{kl}^{el} - e_{kij}E_k;\ \ D_i = e_{ikl}\varepsilon_{kl}^{el} + \kappa_{ij}E_j;\ \ P_i^{R} = P_i^{E} + P_i^{\sigma} \\ \varepsilon^{P} = \dfrac{3}{2}\varepsilon^{SAT}\dfrac{\| \overrightarrow{P^E} \|}{P^{SAT}}\left(\overrightarrow{e_{PR}} \otimes \overrightarrow{e_{PR}} - \dfrac{1}{3}\mathbf{I}\right);\ \ \dot{P}_i^{\sigma} = \dfrac{P^{SAT}}{2\varepsilon^{SAT}}e_j^{0}\dot{\varepsilon}_{ij}^{F};\ \ T_i = \sigma_{ij}e_j^{P} \end{cases}$$

The total strain and electric displacement present a reversible contribution related to the piezoelectric behavior and remnant contributions related to ferroelectric and ferroelastic domain switching. This assumes an analogy between the mechanical and the electric occurring mechanisms. The remnant strain and polarization derive from depolarization and ferroelasticity characterized, respectively, by the internal variables ε^R and P^E. ε^{SAT} and P^{SAT} are, respectively, the maximum remnant strain and polarization corresponding to a fully polarized material. The evolution laws of these two internal variables are presented in the following equations. They assume an elastic–plastic-like behavior with two loading surfaces, two hardening-like laws and two normality conditions corresponding to the mechanical and electrical parts.

$$\begin{cases} f^{\sigma} = \sqrt{\dfrac{3}{2}\left(s_{ij} - \alpha_{ij}\right)\left(s_{ij} - \alpha_{ij}\right)} - \left(\sigma_c + n e_i^{P} E_i\right);\ \ f^{E} = \sqrt{(E_i - X_i)(E_i - X_i)} - E_c \\ \dot{\alpha}_{ij} = \dfrac{2}{3}h\dot{\varepsilon}_{ij}^{F};\ \ \dot{X}_i = \beta \dot{P}_i^{E};\ \ \sqrt{\dfrac{2}{3}\varepsilon_{ij}^{R}\varepsilon_{ij}^{R}} \le \varepsilon^{sat};\ \ \sqrt{P_i^{R}P_i^{R}} \le P^{sat} \\ \dot{\varepsilon}_{ij}^{F} = \dot{\lambda}^{\sigma}\dfrac{\partial f^{\sigma}}{\partial \sigma_{ij}} = \dot{\lambda}^{\sigma}\dfrac{3}{2}\left(\dfrac{s_{ij} - \alpha_{ij}}{\sigma_c + n e_i^{P} E_i}\right);\ \ \dot{P}_i^{E} = \dot{\lambda}^{E}\dfrac{\partial f^{E}}{\partial E_i} = \dot{\lambda}^{E}\left(\dfrac{E_i - X_i}{E_c}\right) \\ \dot{f}^{\sigma} = \dfrac{\partial f^{\sigma}}{\partial \sigma_{ij}}\dot{\sigma}_{ij} + \dfrac{\partial f^{\sigma}}{\partial \alpha_{ij}}\dot{\alpha}_{ij} + \dfrac{\partial f^{\sigma}}{\partial e_i^{P}}\dot{e}_i^{P} + \dfrac{\partial f^{\sigma}}{\partial E_i}\dot{E}_i = 0;\ \ \dot{f}^{E} = \dfrac{\partial f^{E}}{\partial E_i}\dot{E}_i + \dfrac{\partial f^{E}}{\partial X_i}\dot{X}_i = 0 \end{cases}$$

Finally, two conditions of consistency are applied in order to derive the constitutive equations describing the nonlinear electromechanical behavior of the piezoelectric material. These equations are numerically solved by a classical return mapping method combined with the Newton–Rapson technique [131,136]. Fig. 9.3 shows an example of obtained numerical results issued from the solving of these constitutive equations. They show ferroelectric and ferroelastic behaviors induced by a compression loading under constant electric field.

The second family of models consider a local description of piezoelectricity and of the ferroelectric and ferroelastic domain switching at the microstructure scale. The piezoelectric linear constitutive equations have the same expression as in the case of the phenomenological approach. The nonlinear constitutive equations derive from the aforementioned micromechanical formulation initially developed for plastic slip mechanism and extended to martensite variants and here to ferroelectric and ferroelastic domains. Each domain is characterized by its microstructure properties and a correspondent volume fraction assumed as internal variable. The global remnant strain is the contribution sum of intrinsic switching strain for all ferroelectric and ferroelastic active domains weighted by their self-volume fractions. The assuming of a free energy and dissipation expressions at the microstructure scale allow to apply the first and the second principle of thermodynamics in order to derive the local electromechanical constitutive equations. It will be then possible to deduce the corresponding effective electromechanical behavior at the RVE or macroscopic scale by considering analytical or numerical scale transition techniques as self-consistent Mori–Tanaka or FE^2 homogenization methods. The literature presents a wide range of models based on such an approach [124,131,137–143]. In the following is an example of constitutive equations corresponding to the description of the electromechanical behavior of a piezoelectric single crystal based on a crystalline-plasticity-like approach

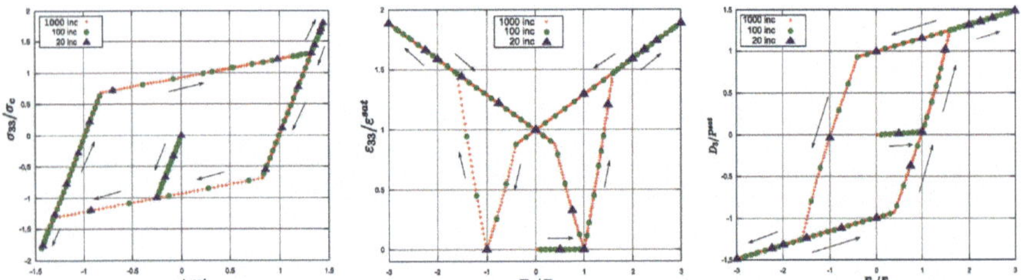

FIGURE 9.3 Ferroelectric and ferroelastic behaviors issued from a phenomenological approach [136]. Source: *Reproduced with permission from W. Zouari, T. Ben Zineb, A. Benjeddou, A FSDT—MITC piezoelectric shell finite element with ferroelectric non-linearity. J. Intell. Mater. Syst. Struct. 20 (2009) 2055–2075. Copyright 2009 Sage publications.*

[131,140]. The Gibbs free energy of this model presents the following expression:

$$\Psi = -\frac{1}{2}\sigma_{ij}(\varepsilon_{ij} - \overline{\varepsilon}^{s}_{ij}) - \frac{1}{2}E^0_n(\overline{D_n} - \overline{P^s_n}) - \frac{1}{2}\sum_{m,n} H^{mn} f^m f^n$$

where $\sigma, \varepsilon, \overline{\varepsilon}^s, E^0_n, \overline{D_n}, \overline{P^s_n}, f^m, H^{mn}$ are, respectively, the stress tensor, the total and spontaneous strain tensors, the electric field, the total and spontaneous polarization vectors, the volume fraction of a given ferroelectric and ferroelastic domain (*m*), and the term of interaction matrix between concerned domains (*n*) and (*m*). Volume fractions of ferroelectric and ferroelastic domains are assumed to be internal variables. As the evolution of a given ferroelectric and ferroelastic domain occurs to the detriment of the other domains, the sum of all the volume fractions is constant and equal to one. The Clausius–Duhem condition allows to determine the driving force for each volume fraction which is compared to a given yield force.

$$F^m = \frac{\partial\Psi}{\partial f^m} = \frac{1}{2}\sigma_{ij}\overline{\varepsilon}^{sm}_{ij} + \frac{1}{2}E^0_n\overline{P}^{sm} - H^{mn}f^n;\ \dot{F}^m = \dot{F}^m_y = \frac{\partial F^m}{\partial\sigma_{ij}}\dot{\sigma}_{ij} + \frac{\partial F^m}{\partial E^0_i}\dot{E}^0_i + \sum_n \frac{\partial F^m}{\partial f^n}\dot{f}^n = 0$$

$$\dot{f}^h = \alpha^h_{ij}\dot{\varepsilon}_{ij} + \beta^h_m\dot{E}^0_m \quad \textit{with} \quad \begin{cases} \alpha^h_{ij} = (H^{gh} + \varepsilon^{sg}_{ij}C_{ijkl}\varepsilon^{sh}_{ij})^{-1}C_{ijkl}\varepsilon^{sg}_{kl} \\ \beta^h_m = (H^{gh} + \varepsilon^{sg}_{ij}C_{ijkl}\varepsilon^{sh}_{ij})^{-1}(\overline{p^{sg}_m} - \varepsilon^{sg}_{ij}e_{mij}) \end{cases}$$

Combining such equations leads to the following constitutive equations:

$$\begin{cases} \dot{\sigma}_{ij} = \left(C_{ijkl} - C_{ijop}\varepsilon^{sh}_{op}\alpha^h_{kl}\right)\dot{\varepsilon}_{kl} - \left(C_{ijkl}\varepsilon^{sh}_{kl}\beta^h_n + e_{nij}\right)\dot{E}^0_n \\ \dot{D}_m = \left(e_{mkl} - e_{mop}\varepsilon^{sh}_{op}\alpha^h_{kl} + \alpha^h_{kl}p^{sh}_m\right)\dot{\varepsilon}_{kl} + \left(-e_{mkl}\varepsilon^{sh}_{kl}\beta^h_n + K_{mn} + \beta^h_n p^{sh}_m\right)\dot{E}^0_n \end{cases}$$

Fig. 9.4 shows an example of simulations issued from the solving of these constitutive equations. They show, for a single crystal, ferroelectric and ferroelastic behaviors induced by electromechanical loading.

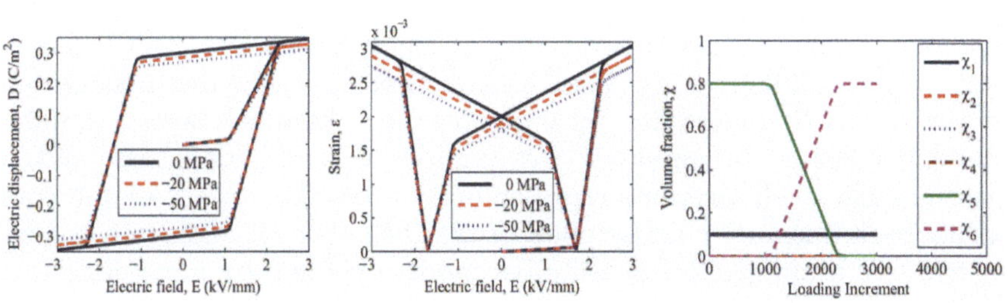

FIGURE 9.4 Ferroelectric and ferroelastic behaviors issued from a micromechanical approach [140]. Source: *Reproduced with permission from R. Jayendiran, Finite element analysis of switching domains using ferroelectric and ferroelastic micromechanical model for single crystal piezoceramics. Ceram. Int. (2016) 15. Copyright 2016 Elsevier publications.*

9.2.3 Modeling of the thermoelectromechanical response of hybrid shape memory alloys/piezo composites

SMA and piezoelectric materials present an interesting multifunctional property leading to innovative applications, particularly in biomedical and aeronautic areas. Most of these applications consider SMAs or piezoelectric materials. Combining properties of these two materials through a hybrid composite could lead to further innovative application as harvester systems adapted to small devices as MEMS [15,144]. It consists of the harvesting of the waste heat related to the Joule effect and converting it to electric power via double multiphysical coupling [145,146]. The SMA layers convert the thermal loading coming from the waste heat into a mechanical deformation via the reverse martensitic transformation. The piezoelectric material, initially polarized, converts the mechanical deformation into an electrical potential via the piezoelectric effect [147,148]. The design of such applications needs the use of numerical tools, mainly based on the finite element method (FEM) and able to predict the global multiphysical response of the composite and local state of all the physical quantities at each material point. Various solutions allow to derive the effective behavior modeling starting from the behavior models of each material layer. The basic solution consists of developing a 3D finite element model for which each layer of the hybrid composite is meshed by 3D continuum solid element with a tetrahedral and/or hexahedral shape [34]. All elements of a given layer have the same multiphysical behavior model well adapted to the corresponding material (SMA or piezoelectric material). This technique needs a high number of elements in order to mesh all the parts, particularly when the surface dimensions are much higher than the thickness and/or when the number of layers through the thickness is important. In addition, it will be mandatory to consider coupled displacements–temperature–electric potential degrees of freedom (dof) for the adopted finite elements. An alternative solution consists of meshing the composite with only one element in the thickness direction. The effective behavior through the thickness derives from a numerical integration with a given number of integration points for each layer. The main drawback resides in consideration of a special element adapted to multilayer composites with different dof and the management of the stress discontinuities through interfaces between layers. Another solution consists of meshing the composite with triangular and/or quadrangular shell elements representing the middle surface of the composite. Each integration gauss point of the shell element takes into account an effective multiphysical constitutive model obtained by homogenization through the thickness with scale transition techniques such as self-consistent or Mori–Tanaka methods [149]. This solution requires the development of a special shell finite element with coupled displacement–temperature–electrical potential dof.

The above section gives a detailed way of finite element modeling of some available models for the thermomechanical behavior of the SMA, for the electromechanical behavior of the piezoelectric layer, and finally for the hybrid composite multiphysical behaviors. The next section deals with state-of-the-art hybrid SMA/piezoelectric composites' manufacturing and characterization, as well as the new developments and efficient energy-harvesting prototypes for the multilayer manufacturing of the composites for the energy-harvesting application.

9.3 Multilayer manufacturing and characterization

The combined effect of smart materials is a promising approach for improved functionalities. A hybrid system composed of shape memory alloy (SMA) and piezoelectric materials provides significant enhancement of the characteristics of the composite system. Typically, SMA possesses two phases called austenite and martensite. It has thermomechanical characteristics and the phase transformation occurs either by stress or temperature [9]. Piezoelectric materials generate an electric response when subjected to mechanical stress. Moreover, they can convert the temperature variation into electrical energy via the pyroelectric effect [13]. The coupling of different properties of the SMAs and piezoelectric materials are presented in Fig. 9.5.

The hybrid device functions by coupling each of the effects in a synergetic way. SMAs depend on certain characteristic temperature and mechanical loading to attain the shape memory effect and the superelasticity. Whereas piezoelectric materials show a change in polarization by

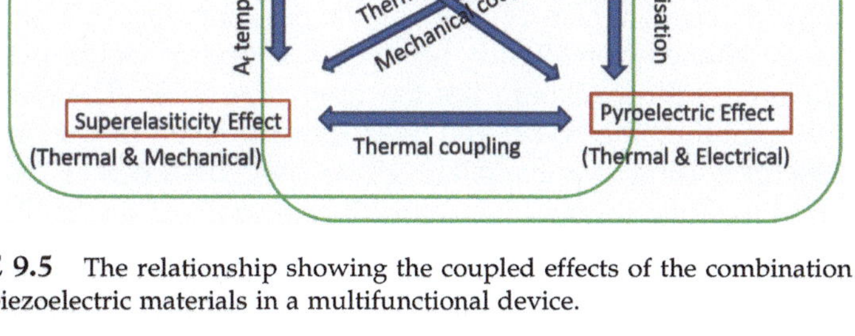

FIGURE 9.5 The relationship showing the coupled effects of the combination of SMAs and the piezoelectric materials in a multifunctional device.

the action of mechanical stress, electric field, or temperature and induce piezoelectric and pyroelectric effects, respectively.

As an example, the SMA with the shape memory effect is subjected to mechanical deformation and applying the thermal loading above the transformation temperature can attain the phase transformation from austenite to martensite and vice versa. The heating induces the pyroelectric effect and the mechanical deformation results in the piezoelectric effect. Hence, the electric charge is created on the surface of the piezoelectric material. Taking into account all these factors and the coupling effects, the design and manufacturing of such a hybrid system is more important for fabrication.

9.3.1 First devices

In 2011 Lebedev et al. [150] experimentally demonstrated thermal energy harvesting by using SMA/piezoelectric laminated composite. The laminated composite composed of 30 μm thick $Ti_{50}Ni_{25}Cu_{25}$ ribbons as an SMA layer with a first-order martensitic transformation and the Micro Fiber Composite (MFC) based on a PZT fiber layer as a piezoelectric material. Fig. 9.6A represents the fabricated thermal energy harvester. The final prototype is made by gluing the SMA ribbon onto the MFC using cyanoacrylate. By heating current modulation, the composite's output voltage measurements were taken at different frequencies. The composite produced a maximum energy density of 30 mJ/cm^3 per phase transition. The same year Namli and Taya [152] designed both experimentally and analytically a thermal energy harvester composed of hybrid SMA/piezoelectric material. The PZT layer and SMA fiber were connected in series and subjected to fluctuating temperature. The authors reported that an electrical signal is obtained starting from a thermal loading with a frequency of 0.05 or 0.1 Hz and with temperature variation between −30°C and 110°C. The experimental results showed an output power of 8.6 μW at 0.1 Hz. However, information such as the dimension of the harvester and the energy per cycle is missing in this study.

Zakharov et al. [151] demonstrated an original laminated NiTiCu/PZT composite for energy harvesting from small temperature variations. The SMA part composed of 40-μm-thick $Ti_{50}Ni_{25}Cu_{25}$ ribbons with a phase transformation temperature range of 50°C−80°C. The commercial MFC is the piezoelectric layer, which consists of 100 μm diameter PZT fibers. The PZT fibers are aligned horizontally between the top and bottom electrodes and it is covered with insulating Kapton films (Fig. 9.6B). The harvester is made by assembling the SMA and piezoelectric layer in the form of a bilayer structure using cyanoacrylate glue. They designed a mechanical structure that uses the thermally induced linear strain of the SMA to bend the PZT plate. The "machine" with an active material of 0.2 cm^3 is able to harvest 90 and 60 μJ during heating and cooling respectively by 35°C. The schematic representation of the machine harvester is shown in Fig. 9.6C.

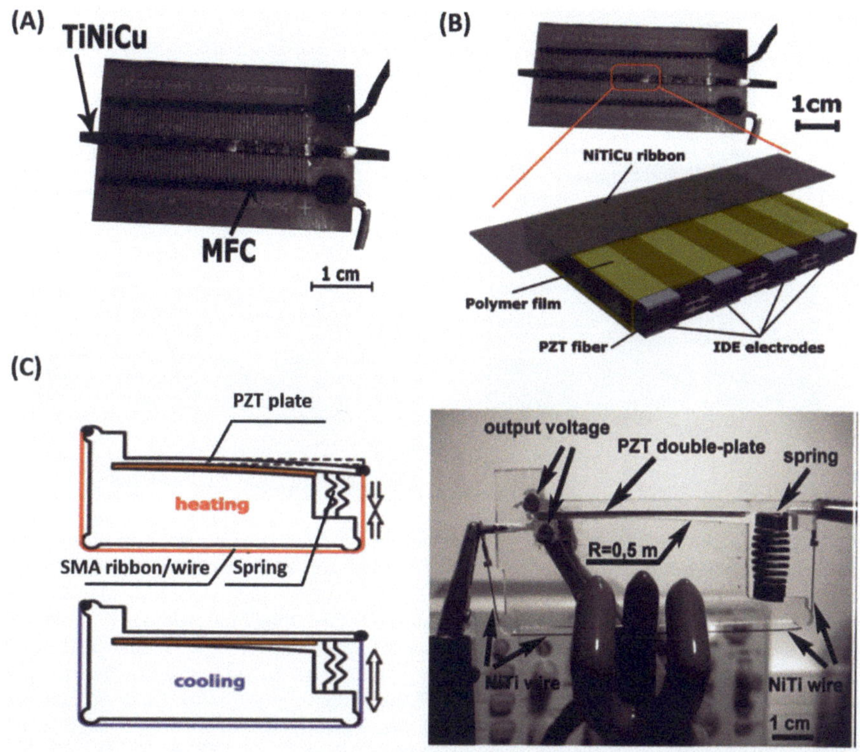

FIGURE 9.6 (A). Photographic image of the NiTiCu-MFC composite thermal energy harvester [150]. (B) NiTiCu ribbon/PZT laminated composites with Kapton covering and the SMA ribbons are glued onto the MFC. (C) Schematic view (left) and the photo (right) of the experimental harvester demonstrator, showing the SMA wire is connected at its ends to the PZT plate [151]. Source: *Reproduced with permission from G.A. Lebedev, B.V. Gusarov, B. Viala, et al., Thermal energy harvesting using shape memory/piezoelectric composites, in: 16th International Solid-State Sensors, Actuators and Microsystems Conference. Beijing, China: IEEE, 2011, pp. 669–670. Copyright 2011IEEE; D. Zakharov, G. Lebedev, O. Cugat, et al. Thermal energy conversion by coupled shape memory and piezoelectric effects. J. Micromech. Microeng. 22 (2012) 094005. Copyright 2012 IOP publications, ff10.1088/0960-1317/22/9/094005ff. hal-00814609f.*

In 2013 the same group proposed an enhanced method for thermal energy harvesting by combined piezoelectric, pyroelectric, and shape memory effect [147]. The generated voltage from the composite is measured by dipping in hot and cold water. The voltage measurements are done using an electrostatic noncontact voltmeter. The results exhibited a 50% of pyroelectric effect increase in a laminated structure. Avirovik et al. [32] developed a novel device design consisting of a piezoelectric bimorph and SMA wire heated remotely using a laser, which shows dual functionality of both as an energy harvester and an actuator (Fig. 9.7A). The piezoelectric bimorph in a cantilever beam configuration

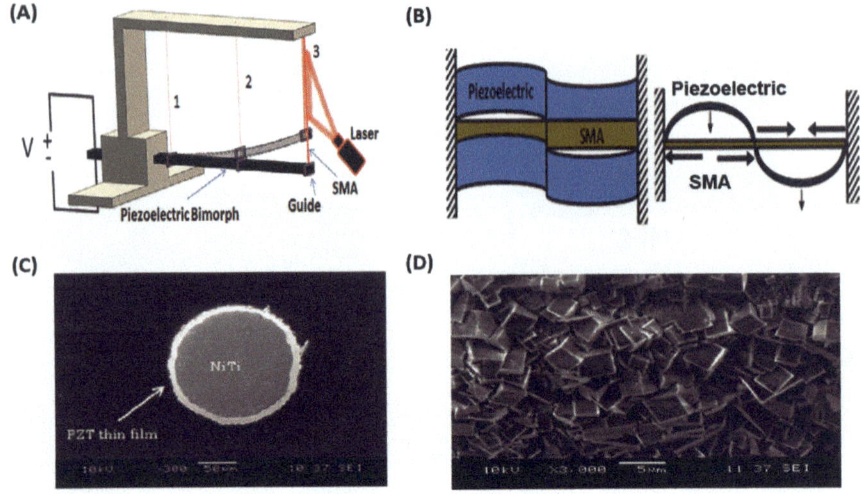

FIGURE 9.7 (A) Schematic diagram showing the piezoelectric bimorph in a cantilever beam configuration and SMA wire is attached to the beam at three different positions and a laser [32]. (B) Schematic of a hybrid SMA/PZT harvester [33]. (C) SEM image of the cross-section of the NiTi/PZT hybrid composite. (D) PZT thin film surface [153]. Source: *Reproduced with permission from Avirovik, D., Kumar, A., Bodnar, R.J., et al. Remote light energy harvesting and actuation using shape memory alloy—piezoelectric hybrid transducer. Smart Mater. Struct. 22 (2013) 052001. Copyright 2013 IOP publications; J.S. Gosliga, D.O.A. Ganilova, Energy harvesting based on the hybridisation of two smart materials, in: EACS 2016 – 6th European Conference on Structural Control, Sheffield, England, 11–13 July 2016; H. Sato, Multifunctional devices combining shape-memory alloy and piezoelectric materials, in: N.C. Goulbourne, H.E. Naguib (Eds.), San Diego, CA, p. 905811.*

is coupled with an SMA wire of diameter 100 μm. The SMA wire is attached to the beam at three different locations. Through this study, the authors showed the dual functionality of the system both as an actuator and energy harvester. In comparison with the reported energy harvesters, the energy conversion efficiency seems lower. However, the applicability as an actuator is proved.

Sato et al. [153] developed a novel hybrid composite with multifunctional properties using PZT thin film and NiTi SMA wire (150 μm diameter). The PZT thin films are directly deposited on the surfaces of the NiTi SMA wires by employing the hydrothermal synthesis method. The piezoelectric thin films can withstand large deformation up to 5% because the thin films are formed by the layering of several crystals. Fig. 9.7C and D show the SEM images of the cross-section of NiTi wire with uniform film formation along the NiTi surface and the surface of the PZT thin layers.

Gosliga et al. [33] developed both finite element and analytical models to describe the dynamic behavior of the hybrid device combining the piezoelectric and shape memory effect (Fig. 9.7B). The FE model predicted electrical energy of 1.19 mJ for a single cycle of the output frequency of 494 Hz. Most of the studies reported so far are dealing with ceramic PZT-based SMA piezoelectric composite.

Gusarov et al. [43] first made the composite by using 110-μm-thick PVDF and two 40-μm-thick NiTiCu ribbons and conductive epoxy as electrodes with patterned Cu tape for electrical contact (Fig. 9.8A and B). The authors showed that the postage stamp size (27.5 mm^3) harvester

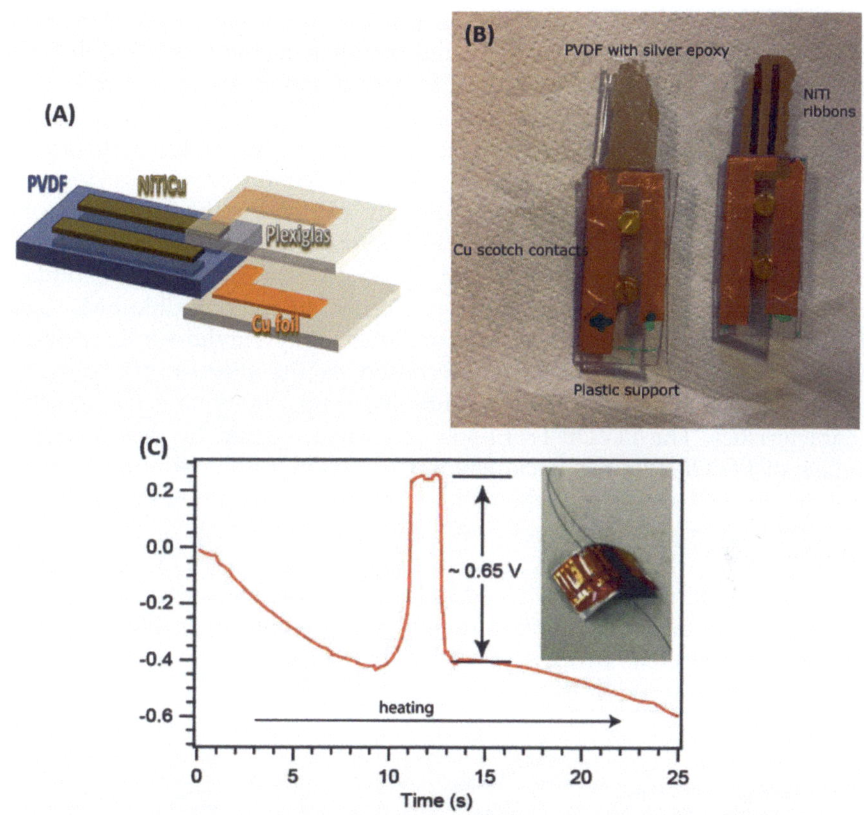

FIGURE 9.8 (A) Schematic representation of the composite. (B) Photographic image of the SMA/PVDF composite, the reference PVDF sample alone (left) and the PVDF (110-mm-thick) + two NiTiCu (40-mm-thick) ribbons on each side for patterned Cu tape [43]. (C) Thermal energy harvesting from the NiTi/ZnO composite [154]. Source: *Reproduced with permission from B. Gusarov, E. Gusarova, B. Viala, et al., Thermal energy harvesting by piezoelectric PVDF polymer coupled with shape memory alloy. Sens. Actuators Phys. 243 (2016) 175–181. Copyright 2013 Elsevier publications; H. Radousky, F. Qian, Y. An, et al., Harvesting mechanical and thermal energy by combining ZnO Nanowires and NiTi shape. Memory Alloy (2017) 8.*

can achieve an energy density of 0.41 mJ/cm^3 with a temperature variation of 20°C. Another related study was done by Radousky et al. [154] to harvest thermal and mechanical energy by combining ZnO nanowires and NiTi SMA. The thermoelectric device yielded an output voltage of ~0.65 V at room temperature bending and heating of the NiTi/ZnO device at 50°C (Fig. 9.8C).

9.3.2 Processing of the shape memory alloys/poly(vinylidene fluoride-trifluoroethylene) hybrid composite

To the best of our knowledge, Gusarov et al. were the only authors who have experimentally demonstrated the coupling of the SMA/PVDF composite. Hence, this work aims to go further to demonstrate that the smart materials arranged in layered composites have the potential to significantly enhance energy conversion and to outperform the state-of-the-art of energy-harvesting devices. Consequently, this section presents a development of a microenergy harvester system by addressing the fabrication of a novel layered composite with stacking layers composed of conventional SMAs and piezoelectric polymer thin layer. A fabricated novel multilayered hybrid composite structure is composed of fully flexible SMA film and a thin layer of the piezoelectric polymer. Poly(vinylidene fluoride-trifluoroethylene) [P(VDF-TrFE)] was chosen as the piezoelectric polymer thin layer due to its excellent flexibility, biocompatibility, and piezoelectric characteristics. The P(VDF-TrFE) solution was prepared by dissolving the pellets of P(VDF-TrFE) in cyclopentanone solvent with continuous stirring at 80°C. Fig. 9.9 shows the schematic diagram of the solution preparation.

A thin layer polymer layer was deposited on the top of the classical NiTi SMA film by a spin-coating technique (Fig. 9.10). On top of the P (VDF-TrFE) thin layer, the Cr/Al electrode was deposited. The schematic representation of the final device is shown in Fig. 9.10B.

The piezoelectric properties of the composite were tested. The P (VDF-TrFE) polymer was subjected to poling to orient the polarization

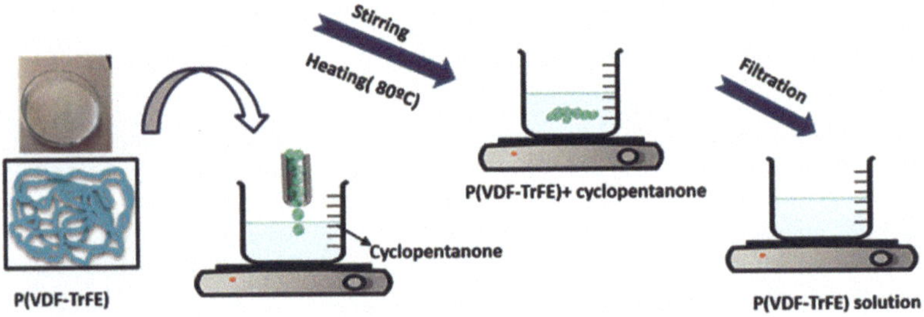

FIGURE 9.9 Schematic diagram of the P(VDF-TrFE) solution preparation (article under review).

of the ferroelectric domains and activate piezoelectric properties. The electric polarization is achieved by the direct contact method. The electric poling of the P(VDF-TrFE) film on NiTi SMA substrates was done by the application of an electric field of 100 V/μm to align the dipoles in the film with the thickness direction. The polarization versus the electric field (P-E loop) (voltage normalized in voltage per micrometer) for P (VDF-TrFE) on SMA is shown in Fig. 9.12A. The polymer exhibits a P-E loop with a saturation polarization (Ps) of 4.9 $\mu C/cm^2$. The obtained d_{33} values for the SMA/P(VDF-TrFE) device is 20.8 pC/N.

In order to improve the adhesion between the layers, a new composite structure was developed. The polymer layer was deposited on a thin flexible polyethylene naphthalate (PEN) substrate and stuck with SMA using a thin layer of flexible epoxy glue (Fig. 9.11). This composite shows a better performance and a good adhesion strength. The electric poling of the P(VDF-TrFE) on the PEN sheet was done in the same way

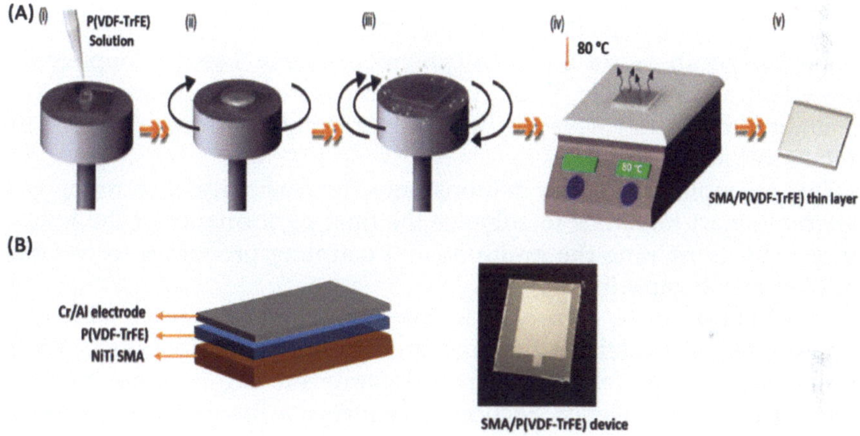

FIGURE 9.10 SMA/Polymer composite film preparation: (A) different steps of spin coating to form the polymer layer on SMA; (B) schematic illustration of the SMA/P(VDF-TrFE) composite and the photographic image (article under review).

FIGURE 9.11 SMA/PEN/P(VDF-TrFE) hybrid multilayer structure with epoxy as an interface bonding layer and photographic image of the final device (article under review).

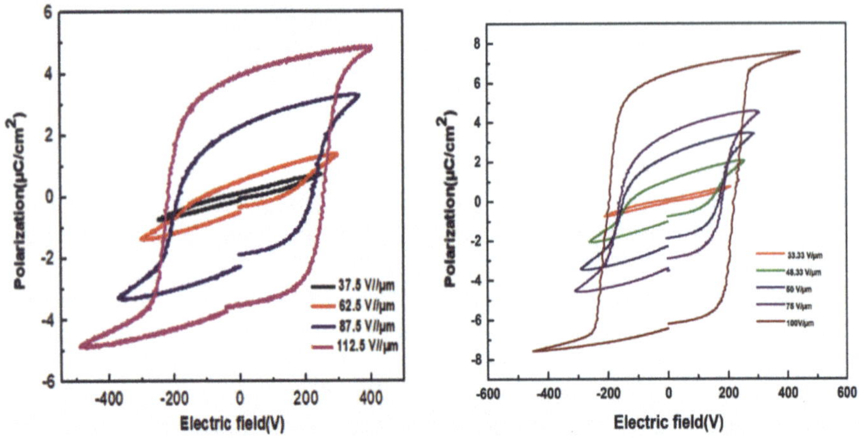

FIGURE 9.12 (A) Polarization hystersis loop of SMA/P(VDF-TrFE) composite. (B) Polarization hystersis loop of P(VDF-TrFE) on PEN (article under review).

as mentioned above for the SMA/polymer system. The P-E loop for the material is shown in Fig. 9.12B. It shows the saturation polarization (P_s) of about 7.6 $\mu C/cm^2$, which is in good agreement with the reported literature values [155,156].

This example of a device demonstrates the feasibility of coupling two different smart materials to enhance the final performance of the hybrid system, by combining the multiphysical coupling properties to enhance the conversion capacity.

The hybrid SMA/piezoelectric polymer composite was fabricated experimentally. Understanding and predicting the distribution of stress, strain, and electric field of the hybrid composite can be done by finite element modeling. Hence, Section 9.1.4 presents the global response of the harvester response and design.

9.4 Finite element analysis of shape memory alloys/piezo composite response for energy harvesting

The idea of combining SMA and piezoelectric layers in hybrid composites for energy harvesting was initially proposed by Lagoudas and Bo [157,158]. They proposed a simplified analytical model that predicts the response of a hybrid composite with an SMA and piezoelectric layers subjected to a cylindrical bending. Shen, Han, Lee, Namli, and Taya [152,159–161] extended such development to the case of thermal loading in order to show that it is possible to deliver an electric voltage from an SMA/piezo composite when subjected to a thermal loading.

They considered the system described by the Fig. 9.13A containing an SMA and piezo layers clamped inside a rigid frame subjected to a heating–cooling cycle under constant compression stress. Fig. 9.13B shows the obtained electric signal function of the applied thermal loading and its frequency. The study contains a simplified analytical modeling of the shape memory effect of the SMA layer and the linear piezoelectric behavior of the piezoelectric layer. It contains also results of an experimental study showing that it is possible to harvest an electric power range between 1 and 50 mW depending on the applied thermal frequency.

Garcia and Lobontiu [162], Zakharov et al. [151], and Gusarov et al. [43] experimentally studied the same kind of hybrid composite and showed that is possible to deliver a given voltage starting from a thermal heating–cooling loading. All available studies experimentally investigated the feasibility of thermal waste harvesting and its conversion to electric power using hybrid SMA/piezo composites. They completed these studies with simplified analytical modeling. Another characterization of such electrothermomechanical response is possible with the finite element method. It consists of meshing the bilayer hybrid composite by solid hexahedral continuum elements with quadratic interpolation (labeled C3D20 in Abaqus). A constitutive model adapted to SMAs, proposed by Chemisky et al. [99], and implemented in Abaqus via Umat, describes the thermomechanical behavior of the SMA layer, and a classical and linear piezoelectric constitutive model describes the electromechanical behavior of the piezoelectric layer. The boundary conditions and loading present two steps. The first step represents the cooling under the M_f temperature followed by a tension preloading applied to only the SMA layer. The second step considers a heating–cooling cycle of the hybrid composite between a minimal temperature lower than M_f and

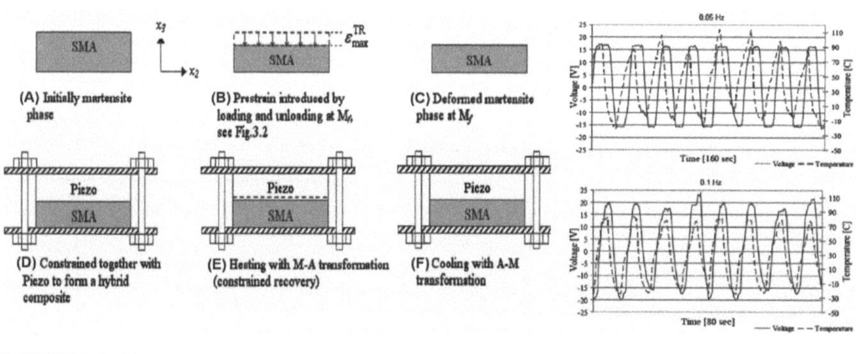

FIGURE 9.13 (A) Proposed device for SMA/piezo hybrid composite [160]; (B) obtained voltage versus temperature [152]. Source: *Reproduced with permission from O.C. Namli, M. Taya, Design of piezo-SMA composite for thermal energy harvester under fluctuating temperature. J. Appl. Mech. 78 (2011) 031001. Copyright 2011 ASME Publications.*

maximum temperature higher than A_f. Fig. 9.14 presents the studied hybrid composite and the corresponding finite element model. The mesh presents four elements through the thickness for the SMA layer with a thickness of 125 mm and two elements for the Piezoelectric element with a variable thickness between 1 and 4 mm. Tables 9.1 and 9.2 present the adopted material parameters for the SMA and piezoelectric layers.

This finite element modeling allowed to predict the distribution of stress, strain, and electric field through the hybrid composite and the global response of voltage function applied temperature.

Fig. 9.15A presents the evolution of the electrical potential function of temperature. During the cooling loading, the forward martensitic transformation occurs in the SMA layer, also inducing the deformation of the piezoelectric layer. The electromechanical coupling induced the increasing of the voltage. During heating, the activation of the reverse transformation recovers the transformation strain in the SMA layer. The piezoelectric layer also recovers its shape and the electric voltage decreases. As the thickness of the layer increases, the induced electrical potential increases. The increasing and decreasing of the electric voltage occurs with hysteresis related to the switching hysteresis between forward and reverse transformation in the SMA layer. The evolution of the maximum and the

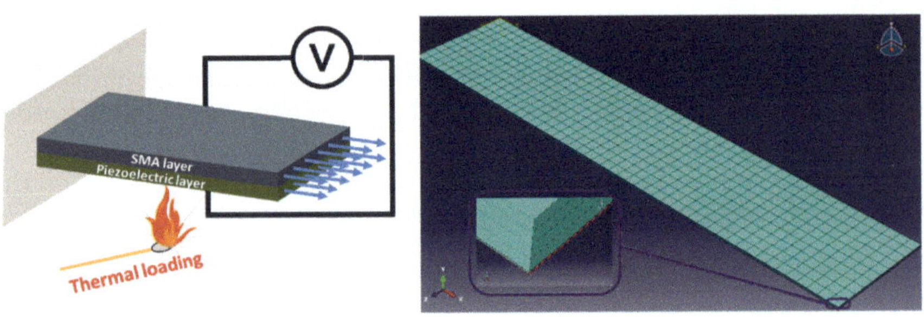

FIGURE 9.14 (A) SMA/piezo hybrid composite subjected to a heating–cooling loading under constant tension preloading, (B) finite element mesh.

TABLE 9.1 Material parameters for the SMA layer.

E(MPa)	ν	$\alpha(°C)^{-1}$	$\varepsilon^T_{tracmar}$	$\varepsilon^{TFA}_{tractmar}$	$\varepsilon^T_{Compmar}$
70,000	0.3	8.10^{-6}	0.05	0.04	0.04
b_d(MPa/°C)	b_r(MPa/°C)	M_S(°C)	A_f(°C)	r_f	
5	6	20	60	0,6	
F_ε(MPa)	H_f(MPa)	$H_{\varepsilon T}$(MPa)	H_{twin}(MPa)	H_s(MPa)	
100	4	1000	40,000	50	

TABLE 9.2 Material parameters for the piezoelectric layer.

Property	PVDF	Units
Relative permittivity	12	$\varepsilon/\varepsilon_0$
Piezoelectric strain coefficient (d_{31})	15	10^{-2} C/N
Piezoelectric strain coefficient (d_{33})	23	pC/N
Electromechanical coupling constant (k_{31})	12	% @1 kHz
Young's modulus	15	GPa
Poisson ratio	0.3	

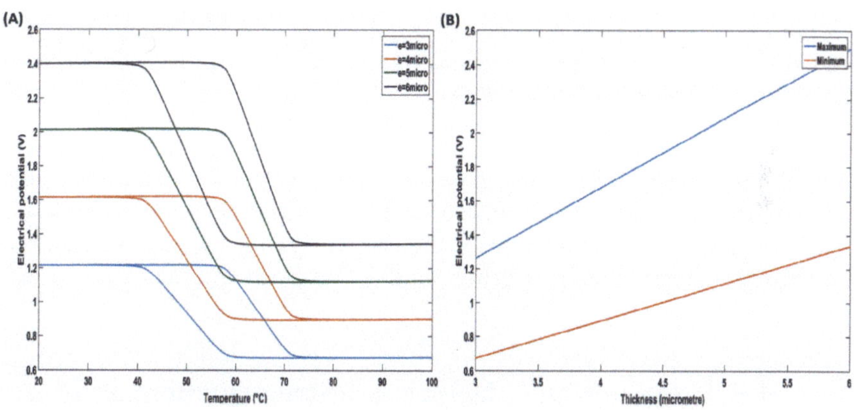

FIGURE 9.15 (A) Electric voltage evolution function of temperature. (B) Electric potential evolution function of piezoelectric layer thickness.

minimum electrical potential created during thermal cycles increases with the increasing of the thickness of the piezoelectric layer, Fig. 9.15B.

These obtained numerical results inspired the development of a waste heat harvester and its conversion to electrical power based on a hybrid SMA/piezoelectric composite subjected to a heating–cooling loading under a constant bending preloading in order to amplify the recovered electric potential. The harvester has a circular shape and is constituted of one or many SMA/piezo sequence layers. It is clamped in its circumference and is subjected to uniform pressure (Fig. 9.16). This solution is under investigation with the development of an experimental device in order to apply the thermal and mechanical loadings and measure the induced electric voltage, the deflection, and the real temperature of the hybrid composite. It is detailed in the next section.

The subsequent section is more focused on the instrumentation and performance of the hybrid composite. The generated electric response and the new experimental bench for the electrothermomechanical

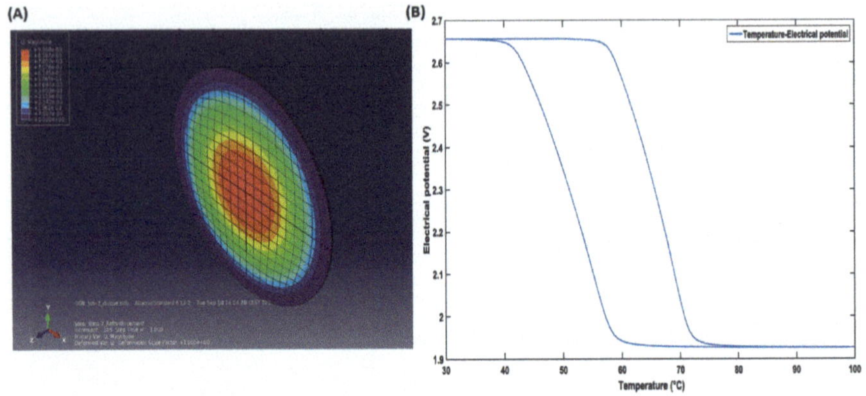

FIGURE 9.16 (A) Finite element model with displacement distribution. (B) Electric potential evolution function of temperature.

characterization of the device is illustrated. Also, the most important part of the energy-harvesting device is the electronic part. The power conversion efficiency of the whole harvester relies on an electronic circuit. Therefore the final section of the chapter gives a brief discussion of the electronic circuit for powering small-scale electronic devices.

9.5 Harvester manufacturing, instrumentation, and performance analysis

9.5.1 Energy harvesting from hybrid composite (shape memory alloys/piezo)

The hybrid SMA/piezo device comprises five layers with Cr/Al as the top and the bottom electrode, two layers of P(VDF-TrFE) piezoelectric thin film, and the SMA substrate. SMA/PEN/P(VDF-TrFE) composite structure using flexible epoxy as an interface bonding layer was also fabricated. All the materials, such as Cr/Al for electrodes and P(VDF-TrFE) film for the active area, were found to adhere well onto the SMA substrate.

To study the performance of the SMA/P(VDF-TrFE) composite, several geometries and configurations were tested. To assemble SMA thin-film sheet on P(VDF-TrFE), a flexible epoxy glue was used as an interface. The SMA film of 125 μm thickness and an area of 1.8 cm × 2.5 cm was cleaned with acetone and isopropanol; a spin coating technique was used to homogenize the thickness of the interface epoxy layer. After that, the PEN/Al/P(VDF-TrFE)/Al device was glued together, followed by at least 24 hours drying at room temperature. The epoxy layer functions as a bonding layer that also allows to follow the deformation

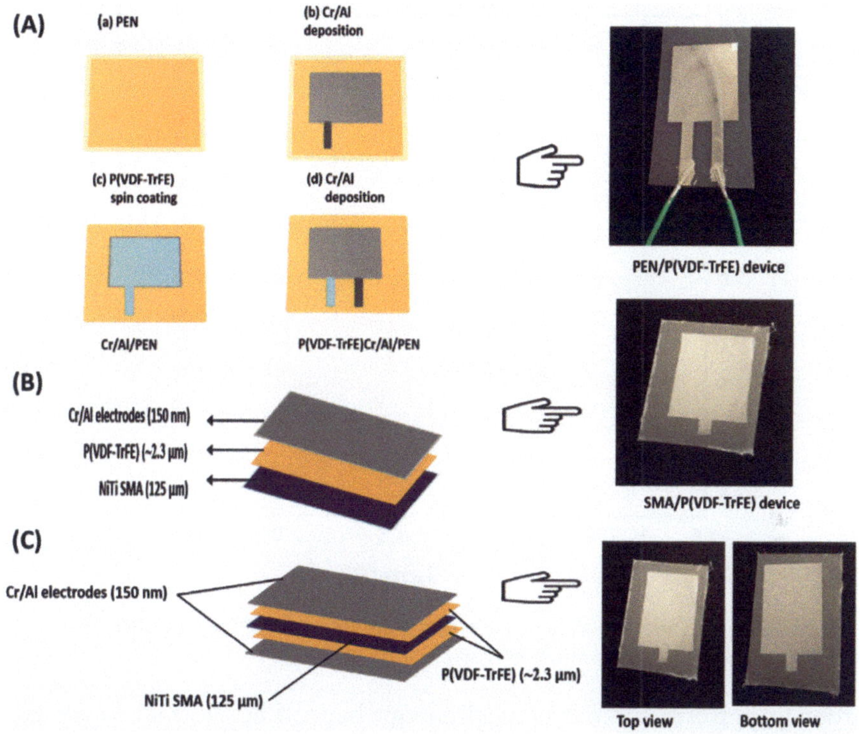

FIGURE 9.17 Schematic diagram of the fabrication process of the flexible piezoelectric energy harvesters. (A) Different steps of the processing of the PEN/P(VDF-TrFE) device and the photographic image of the original device. (B) SMA/P(VDF-TrFE) device structure and original image of the one layered device with Cr/Al to electrode. (C) SMA/P(VDF-TrFE) multilayer device structure and photographic image of the device top view and bottom layer view on the right (article under review).

made by SMA and effectively transfers the stress to the polymer layer (Fig. 9.17). Nevertheless, it can affect the coupling between the two active layers and the final output energy-harvesting performance.

9.5.2 Thermal–mechanical–electrical energy harvesting

The energy harvesting on SMA/P(VDF-TrFE) device, could be highlighted by doing the normal finger impression. It was observed that the charge created on the piezopolymer layer disappeared quickly due to the conducting nature of the SMA substrate. Therefore, to overcome this problem, another device was made using an interface layer using flexible epoxy. SMA/PEN/P(VDF-TrFE) shows good flexibility and the whole device follow easily the deformation made by the SMA layer. In

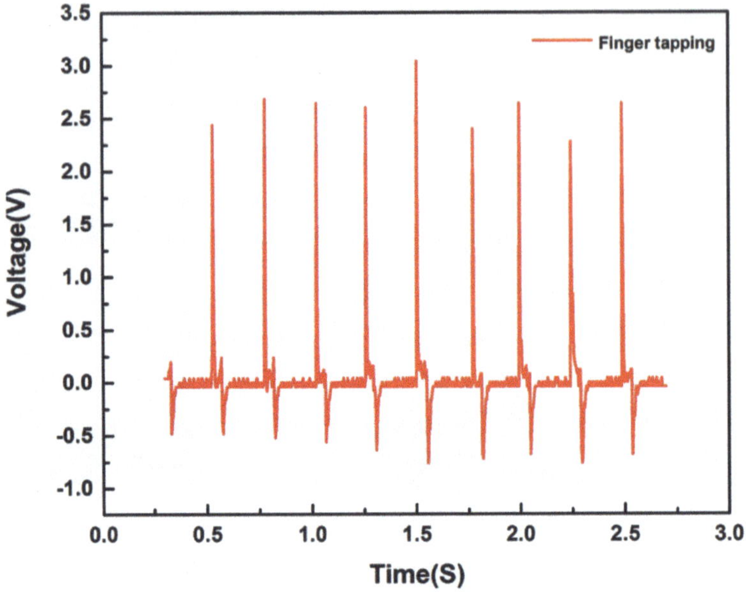

FIGURE 9.18 Output voltage generated from the SMA/PEN/P(VDF-TrFE) device under finger tapping (article under review).

order to perform the mechanical energy harvesting, normal finger tapping was carried out on the whole six-layer assembled sample. The results showed good agreement with the PEN/P(VDF-TrFE) device. The device produces an average output voltage of 2.5–3 V, current of 1 μA, and power output of 3 μW during the finger tapping (Fig. 9.18).

To demonstrate the thermal energy harvesting from the device, the whole device was bent at room temperature. The bent device was placed in a silicon oil bath where the temperature was progressively heated up to 75°C. Above 62°C, the NiTi SMA underwent a reverse martensitic transformation allowing it to recover its original flat shape. Accordingly, the P(VDF-TrFE) device also follows a drastic shape change, which yields an output voltage. An oscilloscope was used to measure the voltage from the device. An output voltage of 1.98 V was obtained during the phase transformation of the SMA (Fig. 9.19). The stress was created in the piezoelectric layer when the device was in a bent shape. During heating, the reverse transformation of the SMA induces the deformation of the PVDF that exhibits a piezoelectric voltage. Therefore this result gives the feasibility of the device to successfully transform the thermal to mechanical to electrical energy. This result corresponds to only one applied cycle of heat and cooling, but several cycles between cooling and heating state will allow one to harvest maximum energy from the device. This system represents a new

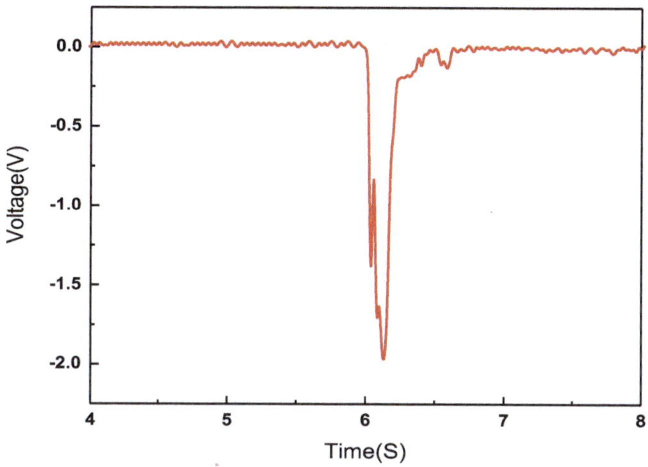

FIGURE 9.19 Output voltage generated from the SMA/P(VDF-TrFE) device during one thermal loading cycle (article under review).

approach for developing a feasible advanced energy material solution for low-power smart micro/nanoelectronics devices.

9.5.3 Electrothermomechanical characterization bench

The electrothermomechanical characterization of the SMA/PVDF hybrid composite could be carried out with a specific bench allowing the application of heating–cooling cycles and measuring the induced electric voltage. The hybrid composite SMA/P(VDF-TrFE) has two specific characteristics: at first, the shape memory effect will take place under thermal constraints. The second is the piezoelectric effect of the PVDF will take place following the mechanical deformation. The design of this experimental test bench should allow to control the thermal loading (measurement and control of current temperature) and measure the induced deflection and electric voltage.

The applied thermal cycles derive from one heating and one cooling source allowing the controlled increasing and the decreasing of the composite temperature. The control system is ensured with the software LabView. The heat source is a laser with a power of 5 W. Usage of a light beam is a noncontact method. The light intensity can be tailored from a remote location to modulate the temperature distribution and hence the actuation characteristics of SMA. The cooling source is a liquid nitrogen-based system. The resulting deformation can be converted into electricity through a piezoelectric bimorph. The simplified model is shown in Fig. 9.20.

The support (marked 1) clamps the hybrid composite on its circumference, a laser (marked 3) of 5 W power is used to heat this sample following a focusing of the beam with two lenses (marked 4), so that the

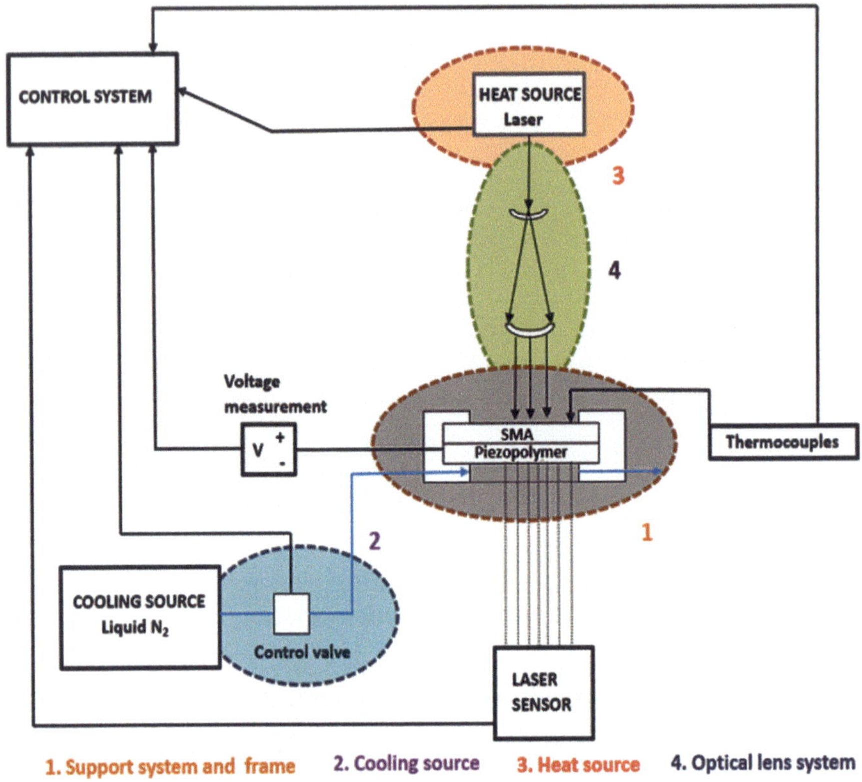

FIGURE 9.20 Experimental bench to energy harvesting from the hybrid composite (SMA/piezo).

laser beam focuses on the entire surface of the hybrid composite leading to an homogeneous heating.

The cooling system consist of a nitrogen reservoir controlled with an electrovalve system that injects doses of liquid nitrogen. This heating/cooling will be simultaneously applied thanks to a movement sensor that manages the triggering of the two subsystems (laser/nitrogen injection). This sensor measures in real-time the deformation of the sample, which is the fundamental behavior of the SMA layer.

The voltage recovered is visualized in the form of a curve that can be obtained in the form of CSV. All of this system can be controlled by a PC through a command interface.

9.5.4 Electronic circuits for piezoelectric energy harvesting

A final important aspect of the energy recovery is the electronic harvesting circuit, which is a major part of piezoelectric energy-harvesting

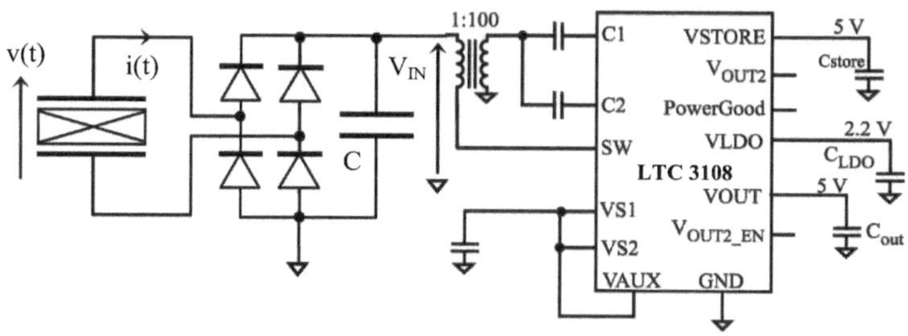

FIGURE 9.21 LTC3108 Boost converter.

systems. This will play a significant role in the power conversion efficiency of the whole system. The main problem in this case is to convert the electronic charge into electric power because there are few charges generated by the piezoelectric effect. Mostly, the piezoelectric energy-harvesting system interface circuit contains three key elements. They are an AC–DC rectifier that rectifies the AC voltage generated from the piezoelectric element into DC. Another one is the voltage regulator that regulates the DC power supply to the storage device like a capacitor. The third important element is the storage device which stores the harvested energy [123,163].

The harvesting of energy from this hybrid composite SMA/P(VDF-TrFE) would be very useful for many applications, such as in micro or nano piezoelectric generators. However, the voltage collected is too low to function for a buck regulator converter. In this case, Zhao et al. [164] suggest adding an adaptive DC–DC boost converter after the rectifier. In this study, the optimum choice is to use the LTC3108 integrated converter step-up, as shown in Fig. 9.21. This converter can work with very low VIN input levels (20 mV).

LTC 3108 is intended for applications that require low average power (in the order of milliwatts or less) and has been designed to be powered by any source that can provide a minimum input voltage of approximately 20 mV, with low source resistance (ideally 3Ω or less). This could be a continuously operating microcontroller or a wireless sensor/transmitter operating at a low duty cycle. Therefore, this converter seems perfectly adapted to the characteristic variations of the SMA/P(VDF-TrFE) device. The integration of this electronic stage in the device is in progress.

9.6 Conclusion

Smart materials with the combined effect of different properties are a promising approach for improved functionalities. A hybrid system composed of shape memory alloy (SMA) and piezoelectric materials provides significant enhancement of the characteristics of the composite system. Therefore the hybrid device allows the coupling of various properties such as thermomechanical and electromechanical behavior in a synergetic way. These hybrid systems have found application in various fields, including sensors, actuators, and energy harvesting. Also, small-scale energy harvesting to power self-powered electronic devices is increasing. In this context, the ability to combine thermal and mechanical harvesting using smart materials requires further study. So far most of the studies have reported on the combined effect of SMA and ceramic composite. Hence, it will be interesting to focus on the development of a hybrid system composed of SMA and piezoelectric polymer P(VDF-TrFE) film, because the excellent flexibility and the high piezoelectric coefficient of the polymer is well suited for the fabrication of devices with SMA and PVDF layers.

This book chapter presents the design, fabrication, and characterization of a novel hybrid composite composed of a fully flexible shape memory alloy film and a thin layer of the piezoelectric polymer. The modeling of the SMA and piezoelectric layers and the whole hybrid composite, the experimental synthesis of the polymer, the deposition of the thin layer of the polymer on the SMA, and the piezoelectric properties of the composite were detailed. The hybrid SMA/piezoelectric polymer composite can simultaneously harvest both mechanical and thermal energy. The developed hybrid device can combine the multiphysical coupling properties such as the shape memory effect and piezoelectric effect of two different smart materials.

The finite-element simulation of the hybrid SMA/piezoelectric polymer composite allowed us to predict the delivered voltage for a given heating–cooling loading. Through this model prediction, the applied thermal cycles generate the electrical response from the SMA and the piezoelectric polymer layers. It is an interesting predictive numerical tool that could help to enhance the harvesting performances of such hybrid composites. In order to characterize the hybrid composite experimentally, we have developed an electrothermomechanical bench. In the setup, the cooling can be done by the circulation of liquid nitrogen and temperature increase by laser heating. Thus the experimental setup is able to perform the fast heating and cooling of the hybrid device to generate more electrical response from the device for temperature variations up to 70°C. The numerical finite element tool allows to predict and optimize the harvesting process considering such hybrid composites by determining the adapted number of layers and their thicknesses, the shape of the neutral surface, and the appropriate applied mechanical preloading.

The electronic circuit carries out an important role in the energy-harvesting system. The power conversion efficiency of the whole system strongly depends on the electronic circuit. The main problem with such a kind of system is to convert the electronic charge into electric power because there are few charges generated by the piezoelectric thin layer. In order to harvest the energy efficiently from the device, a perspective is to implement an integrated converter step-up like the LTC3108, which will make it possible to store the produced energy in a battery for future usage. Thus the device will be able to convert electrical energy from the ambient environment and thereby power small electronics, such as wireless sensors, MEMS, and biomedical devices in an autonomous way.

References

[1] S. Bodkhe, P. Ermanni, 3D printing of multifunctional materials for sensing and actuation: merging piezoelectricity with shape memory, Eur. Polym. J. 132 (2020) 109738.

[2] K.J. Narayana, R. Gupta Burela, A review of recent research on multifunctional composite materials and structures with their applications, Mater. Today Proc. 5 (2018) 5580–5590.

[3] S. Sukumaran, S. Chatbouri, D. Rouxel, et al., Recent advances in flexible PVDF based piezoelectric polymer devices for energy harvesting applications. J. Intell. Mater. Syst. Struct.; 1045389X20966058.

[4] H. He, X. Lu, E. Hanc, et al., Advances in lead-free pyroelectric materials: a comprehensive review, J. Mater. Chem. C. 8 (2020) 1494–1516.

[5] P. Costa, J. Nunes-Pereira, N. Pereira, et al., Recent progress on piezoelectric, pyroelectric, and magnetoelectric polymer-based energy-harvesting devices, Energy Technol. 7 (2019) 1800852.

[6] G. Costanza, M.E. Tata, Shape memory alloys for aerospace, recent developments, and new applications: a short review, Materials 13 (2020) 1856.

[7] S. Zareie, A.S. Issa, R.J. Seethaler, et al., Recent advances in the applications of shape memory alloys in civil infrastructures: a review, Structures 27 (2020) 1535–1550.

[8] L. Sun, W.M. Huang, Z. Ding, et al., Stimulus-responsive shape memory materials: a review, Mater. Des. 33 (2012) 577–640.

[9] I. Stachiv, E. Alarcon, M. Lamac, Shape memory alloys and polymers for MEMS/NEMS applications: review on recent findings and challenges in design, preparation, and characterization, Metals 11 (2021) 415.

[10] M. Mehrpouya, H.C. Bidsorkhi, MEMS applications of NiTi based shape memory alloys: a review, Micro Nanosyst. 8 (2017) 79–91.

[11] H. Sun, M. Yin, W. Wei, et al., MEMS based energy harvesting for the Internet of Things: a survey, Microsyst. Technol. 24 (2018) 2853–2869.

[12] J. Iannacci, Microsystem based Energy Harvesting (EH-MEMS): powering pervasivity of the Internet of Things (IoT) – a review with focus on mechanical vibrations, J. King Saud. Univ. Sci. 31 (2019) 66–74.

[13] C. Covaci, A. Gontean, Piezoelectric energy harvesting solutions: a review, Sensors 20 (2020) 3512.

[14] N. Sezer, M. Koç, A comprehensive review on the state-of-the-art of piezoelectric energy harvesting, Nano Energy 80 (2021) 105567.

[15] A. Adeodato, B.T. Duarte, L.L.S. Monteiro, et al., Synergistic use of piezoelectric and shape memory alloy elements for vibration-based energy harvesting, Int. J. Mech. Sci. 194 (2021) 106206.

[16] S. Khanna, P. Marathey, Utsav, et al., Unravelling camphor mediated synthesis of TiO_2 nanorods over shape memory alloy for efficient energy harvesting, Appl. Surf. Sci. 541 (2021) 148489.
[17] Shape Memory Alloys, Boston, MA: Springer US. Epub ahead of print 2008. Available from: https://doi.org/10.1007/978-0-387-47685-8.
[18] Molecular dynamics simulation of shape-memory behavior, INTECH Open Access Publisher, http://www.intechopen.com/articles/show/title/molecular-dynamics-simulation-of-shape-memory-behavior, 2010 (accessed 17.09.19).
[19] Achieving ultra-large elastic strains in Nb thin films on NiTi phase-transforming substrate by the principle of lattice strain matching, Elsevier Enhanced Reader. Available from: https://doi.org/10.1016/j.matdes.2020.109257.
[20] A. Hubert, N.R. Calchand, Y.L. Gorrec, et al., Magnetic shape memory alloys as smart materials for micro-positioning devices. 2012. Epub ahead of print 2012. Available from: https://doi.org/10.7716/aem.v1i2.10.
[21] T. Sawaguchi, T. Maruyama, H. Otsuka, et al., Design concept and applications of Fe–Mn–Si-based alloys—from shape-memory to seismic response control, 2016. Epub ahead of print 2016. Available from: https://doi.org/10.2320/matertrans.mb201510.
[22] Z. Stošić, D. Manasijević, L. Balanović, et al., Effects of composition and thermal treatment of Cu-Al-Zn alloys with low content of al on their shape-memory properties, Mater. Res. 20 (2017) 1425–1431.
[23] S. Sathish, U.S. Mallik, T.N. Raju, Microstructure and shape memory effect of Cu-Zn-Ni shape memory alloys, J. Min. Mater. Charact. Eng. 2 (2014) 71–77.
[24] J. Mohd Jani, M. Leary, A. Subic, et al., A review of shape memory alloy research, applications and opportunities, Mater. Des. 1980–2015 56 (2014) 1078–1113.
[25] M.A. Jithin, L.G. Kolla, G.N.V.R. Vikram, et al., Pulsed DC magnetron sputtered titanium nitride thin films for localized heating applications in MEMS devices, Sens. Actuators Phys. 272 (2018) 199–205.
[26] S. Jayachandran, S.S. Mani Prabu, M. Manikandan, et al., Exploring the functional capabilities of NiTi shape memory alloy thin films deposited using electron beam evaporation technique, Vacuum 168 (2019) 0960-131726.
[27] A. Ishida, V. Martynov, Sputter-deposited shape-memory alloy thin films: properties and applications, MRS Bull. 27 (2002) 111–114.
[28] P. Šittner, R. Stalmans, Developing hybrid polymer composites with embedded shape-memory alloy wires, JOM 52 (2000) 15–20.
[29] M. Merlin, M. Scoponi, C. Soffritti, et al., On the improved adhesion of NiTi wires embedded in polyester and vinylester resins, Frat. (Ed.) Integrità Strutt. 9 (2014) 127–137.
[30] A. Cohades, V. Michaud, Shape memory alloys in fibre-reinforced polymer composites, Adv. Ind. Eng. Polym. Res. 1 (2018) 66–81.
[31] K. Neuking, A. Abu-Zarifa, G. Eggeler, Surface engineering of shape memory alloy/polymer-composites: improvement of the adhesion between polymers and pseudoelastic shape memory alloys, Mater. Sci. Eng. A (2008) 481–482. 606–611.
[32] D. Avirovik, A. Kumar, R.J. Bodnar, et al., Remote light energy harvesting and actuation using shape memory alloy—piezoelectric hybrid transducer, Smart Mater. Struct. 22 (2013) 052001.
[33] J.S. Gosliga, D.O.A. Ganilova, Energy harvesting based on the hybridisation of two smart materials, in: EACS 2016 – 6th European Conference on Structural Control, Sheffield, England: 11–13 July 2016.
[34] R. Xu, C. Bouby, H. Zahrouni, et al., 3D modeling of shape memory alloy fiber reinforced composites by multiscale finite element method, Compos. Struct. 200 (2018) 408–419.
[35] T. Merzouki, Finite Element analysis of a shape memory alloy actuator for a micropump, Simul. Model Pract. Theory 15 (2012).

[36] J. Chen, Q.C. Xu, M. Blaszkiewicz, et al., Lead zirconate titanate films on nickel-titanium shape memory alloys: SMARTIES, J. Am. Ceram. Soc. 75 (1992) 2891–2892.

[37] K. Sappati, S. Bhadra, Piezoelectric polymer and paper substrates: a review, Sensors 18 (2018) 3605.

[38] M. Safaei, H.A. Sodano, S.R. Anton, A review of energy harvesting using piezoelectric materials: state-of-the-art a decade later (2008–2018), Smart Mater. Struct. 28 (2019) 113001.

[39] X. Chen, X. Han, Q.-D. Shen, PVDF-based ferroelectric polymers in modern flexible electronics, Adv. Electron. Mater. 3 (2017) 1600460.

[40] C. Liu, Recent developments in polymer MEMS, Adv. Mater. 19 (2007) 3783–3790.

[41] M.T. Chorsi, E.J. Curry, H.T. Chorsi, et al., Piezoelectric biomaterials for sensors and actuators, Adv. Mater. 31 (2019) 1802084.

[42] Q. Jing, S. Kar-Narayan, Nanostructured polymer-based piezoelectric and triboelectric materials and devices for energy harvesting applications, J. Phys. Appl. Phys. 51 (2018) 303001.

[43] B. Gusarov, E. Gusarova, B. Viala, et al., Thermal energy harvesting by piezoelectric PVDF polymer coupled with shape memory alloy, Sens. Actuators Phys. 243 (2016) 175–181.

[44] R. Augustine, F. Sarry, N. Kalarikkal, et al., Surface acoustic wave device with reduced insertion loss by electrospinning P (VDF–TrFE)/ZnO nanocomposites, Nano-Micro Lett. 8 (2016) 282–290.

[45] R. Hadji, V.S. Nguyen, B. Vincent, et al., Preparation and characterization of P (VDF-TrFE)/Al_2 O_3 nanocomposite, IEEE Trans. Ultrason. Ferroelectr. Freq. Control. 59 (2012) 163–167.

[46] K.S. Ramadan, D. Sameoto, S. Evoy, A review of piezoelectric polymers as functional materials for electromechanical transducers, Smart Mater. Struct. 23 (2014) 033001.

[47] L.C. Brinson, One-dimensional constitutive behavior of shape memory alloys: thermomechanical derivation with non-constant material functions and redefined martensite internal variable, J. Intell. Mater. Syst. Struct. 4 (1993) 229–242.

[48] C. Liang, C.A. Rogers, One-dimensional thermomechanical constitutive relations for shape memory materials, J. Intell. Mater. Syst. Struct. 1 (1990) 207–234.

[49] C. Cisse, W. Zaki, T. Ben Zineb, A review of constitutive models and modeling techniques for shape memory alloys, Int. J. Plast. 76 (2016) 244–284.

[50] D.C. Lagoudas, Z. Bo, M.A. Qidwai, A unified thermodynamic constitutive model for SMA and finite element analysis of active metal matrix composites, Mech. Compos. Mater. Struct. 3 (1996) 153–179.

[51] E. Patoor, A. Eberhardt, M. Berveiller, Thermomechanical behavior of shape memory alloys, ESOMAT 1989—*Ist European Symposium on Martensitic Transformations in Science and Technology*, EDP Sciences, Bochum, Germany, 1989, pp. 133–140.

[52] R. Ahluwalia, T. Lookman, A. Saxena, et al., Landau theory for shape memory polycrystals, Acta Mater. (2004) 10.

[53] D.R. Mahapatra, R.V.N. Melnik, A dynamic model for phase transformations in 3d samples of shape memory alloys, in: V.S. Sunderam, G.D. van Albada, P.M.A. Sloot, et al. (Eds.), Computational Science – ICCS 2005, Springer, Berlin, Heidelberg, 2005, pp. 25–32.

[54] F. Falk, Ginzburg-Landau theory of static domain walls in shape-memory alloys, Z. Fr. Phys. B Condens. Matter 51 (1983) 177–185.

[55] F. Falk, Model free energy, mechanics, and thermodynamics of shape memory alloys, Acta Metall. 28 (1980) 1773–1780.

[56] R. Abeyaratne, J.K. Knowles, On the driving traction acting on a surface of strain discontinuity in a continuum, J. Mech. Phys. Solids 38 (1990) 345–360.

[57] J.M. Ball, R.D. James, Fine phase mixtures as minimizers of energy, in: S.S. Antman, H. Brezis, B.D. Coleman, et al. (Eds.), Analysis and Continuum Mechanics: A

Collection of Papers Dedicated to J. Serrin on His Sixtieth Birthday. Springer, Berlin, Heidelberg, pp. 647–686.

[58] G.R. Barsch, J.A. Krumhansl, Nonlinear and nonlocal continuum model of transformation precursors in martensites, Metall. Trans. A 19 (1988) 761–775.

[59] R.P. Dhote, R.N.V. Melnik, J. Zu, Dynamic multi-axial behavior of shape memory alloy nanowires with coupled thermo-mechanical phase-field models, Meccanica 49 (2014) 1561–1575.

[60] V.I. Levitas, D.L. Preston, Three-dimensional Landau theory for multivariant stress-induced martensitic phase transformations. I. Austenite $\leftrightarrow$ Martensite, Phys. Rev. B 66 (2002) 134206.

[61] V.I. Levitas, D.L. Preston, Three-dimensional Landau theory for multivariant stress-induced martensitic phase transformations. II. Multivariant phase transform. stress. space analysis, Phys. Rev. B 66 (2002) 134207.

[62] L.X. Wang, R.V.N. Melnik, Thermo-mechanical wave propagations in shape memory alloy rod with phase transformations, Mech. Adv. Mater. Struct. 14 (2007) 665–676.

[63] B.J. Alder, T.E. Wainwright, Studies in molecular dynamics. I. General method, J. Chem. Phys. 31 (1959) 459–466.

[64] A. Rahman, Correlations in the motion of atoms in liquid argon, Phys. Rev. 136 (1964) A405–A411.

[65] M. Parrinello, A. Rahman, Polymorphic transitions in single crystals: a new molecular dynamics method, J. Appl. Phys. 52 (1981) 7182–7190.

[66] S. Ozgen, O. Adiguzel, Investigation of the thermoelastic phase transformation in a NiAl alloy by molecular dynamics simulation, J. Phys. Chem. Solids 65 (2004) 861–865.

[67] T. Suzuki, M. Shimono, A simple model for martensitic transformation, J. Phys. IV Proc. 112 (2003) 129–132.

[68] M.S. Daw, M.I. Baskes, Embedded-atom method: derivation and application to impurities, surfaces, and other defects in metals, Phys. Rev. B 29 (1984) 6443–6453.

[69] P. Blanc, C. Lexcellent, Micromechanical modelling of a CuAlNi shape memory alloy behaviour, Mater. Sci. Eng. A 378 (2004) 465–469.

[70] M. Huang, X. Gao, L.C. Brinson, A multivariant micromechanical model for SMAs Part 2. Polycrystal model, Int. J. Plast. 16 (2000) 1371–1390.

[71] V.I. Levitas, I.B. Ozsoy, Micromechanical modeling of stress-induced phase transformations. Part 1. Thermodynamics and kinetics of coupled interface propagation and reorientation, Int. J. Plast. 25 (2009) 239–280.

[72] A. Sadjadpour, K. Bhattacharya, A micromechanics inspired constitutive model for shape-memory alloys: the one-dimensional case, Smart Mater. Struct. 16 (2007) S51–S62.

[73] Q.P. Sun, K.C. Hwang, Micromechanics modelling for the constitutive behavior of polycrystalline shape memory alloys—I. Derivation of general relations, J. Mech. Phys. Solids 41 (1993) 1–17.

[74] Y. Kudoh, M. Tokonami, S. Miyazaki, et al., Crystal structure of the martensite in Ti-49.2 at.%Ni alloy analyzed by the single crystal X-ray diffraction method, Acta Metall. 33 (1985) 2049–2056.

[75] Q.P. Sun, K.C. Hwang, S.W. Yu, A micromechanics constitutive model of transformation plasticity with shear and dilatation effect, J. Mech. Phys. Solids 39 (1991) 507–524.

[76] X.M. Wang, B.X. Xu, Z.F. Yue, Micromechanical modelling of the effect of plastic deformation on the mechanical behaviour in pseudoelastic shape memory alloys, Int. J. Plast. 24 (2008) 1307–1332.

[77] C. Yu, G. Kang, D. Song, et al., Micromechanical constitutive model considering plasticity for super-elastic NiTi shape memory alloy, Comput. Mater. Sci. 56 (2012) 1–5.

[78] Z.K. Lu, G.J. Weng, A self-consistent model for the stress–strain behavior of shape-memory alloy polycrystals, Acta Mater. 46 (1998) 5423–5433.

[79] P. Šittner, V. Novák, Anisotropy of martensitic transformations in modeling of shape memory alloy polycrystals, Int. J. Plast. 16 (2000) 1243–1268.

[80] P. Thamburaja, L. Anand, Polycrystalline shape-memory materials: effect of crystallographic texture, J. Mech. Phys. Solids 49 (2001) 709–737.

[81] Batdorf: A mathematical theory of plasticity based. Google Scholar, <https://scholar.google.com/scholar_lookup?title = A%C2%A0Mathematical%20Theory%20of%20Plasticity%20Based%20on%20the%20Concept%20of%20Slip&publication_year = 1949&author = S.B.%20Batdorf&author = B.%20Budiansky> (accessed 8.04.21).

[82] Bažant: Microplane model for strain controlled inelastic. Google Scholar, <https://scholar.google.com/scholar_lookup?title = Microplane%20Model%20for%20Strain%20Controlled%20Inelastic%20Behaviour&publication_year = 1984&author = Z.%20Ba%C5%BEant> (accessed 8.04.21).

[83] C. Collard, T. Ben Zineb, Simulation of the effect of elastic precipitates in SMA materials based on a micromechanical model, Compos. Part. B Eng. 43 (2012) 2560–2576.

[84] C. Niclaeys, T. Ben Zineb, S. Arbab-Chirani, et al., Determination of the interaction energy in the martensitic state, Int. J. Plast. 18 (2002) 1619–1647.

[85] K. Tanaka, S. Nagaki, A thermomechanical description of materials with internal variables in the process of phase transitions, Ing-Arch 51 (1982) 287–299.

[86] F. Auricchio, R.L. Taylor, Shape-memory alloys: modelling and numerical simulations of the finite-strain superelastic behavior, Comput. Methods Appl. Mech. Eng. 143 (1997) 175–194.

[87] A. Bertram, Thermo-mechanical constitutive equations for the description of shape memory effects in alloys, Nucl. Eng. Des. 74 (1983) 173–182.

[88] B. Raniecki, C. Lexcellent, K. Tanaka, Thermodynamic models of pseudoelastic behaviour of shape memory alloys, Memory Alloy (1992) 24.

[89] B. Raniecki, C. Lexcellent, RL-models of pseudoelasticity and their specification for some shape memory solids, Eur. J. Mech. e A/Solids 13 (1994) 21–50.

[90] B. Raniecki, Ch Lexcellent, Thermodynamics of isotropic pseudoelasticity in shape memory alloys, Eur. J. Mech. A Solids 17 (1998) 185–205.

[91] A.C. Souza, E.N. Mamiya, N. Zouain, Three-dimensional model for solids undergoing stress-induced phase transformations, Eur. J. Mech. ASolids 17 (1998) 789–806.

[92] F. Auricchio, L. Petrini, Improvements and algorithmical considerations on a recent three-dimensional model describing stress-induced solid phase transformations, Int. J. Numer. Methods Eng. 55 (2002) 1255–1284.

[93] C. Bouvet, S. Calloch, C. Lexcellent, A phenomenological model for pseudoelasticity of shape memory alloys under multiaxial proportional and nonproportional loadings, Eur. J. Mech. ASolids 23 (2004) 37–61.

[94] V. Evangelista, S. Marfia, E. Sacco, Phenomenological 3D and 1D consistent models for shape-memory alloy materials, Comput. Mech. 44 (2009) 405.

[95] C. Lexcellent, M.L. Boubakar, C. Bouvet, S. Calloch, About modelling the shape memory alloy behaviour based on the phase transformation surface identification under proportional loading and anisothermal conditions, Int. J. Solids Struct. 43 (2006) 613–626.

[96] L. Saint-Sulpice, S. Arbab-Chirani, S. Calloch, Thermomechanical cyclic behavior modeling of Cu-Al-Be SMA materials and structures, Int. J. Solids Struct. 49 (2012) 0960-1317–1102.

[97] F. Thiebaud, C. Lexcellent, M. Collet, et al., Implementation of a model taking into account the asymmetry between tension and compression, the temperature effects in

a finite element code for shape memory alloys structures calculations, Comput. Mater. Sci. 41 (2007) 208–221.

[98] J.G. Boyd, D.C. Lagoudas, A thermodynamical constitutive model for shape memory materials. Part I. The monolithic shape memory alloy, Int. J. Plast. 12 (1996) 805–842.

[99] Y. Chemisky, A. Duval, E. Patoor, et al., Constitutive model for shape memory alloys including phase transformation, martensitic reorientation and twins accommodation, Mech. Mater. 43 (2011) 361–376.

[100] S. Leclercq, C. Lexcellent, A general macroscopic description of the thermomechanical behavior of shape memory alloys, J. Mech. Phys. Solids 44 (1996) 953–980.

[101] Z. Moumni, W. Zaki, Q.S. Nguyen, Theoretical and numerical modeling of solid–solid phase change: application to the description of the thermomechanical behavior of shape memory alloys, Int. J. Plast. 24 (2008) 614–645.

[102] B. Peultier, T. Ben Zineb, E. Patoor, Macroscopic constitutive law of shape memory alloy thermomechanical behaviour. Application to structure computation by FEM, Mech. Mater. 38 (2006) 510–524.

[103] F. Auricchio, S. Marfia, E. Sacco, Modelling of SMA materials: training and two way memory effects, Comput. Struct. 81 (2003) 2301–2317.

[104] A. Duval, M. Haboussi, T. Ben Zineb, Modelling of localization and propagation of phase transformation in superelastic SMA by a gradient nonlocal approach, Int. J. Solids Struct. 48 (2011) 1879–1893.

[105] S. Hazar, W. Zaki, Z. Moumni, et al., Steady state crack growth in shape memory alloys. American society of mechanical engineers digital collection, Epub ahead of print (2014). Available from: https://doi.org/10.1115/SMASIS2013-3071. 20 February.

[106] C. Morin, A comprehensive approach for fatigue analysis of shape memory alloys. PhD thesis, Ecole Polytechnique X, <https://pastel.archives-ouvertes.fr/pastel-00608205>, 2011, (accessed 8.04.21).

[107] A.P. Stebner, L.C. Brinson, Explicit finite element implementation of an improved three dimensional constitutive model for shape memory alloys, Comput. Methods Appl. Mech. Eng. 257 (2013) 17–35.

[108] W. Zaki, An efficient implementation for a model of martensite reorientation in martensitic shape memory alloys under multiaxial nonproportional loading, Int. J. Plast. 37 (2012) 72–94.

[109] L. Anand, M.E. Gurtin, Thermal effects in the superelasticity of crystalline shape-memory materials, J. Mech. Phys. Solids 51 (2003) 1015–1058.

[110] J. Arghavani, F. Auricchio, R. Naghdabadi, et al., An improved, fully symmetric, finite-strain phenomenological constitutive model for shape memory alloys, Finite Elem. Anal. Des. 47 (2011) 166–174.

[111] J. Arghavani, F. Auricchio, R. Naghdabadi, et al., On the robustness and efficiency of integration algorithms for a 3D finite strain phenomenological SMA constitutive model, Int. J. Numer. Methods Eng. 85 (2011) 107–134.

[112] D. Christ, S. Reese, A finite element model for shape memory alloys considering thermomechanical couplings at large strains, Int. J. Solids Struct. 46 (2009) 3694–3709.

[113] C.H. Müller, O.T. Bruhns, A thermodynamic finite-strain model for pseudoelastic shape memory alloys, Int. J. Plast. 22 (2006) 1658–1682.

[114] S. Reese, D. Christ, Finite deformation pseudo-elasticity of shape memory alloys – Constitutive modelling and finite element implementation, Int. J. Plast. 24 (2008) 455–482.

[115] S. Stupkiewicz, H. Petryk, Finite-strain micromechanical model of stress-induced martensitic transformations in shape memory alloys, Mater. Sci. Eng. A 438–440 (2006) 126–130.

[116] P. Thamburaja, A finite-deformation-based phenomenological theory for shape-memory alloys, Int. J. Plast. 26 (2010) 1195–1219.
[117] A. Ziółkowski, Three-dimensional phenomenological thermodynamic model of pseudoelasticity of shape memory alloys at finite strains, Contin. Mech. Thermodyn. 19 (2007) 379–398.
[118] A. Bhattacharyya, D.C. Lagoudas, A stochastic thermodynamic model for the gradual thermal transformation of SMA polycrystals, Smart Mater. Struct. 6 (1997) 235–250.
[119] M. Fischlschweiger, E.R. Oberaigner, Kinetics and rates of martensitic phase transformation based on statistical physics, Comput. Mater. Sci. 52 (2012) 189–192.
[120] M.L. Bellac, M.L. Bellac, F. Mortessagne, et al., Equilibrium and Non-Equilibrium Statistical Thermodynamics, Cambridge University Press, 2004.
[121] I. Müller, Pseudo-elastic hysteresis in shape memory alloys, Phys. B Condens. Matter 407 (2012) 1314–1315.
[122] I. Müller, S. Seelecke, Thermodynamic aspects of shape memory alloys, Math. Comput. Model. 34 (2001) 1307–1355.
[123] Y. Wu, A. Badel, F. Formosa, et al., Self-powered optimized synchronous electric charge extraction circuit for piezoelectric energy harvesting, J. Intell. Mater. Syst. Struct. 25 (2014) 2165–2176.
[124] M. Kamlah, Ferroelectric and ferroelastic piezoceramics – modeling of electromechanical hysteresis phenomena, Contin. Mech. Thermodyn. 13 (2001) 219–268.
[125] P.J. Chen, Three dimensional dynamic electromechanical constitutive relations for ferroelectric materials, Int. J. Solids Struct. 16 (1980) 1059–1067.
[126] P.J. Chen, S.T. Montgomery, A macroscopic theory for the existence of the hysteresis and butterfly loops in ferroelectricity, Ferroelectrics 23 (1980) 199–207.
[127] E. Bassiouny, A.F. Ghaleb, G.A. Maugin, Thermodynamical formulation for coupled electromechanical hysteresis effects—I. Basic equations, Int. J. Eng. Sci. 26 (1988) 1279–1295.
[128] E. Bassiouny, G.A. Maugin, Thermodynamical formulation for coupled electromechanical hysteresis effects—III. Parameter identification, Int. J. Eng. Sci. 27 (1989) 975–987.
[129] A.C.F. Cocks, R.M. Mcmeeking, A phenomenological constitutive law for the behaviour of ferroelectric ceramics, Ferroelectrics 228 (1999) 219–228.
[130] M. Elhadrouz, T.B. Zineb, E. Patoor, Constitutive law for ferroelectric and ferroelastic single crystals: a micromechanical approach, Comput. Mater. Sci. 32 (2005) 355–359.
[131] M. Elhadrouz, T.B. Zineb, E. Patoor, Constitutive law for ferroelastic and ferroelectric piezoceramics, J. Intell. Mater. Syst. Struct. 16 (2005) 221–236.
[132] M. Kamlah, U. Böhle, Finite element analysis of piezoceramic components taking into account ferroelectric hysteresis behavior, Int. J. Solids Struct. 38 (2001) 605–633.
[133] M. Kamlah, C. Tsakmakis, Phenomenological modeling of the non-linear electromechanical coupling in ferroelectrics fn2, Int. J. Solids Struct. 36 (1999) 669–695.
[134] S. Klinkel, A phenomenological constitutive model for ferroelastic and ferroelectric hysteresis effects in ferroelectric ceramics, Int. J. Solids Struct. 43 (2006) 7197–7222.
[135] R.M. McMeeking, C.M. Landis, A phenomenological multi-axial constitutive law for switching in polycrystalline ferroelectric ceramics, Int. J. Eng. Sci. 40 (2002) 1553–1577.
[136] W. Zouari, T. Ben Zineb, A. Benjeddou, A FSDT—MITC piezoelectric shell finite element with ferroelectric non-linearity, J. Intell. Mater. Syst. Struct. 20 (2009) 2055–2075.
[137] S.C. Hwang, J.E. Huber, R.M. McMeeking, et al., The simulation of switching in polycrystalline ferroelectric ceramics, J. Appl. Phys. 84 (1998) 1530–1540.

[138] S.C. Hwang, R.M. McMeeking, A finite element model of ferroelectric polycrystals, Ferroelectrics 211 (1998) 177–194.

[139] S.C. Hwang, R.M. McMeeking, A finite element model of ferroelastic polycrystals, Int. J. Solids Struct. 36 (1999) 1541–1556.

[140] R. Jayendiran, Finite element analysis of switching domains using ferroelectric and ferroelastic micromechanical model for single crystal piezoceramics, Ceram. Int. 15 (2016).

[141] M. Kamlah, A.C. Liskowsky, R.M. McMeeking, et al., Finite element simulation of a polycrystalline ferroelectric based on a multidomain single crystal switching model, Int. J. Solids Struct. 42 (2005) 2949–2964.

[142] R.M. Mcmeeking, S.C. Hwang, On the potential energy of a piezoelectric inclusion and the criterion for ferroelectric switching, Ferroelectrics 200 (1997) 151–173.

[143] T. Michelitsch, W.S. Kreher, A simple model for the nonlinear material behavior of ferroelectrics, Acta Mater. 46 (1998) 5085–5094.

[144] C.R. Knick, D.J. Sharar, A.A. Wilson, et al., High frequency, low power, electrically actuated shape memory alloy MEMS bimorph thermal actuators, J. Micromech. Microeng. 29 (2019) 075005.

[145] J. Joseph, M. Ohtsuka, H. Miki, et al., Upscaling of thermomagnetic generators based on Heusler alloy films, Joule 4 (2020) 2718–2732.

[146] S. Rouabah, B. Vincent, A. Chaabi, et al., Brillouin spectroscopy study of the radiation resistance of ferroelectric copolymer filled with Al_2 O_3 and ZnO nanoparticles, Eur. Phys. J. Appl. Phys. 77 (2017) 30701.

[147] D. Zakharov, B. Gusarov, E. Gusarova, et al., Combined pyroelectric, piezoelectric and shape memory effects for thermal energy harvesting, J. Phys. Conf. Ser. 476 (2013) 012021.

[148] A. Rami Reddy, M. Umapathy, D. Ezhilarasi, et al., Piezoelectric energy harvester with shape memory alloy actuator using solar energy, IEEE Trans. Sustain. Energy 6 (2015) 1409–1415.

[149] H. Tan, Y. Huang, C. Liu, et al., The Mori–Tanaka method for composite materials with nonlinear interface debonding, Int. J. Plast. 21 (2005) 1890–1918.

[150] G.A. Lebedev, B.V. Gusarov, B. Viala et al., Thermal energy harvesting using shape memory/piezoelectric composites, in: 2011 16th International Solid-State Sensors, Actuators and Microsystems Conference, Beijing, China, IEEE, pp. 669–670.

[151] D. Zakharov, G. Lebedev, O. Cugat, et al., Thermal energy conversion by coupled shape memory and piezoelectric effects, J. Micromech. Microeng. 22 (2012) 094005.

[152] O.C. Namli, M. Taya, Design of piezo-SMA composite for thermal energy harvester under fluctuating temperature, J. Appl. Mech. 78 (2011) 031001.

[153] H. Sato, Multifunctional devices combining shape-memory alloy and piezoelectric materials, in: N.C. Goulbourne, H.E. Naguib (Eds.), San Diego, CA, p. 90581I.

[154] H. Radousky, F. Qian, Y. An, et al., Harvesting mechanical and thermal energy by combining ZnO Nanowires and NiTi shape. Memory Alloy 2017;8.

[155] M. Dietze, M. Es-Souni, Structural and functional properties of screen-printed PZT–PVDF-TrFE composites, Sens. Actuators Phys. 143 (2008) 329–334.

[156] N. Neumann, R. Köhler, G. Hofmann, Pyroelectric thin film sensors and arrays based on P (VDF/TrFE), Integr. Ferroelectr. 6 (1995) 213–230.

[157] D.C. Lagoudas, Z. Bo, The cylindrical bending of composite plates with piezoelectric and SMA layers, Smart Mater. Struct. 3 (1994) 309–317.

[158] B. Lester, T. Baxevanis, Y. Chemisky, et al., Review and perspectives: shape memory alloy composite systems, 62.

[159] J.-K. Lee, M. Taya, Modeling for piezoelectric-shape memory alloy composites, Arch. Appl. Mech. 81 (2011) 629–640.

[160] O.C. Namli, J.-K. Lee, M. Taya, Modeling of piezo-SMA composites for thermal energy harvester, Behavior and Mechanics of Multifunctional and Composite Materials, International Society for Optics and Photonics, 2007, p. 65261L.
[161] X. Shen, J.-H. Han, I. Lee, Study of a reduced and internally biased oxide wafer PZT actuator and its integration with shape memory alloy 6.
[162] E. Garcia, N. Lobontiu, Induced-strain multimorphs for microscale sensory actuation design 9.
[163] A. Badel, D. Guyomar, E. Lefeuvre, et al., Piezoelectric energy harvesting using a synchronized switch technique, J. Intell. Mater. Syst. Struct. 17 (2006) 831–839.
[164] Y. Zhao, K. Wang, M. Guan, An adaptive boost converter for low voltage piezoelectric energy harvesting, Ferroelectrics 502 (2016) 107–118.
[165] P. Xolin, C. Collard, M. Engels-Deutsch, et al., Finite element and experimental structural analysis of endodontic rotary file made of Cu-based single crystal SMA considering a micromechanical behavior model, Int. J. Solids Struct. (2021). S0020768321000238.

CHAPTER

10

Designing piezo- and pyroelectric energy harvesters

Sujoy Kumar Ghosh[1], *Luana Persano*[1] *and Biswajit Mahanty*[2]

[1]Laboratorio NEST, Istituto Nanoscienze-CNR and Scuola Normale Superiore, Pisa, Italy, [2]Department of Electronics & Communication Engineering, Ramgarh Engineering College, Ramgarh, India

10.1 Introduction

In the modern lifestyle, the usefulness of Internet of Things (IoT) is increasing rapidly on a daily basis, especially in the area of information and communication technologies (ICTs) where billions of electronic gadgets and sensors are interconnected through the Internet. More important, in the IoT-based healthcare monitoring, sensors are integrated into our regularly used textiles or can be attached to the patient's skin for real-time monitoring of vital signs and physiological signal [1]. All these gadgets and most of the sensors, require constant power supply to operate continuously. In this regard, rechargeable batteries can play an important role. Batteries strongly contributed to the development of modern technologies, enabling the proliferation of portable devices and electric vehicles, with substantial benefits for society. However, the rapid growth of such technologies, which are effectively dependent on batteries, has highlighted the ethical and environmental issues associated with the use of mineral resources for their manufacture and the issues associated with safe use and nonhazardous disposal. Indeed, only a small fraction of batteries are recycled, thus further exacerbating the global supply of strategic element materials. A potential alternative is offered by the development of green, renewable and nonexhaustible

Engineered Polymer Nanocomposites for Energy Harvesting Applications
DOI: https://doi.org/10.1016/B978-0-12-824155-4.00003-X

energy sources. Nanogenerators (NGs), and in particular piezoelectric and pyroelectric NGs, which convert mechanical and thermal energy respectively into electricity, are very promising candidates to achieve these goals [2]. Indeed, they can harvest energy from widely available green environmental energy sources such as acoustic signals, kinetic energy, mechanical energy, and temperature gradient. In particular, the kinetic energy from different vibrations is ubiquitous in our surroundings and it includes different human actions such as running, walking, muscle vibrations, heartbeat, respiration, and structural vibrations such as those related to transport of vehicles, fluid flows, wind, and ocean motions. Furthermore, heat waste from industry and heat generated by mechanical friction among dissimilar surfaces are also widely available sources of energy. An additional advantage is represented by the capability of NGs to power up very small, low-power (μW-mW range) electronic gadgets, without the need for maintenance and the use of highly toxic chemicals and metals. Furthermore, the piezo- and pyroelectric-based devices could be used in a wide range of advanced nano-/microscopic functional systems, such as sensors, actuators, transistors, nonvolatile memory, capacitor, thermal infrared switch, and flexible displays [3], with application fields spanning from self-powered catalysis [4] to autonomous surface-enhanced Raman scattering [5], from temperature stimulated hydrogen generation [6] to pollutants degradation [7] and disinfection of bacteria [8], from implantable electronics to [9] drug delivery [10], to name a few (Fig. 10.1). To date, despite a brittle (low toughness) and rigid (high stiffness) nature, a large number of inorganic and lead-based materials have been used to fabricate piezo- and pyroelectric NGs. On the other hand, the organic piezoelectric materials, for example, poly(vinylidene fluoride) (PVDF)-$(CH_2$-$CF_2)n$- and its copolymers are suitable alternatives to the lead-based materials. These organic piezoelectric materials are getting more popular due to their reduced cost, biocompatibility, flexibility, durability, chemical stability, and high piezoelectric coefficient, although their piezoelectric coefficients are still lower when compared with inorganic materials.

This chapter will discuss recent developments in piezoelectric, pyroelectric, and combined piezopyro-NGs with focus on innovative devices and material design strategies. A final discussion on the importance of further developments for the design of the future electronics based on them is also given.

10.2 Piezoelectric nanogenerator

The name piezoelectricity is derived from the Greek word "piezein," which means to press or squeeze. When a piezoelectric material is under

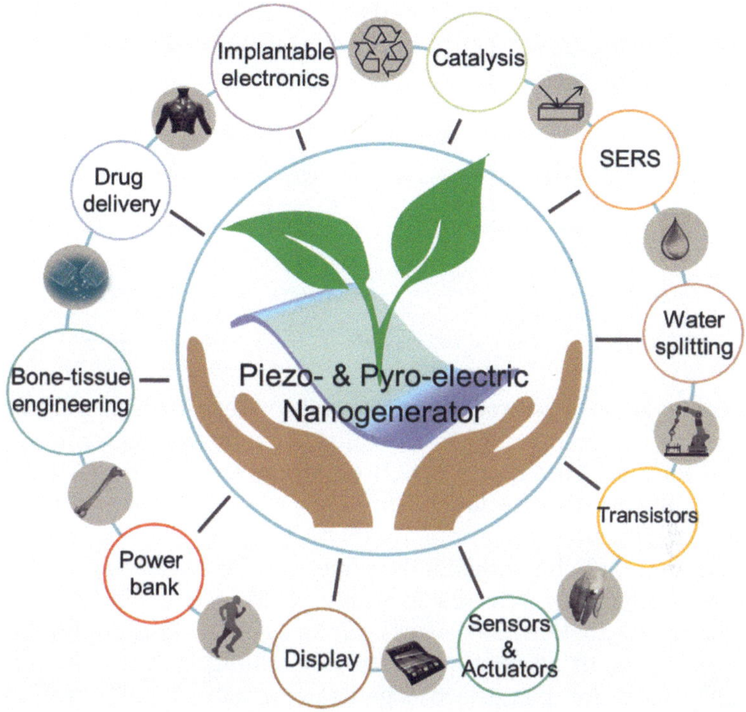

FIGURE 10.1 Fields of application of piezo- and pyroelectric nanogenerators for self-powered electronics.

stress, the electric polarization is achieved within the material. This phenomenon is called the piezoelectric effect which was first observed by Pierre and Jacques Curie in 1880 in a single crystal of quartz. In general there are two types of piezoelectricity, the "direct piezoelectric effect," where an external mechanical stress applied to the material produces an electrical signal, and the "converse piezoelectric effect," where an external electrical signal applied to the material induces a mechanical deformation (Fig. 10.2). Piezoelectric materials are usually noncentrosymmetric crystals, where an equal number of positive and negative charges is the result of charge neutrality existing across the virtual polar axis and is called balanced equilibrium condition. When an external mechanical stress or vibrations is applied to the material, the material is deformed, the equilibrium condition fails and charges are generated across the material causing the formation of an electrical dipole. Finally the formed surface charges can be collected by proper electrodes.

Most piezoelectric materials of technological importance for energy-harvesting possess a well-defined polar axis, and the energy-harvesting performance depends on the direction of the applied strain relative to

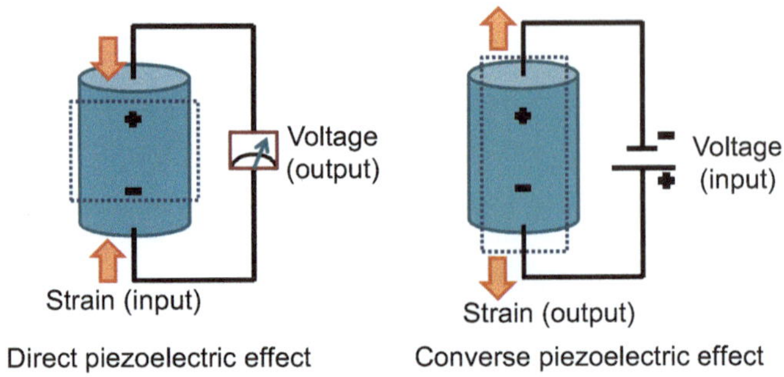

FIGURE 10.2 Scheme of the piezoelectric effect (direct & converse effect).

this polar axis. In a piezoelectric ceramic or ferroelectric polymer the polar axis is the poling direction, whilst for nonferroelectric crystalline materials such as zinc oxide (ZnO) or aluminum nitride (AlN) this is defined by the crystal orientation. In these cases, the polar axis is referred to as the "3" direction.

By symmetry, all directions in the plane at right angles to the polar axis are equivalent and are referred to as the "1" direction. This is typical of most ceramic piezoelectrics. A stress can be applied either in the direction of the polar axis, or at right angles to it, resulting in two configurations commonly used for piezoelectric generators, named "33" and "31" in Fig. 10.3. Other configurations are possible, and the situation is more complex for materials with lower symmetry, but these two configurations cover most situations of practical piezoelectric energy harvesters [11]. The materials selection and structural design are the fundamental key factors for the development of piezoelectric nanogenerators (PENGs).

In the following, PENGs have been classified into three main classes depending on the typology of piezoelectric material used: (1) inorganic piezoelectric materials; (2) organic piezoelectric materials; and (3) biodegradable piezoelectric materials

10.2.1 Inorganic piezoelectric materials

10.2.1.1 Zinc oxide nanowires-based piezoelectric nanogenerators

In 2006, Prof. Z. L. Wang from Georgia Tech reported the first PENGs based on ZnO NWs [12]. The mechanism of the power generator relies on the coupling of piezoelectric and semiconducting properties of ZnO as well as the formation of a Schottky barrier (Fig. 10.4A (i)). Among the known one-dimensional (1D) nanomaterials, ZnO has several advantages. For example, benefitting from the both piezoelectric and semiconducting properties of ZnO, the electromechanically coupled sensors and transducers were realized. In addition, it can be used in different biomedical

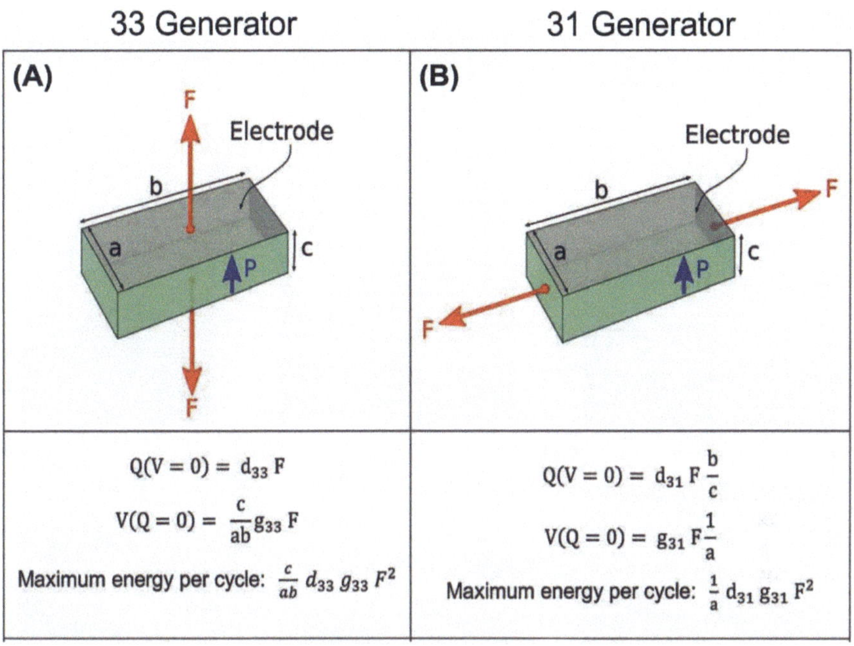

FIGURE 10.3 Conventional energy-harvesting configuration of piezoelectric nanogenerator in (A) longitudinal (33) mode and (B) transverse (31) mode, where in the figure Q represents the generated piezoelectric charge, V corresponds to the generated piezoelectric voltage, F is the applied compressive force, P is direction of polarization, g_{xx} and d_{xx} are the piezoelectric voltage and charge coefficients, respectively. Source: *Reproduced from C.R. Bowen, H.A. Kim, P.M. Weaver, S. Dunn, Energy Environ. Sci. 7 (2014) 25 with permission by Publisher.*

applications due to its biocompatible property. Furthermore, several nanostructures formation capability of ZnO could be used in different application fields according to the need of the platform.

In the first demonstration, aligned vertical ZnO NWs were grown on *c* plane–oriented α-Al_2O_3 substrates, using Au particles as a catalyst, by a vapor liquid-solid (VLS) process (Fig. 10.4A (ii)). NWs exhibit a length of 0.2–0.5 μm and the generated voltage was in the range of 6–9 mV. Later, Yang et al. prepared an array of laterally packaged ZnO NWs by using the same physical vapor deposition method and obtaining NWs with diameters of 3–5 μm and lengths of 200–300 μm [19]. The PENG was prepared by contacting the two opposite sides of a single ZnO NW, which was positioned on a flexible substrate. The generated open circuit output voltage (V_{oc}) and short circuit current (I_{sc}) were 20–50 mV and 400–750 pA respectively under continuous bending – releasing strain of 0.05%–0.1%. However, the power output was not enough to drive a real device. Practical application was demonstrated by the integration of

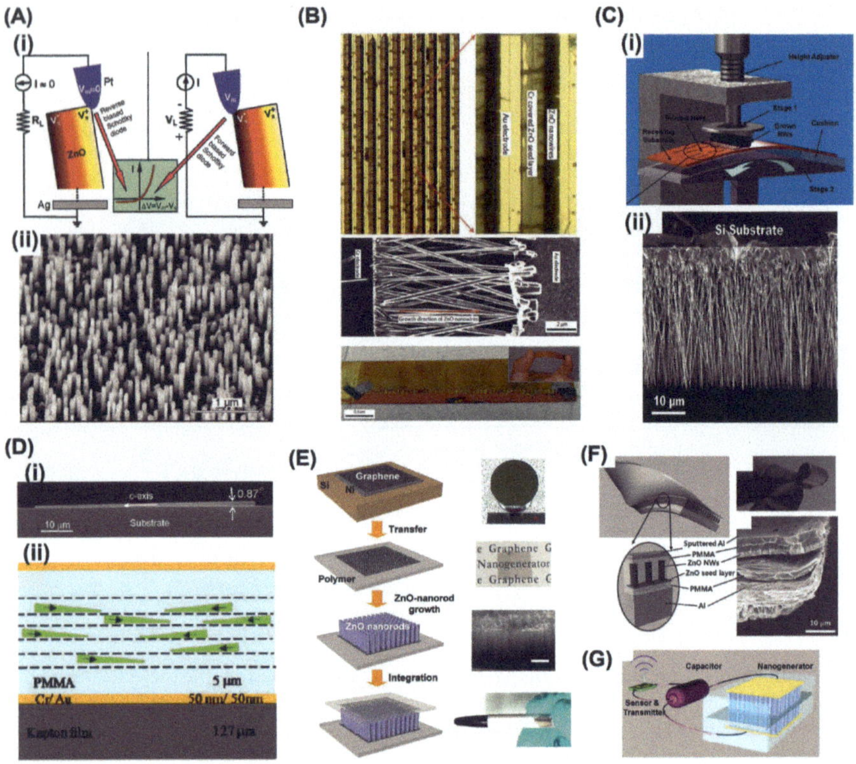

FIGURE 10.4 ZnO NWS array-based PENGs. (A) (i) Power generation mechanism of ZnO NWs under an AFM tip and (ii) Scanning electron microscopy (SEM) micrographs of aligned ZnO NWs grown on α-Al_2O_3 substrate (Source: Reproduced with permission Z.L. Wang, J. Song, Science 312 (2006) 242 [12]). (B) The upper panel shows optical microscopy image of the laterally arranged ZnO nanowire arrays-based PENG, middle panel shows the SEM image of the same PENG and lower panel shows digital photograph of the PENG including flexibility demonstration in the inset (Source: Reproduced with permission S. Xu, Y. Qin, C. Xu, Y. Wei, R. Yang, Z.L. Wang, Nat. Nanotechnol. 5 (2010) 366 [13]). (C) (i) The schematic of the experiment used in the sweeping-printing-method for vertically grown ZnO NWs transfer to flexible substrate and (ii) SEM image of the ZnO NWs grown on Si substrate (Source: Reproduced with permission G. Zhu, R. Yang, S. Wang, Z. L. Wang, Nano Lett. 10 (2010) 3151 [14]). (D) (i) SEM image of the conical shape ZnO nanowire on a flat surface and (ii) schematically represented PMMA/conical shape ZnO NWs composite-based device (Source: Reproduced with permission Y. Hu, Y. Zhang, C. Xu, G. Zhu, Z.L. Wang, Nano Lett. 10 (2010) 5025 [15]). (E) (i) Transparent graphene electrode integrated vertically grown ZnO nanorod-based rollable PENG.(Source: Reproduced with permission D. Choi, M.-Y. Choi, W.M. Choi, H.-J. Shin, H.-K. Park, J.-S. Seo, et al., Adv. Mater. 22 (2010) 2187 [16]). (F) Device schematic and representative images of the super flexible ZnO-based PENG prepared on Al-foil (Source: Reproduced with permission S. Lee, S.-H. Bae, L. Lin, Y. Yang, C. Park, S.-W. Kim, et al., Adv. Funct. Mater 23 (2013) 2445 [17]). (G) Schematic demonstration of an self-powered wireless system powered by ZnO NWs PENG. (Source: Reproduced with permission Y. Hu, Y. Zhang, C. Xu, L. Lin, R. L. Snyder, Z.L. Wang, Nano Lett. 11 (2011) 2572 [18]).

700 rows of ZnO NWs (Fig. 10.4B) which produced a peak voltage of 1.26 V and a power density of 2.7 mW/cm^3 at a strain of 0.19%. The realized array was potentially sufficient to power a nanowire pH sensor and a nanowire UV sensor [13]. In order to meet the demand for high power throughput, further improvements were achieved by the use of a scalable sweeping-printing-method (Fig. 10.4C (i)) where, vertically aligned ZnO NWs (Fig. 10.4C (ii)) were transferred to a receiving substrate to form horizontally aligned arrays and then, parallel stripe type of electrodes were deposited to connect all of the NWs together [14]. With this configuration, the PENG generated $V_{oc} \sim 2.03$ V and a maximum output power density of 11 mW/cm^3 which was sufficient to charge up few capacitors after full wave rectification and to turn on commercial light emitting diode (LED). Further design of PENG was demonstrated by using as-grown ZnO NWs with conical shape (Fig. 10.4D (i)) [15]. A nanogenerator was fabricated by simply dispersing the NWs into a poly (methyl methacrylate) (PMMA) film to form a composite structure (Fig. 10.4D (ii)). The conical geometry induced unipolar assembly of the nanowires in such a composite structure results in the formation of a macroscopic piezoelectric potential across the film upon the application of a mechanical deformation. By applying a compressive strain of 0.11% at a rate of 3.67% s^{-1}, an output voltage up to 2 V (equivalent open circuit voltage of 3.3 V) was generated. However, such devices show poor flexibility and mechanical durability which limits their lifetime and the device performances. This problem was addressed by fabricating heterogeneous 3D nanostructure consisting of vertically aligned 1D ZnO nanorods on a 2D transparent graphene electrode (Fig. 10.4E) [16]. ZnO nanorods where directly synthesized on graphene by using an aqueous solution method at 95°C. The device was then rolled up several times and was capable to generate an output current density of 2 $\mu A/cm^2$. An alternative approach, consists in the fabrication of a super flexible PENG based on 18 μm thick Al foil that acts as both substrate and electrode (Fig. 10.4F) [17]. A 2 μm thick layer of PMMA was coated on the Al foil prior to the growth of the ZnO NWs by a hydrothermal process at 95°C for 5 hours. Finally, a 3 μm thick layer of PMMA was coated on the surface of as-grown ZnO NWs before the deposition of a 40 nm thick Al film, by sputtering. The total thickness of the device was about 25 μm. This device could be used as an energy harvester/scavenger ($V_{oc} \sim 0.2$ V and $I_{sc} \sim 2$ nA) but also as an effective self-powered sensor for detecting wrinkling of a human face. In another work, a self-powered wireless system was powered by a PENG composed of ZnO NWs [18]. The free standing PENG was used to store the generated energy into capacitor after signal rectification which was further used to power up sensor and transmitter. The system transmitted the signal up to 5 − 10 m.

10.2.1.2 Polycrystalline lead zirconate titanate-based piezoelectric nanogenerators

Polycrystalline lead zirconate titanate (PZT) ceramics are among the most widely used piezoelectric materials due to their remarkably higher dielectric constant (PZT 4 (hard PZT) ~1300 and PZT 5H (soft PZT) ~3400) as well as high piezoelectric coefficient (PZT 4 (hard PZT): d_{33} ~289 pC/N and PZT 5H (soft PZT): d_{33} ~593 pC/N) [11]. However, owing to their inherently high stiffness and brittle nature (maximum safe strain ~0.2%), thin films of PZT materials have limited applications in flexible and stretchable electronics. In order to overcome this issue, a "piezorubber" chip (Fig. 10.5A (i)) was

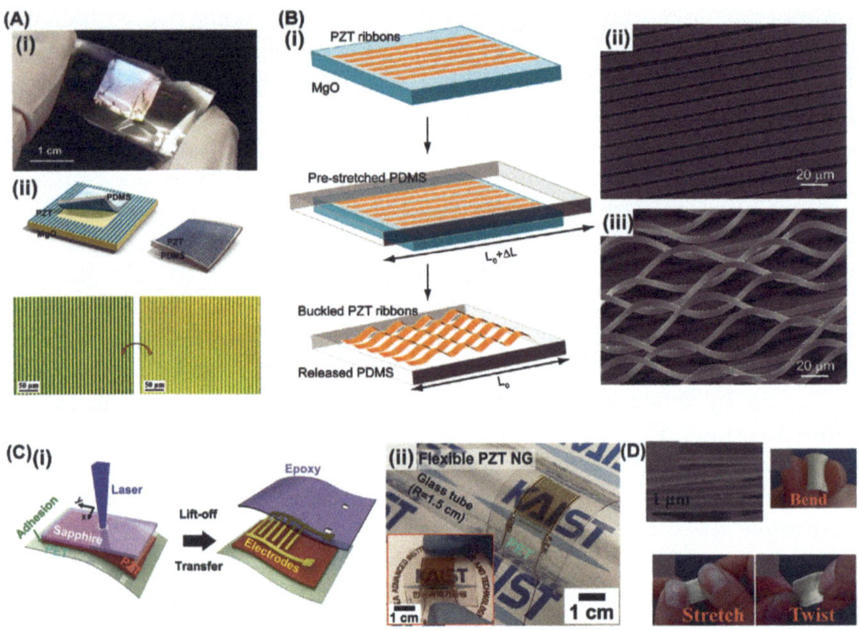

FIGURE 10.5 PZT-based PENGs. (A) (i) Photograph of a rubber chip based on PZT, (ii) scheme and optical microscopy micrograph of PZT ribbons on a PDMS substrate (Source: Reproduced with permission Y. Qi, N.T. Jafferis, K. Lyons, C.M. Lee, H. Ahmad, M.C. McAlpine, Nano Lett. 10 (2010) 524 [20]). (B) (i) Scheme of the process used for the formation of wavy/buckled PZT ribbons. SEM images of PZT ribbons transfer printed onto PDMS surface without (ii) and with prestrain conditions (Source: Reproduced with permission Y. Qi, J. Kim, T.D. Nguyen, B. Lisko, P.K. Purohit, M.C. McAlpine, Nano Lett. 11 (2011) 1331 [21]). (C) (i) Schematic diagram of the fabrication process used to fabricate large area PENG by using the laser lift-off method and (ii) representative photographs of the device attached to a glass tube and bent by human fingers (the inset) (Source: Reproduced with permission K.-I. Park, J. H. Son, G.-T. Hwang, C.K. Jeong, J. Ryu, M. Koo, et al., Adv. Mater. 26 (2014) 2514 [22]). (D) SEM image of aligned electrospun PZT nanofibers and optical photographs of the same under different deformations. (Source: Reproduced with permission L. Gu, N. Cui, L. Cheng, Q. Xu, S. Bai, M. Yuan, et al., Nano Lett. 13 (2013) 91 [23]).

fabricated by patterning PZT ribbon array (500 nm thick and 5 μm wide PZT ribbons with 10 μm center-to-center spacing) on MgO substrate and subsequent transfer printing onto a 2.5–5 mm thick polydimethylsiloxane (PDMS) substrate (Fig. 10.5A (ii)) [20]. The device was characterized by piezoforce microscopy resulting in the demonstration of high electromechanical energy conversion metrics. However, to enable increased device elasticity, wavy/buckled PZT ribbons were fabricated by transfer of PZT ribbons onto prestrained (~8%) PDMS substrate [21] and subsequent release of the strain (Fig. 10.5B (i)). Indeed, in flat PZT ribbons (Fig. 10.5B (ii)), fracture starts at tensile strain <1%, while wavy/buckled PZT ribbons (Fig. 10.5B (iii)) can withstand an applied tensile strains up to >8% and a compressive strains of -1% without fracture. The fabricated PENG, consisting of 10 wavy ribbons can generate output current density of 2.5 $\mu A/mm^2$ under 8% stretching strain. In order to increase the output performance along with the device flexibility, a PENG was developed by using a PZT thin film deposited on a flexible polyethylene terephthalate (PET) substrates by a laser lift-off (LLO) process (Fig. 10.5C (i)) [22]. A single flexible PENG (1.5 × 1.5 cm) (Fig. 10.5C (ii)) generated V_{oc} ~200 V and I_{sc}~ 150 $\mu A/cm^2$ under a bending strain of ~0.386% at a rate of ~2.32% s^{-1}. The energy harvested by using a large-area thin film NG (3.5 × 3.5 cm) was directly used to power over 100 blue LEDs without any external electric source and circuits. An alternative approach, consisting in the realization of vertically aligned ultralong (~ 420 μm) PZT nanofibers bonded with PDMS, was reported by Z. L. Wang et al. (Fig. 10.5D) [23].

The resulting device could be bent, stretched, or twisted to a large extent without breaking and could generate 209 V of output voltage and 53 μA of short circuit current under a high periodic impact pressure >0.53 MPa.

10.2.1.3 Composite-based materials-based piezoelectric nanogenerators

Conventionally, composite material-based PENGs are composed of piezoelectric nanomaterials dispersed into an elastomeric matrix. This is a promising approach for large-scale, flexible energy-harvesting applications. The composite-based PENGs have several advantages, including a simple fabrication process, cost effectiveness and mechanical robustness. The robustness and enhanced electric output broadens their potential application in consumer electronics, environmental monitoring, and sensor networks. For example, a flexible, bendable, and even stretchable large area (13 × 13 cm) composite-based PENG was fabricated by using a coating deposition process (spin-coating or Mylar bar-coating) to deposit barium titanate (BTO) nanoparticles (~100 nm) and universal graphitic carbon within a PDMS matrix (Fig. 10.6A) [24]. Under applying a force of ~200 N, the device (active area of 5 × 7 cm) could generate

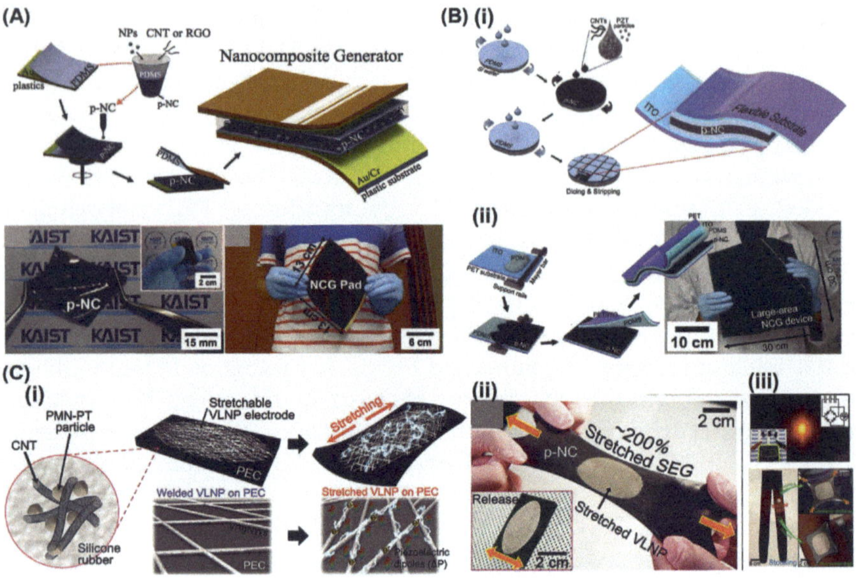

FIGURE 10.6 Composite-based PENGs. (A) Schematic representation of flexible, bendable, and stretchable large area composite PENG based on BTO nanoparticles and universal graphitic carbon in a matrix of PDMS (Source: Reproduced with permission K.-I. Park, M. Lee, Y. Liu, S. Moon, G.-T. Hwang, G. Zhu, et al., Adv. Mater. 24 (2012) 2999 [24]). (B) Schematic diagrams of (i) a flexible composite nanogenerator based on PZT particles and MWCNTs and (ii) of the fabrication process used to fabricate large-area device (30 × 30 cm) (Source: Reproduced with permission K.-I. Park, C.K. Jeong, J. Ryu, G.-T. Hwang, K.J. Lee, Adv. Energy Mater. 3 (2013) 1539 [25]). (C) (i) Schematic illustration of the fabrication process used to fabricate hyperstretchable PENG composed of PMN-PT particles, MWCNTs, and silicone rubber matrix using very long nanowire percolation (VLNP) electrodes. (ii) Photograph of the PENG subject to 200% stretching. The inset shows a photograph of the unstretched PENG. (iii) The upper panel shows consumer electronics operated by the developed PENG which was further integrated in a wearable cloth for human kinematic energy harvesting. Bottom panel. (Source: Reproduced with permission C.K. Jeong, J. Lee, S. Han, J. Ryu, G.-T. Hwang, D.Y. Park, et al., Adv. Mater. 27 (2015) 2866 [26]).

output signals of 1.5 V and 150 nA which were used to drive LEDs and capacitors. Improved performances were achieved by dispersing PZT particles ($\sim 1\ \mu m$) and multiwalled carbon nanotubes (MWCNTs) (diameters of ≈ 15 nm and lengths of $\approx 10\ \mu m$) in a PDMS matrix (Fig. 10.6B (i)) [25]. A PENG device of 30×30 cm was fabricated by using the Mayer bar-coating technique (Fig. 10.6B (ii)). The device output voltage was ≈ 100 V and the current signal up to $\approx 10\ \mu A$. This generated power was used to operate an LCD screen and 12 commercial LED arrays without any external power source and circuits. With a different device design, an highly stretchable composite PENG was fabricated by using ultralong Ag nanowires (maximum length up to

500 μm (average length of 150 μm) and diameter of 100–150 nm) as stretchable electrodes (Fig. 10.6C (i)) [26]. The composite PENG was composed by dispersing lead magnesium niobate–lead titanate (PMN-PT) particles and MWCNTs in a silicone elastomer matrix. This PENG could be stretchade up to 200% strain (Fig. 10.6C (ii)) and generated output signals of 4 V and 500 nA. The output electric power generated from the stretchable PENG was sufficient to operate commercial electronic units, such as LEDs and capacitors (top panel of Fig. 10.6C (iii)). Benefitting from its stretchability, the device was sewn to fabrics to harvest biomechanical stretching motions (lower panel of Fig. 10.6C (iii)). Additional piezoelectric materials used to fabricate composite PENG by dispersing them into PDMS elastomer are: potassium niobate ($KNbO_3$) nanowires ($V_{oc}\sim 16$ V, $I_{sc}\sim 230$ nA, power ~ 3.6 μW, device area ~ 3.5 × 4 cm) [27], sodium niobate ($NaNbO_3$) nanowires ($V_{oc}\sim 3.2$ V, $I_{sc}\sim 72$ nA, power ~ 0.6 mW/cm^3, device thickness ~ 100 μm, compression strain ~ 0.23%) [28], perovskite zinc stannate ($ZnSnO_3$) nanocubes ($V_{oc}\sim 40$ V, $I_{sc}\sim 0.4$ μA, power ~ 10.8 μW/cm^3, electrode area ~ 32 × 26 mm^2, compression stress ~ 12 kPa) [29].

10.2.2 Organic piezoelectric materials

Organic piezoelectric materials, and in particular, PVDF and its copolymers, are easy to process, biocompatible and chemically stable [30]. Compared to inorganic materials, the only constraint for the use of polymers in energy-harvesting applications is their lower piezoelectric coefficient, d_{33}. Strategies so far proposed for improving the piezoelectric performances of these materials include, the formation of porous structures, bioinspired interconnected microstructures and nanostructures.

For example, Mao et al. proposed a sponge-like mesoporous PVDF-based PENG (Fig. 10.7A (i)) [31]. The flexible thin film configuration allows the PENG to be directly attached to electronic devices and to use the device's own weight as the proof mass to amplify the oscillation of the PVDF film. This nanogenerator was used to directly power up a smart phone (Fig. 10.7A (ii)). Additionally, Yang et al. demonstrated ice-templated freeze casting method (Fig. 10.2E (i)) for generating porous PVDF-based ferroelectrets with very high d_{33} values (264 pC/N after poling) [39]. Conventionally, electrical poling is necessary to align dipoles of the polar β-phase (*all-trans configuration*) of PVDF thus enhancing the output power of the piezoelectric devices. However, the poling procedure, often limits the device design due to repetitive electrical breakdown failure taking place during the procedure and, in addition, to form porous structure additional fabrication steps are needed thus increasing the complexity and duration of the whole fabrication process. An effective and efficient in-situ and template free strategy was proposed by Ghosh et al. through the incorporation of a metal salt such as, platinum (Pt)-salt, directly into a solution of P (VDF-HFP) in

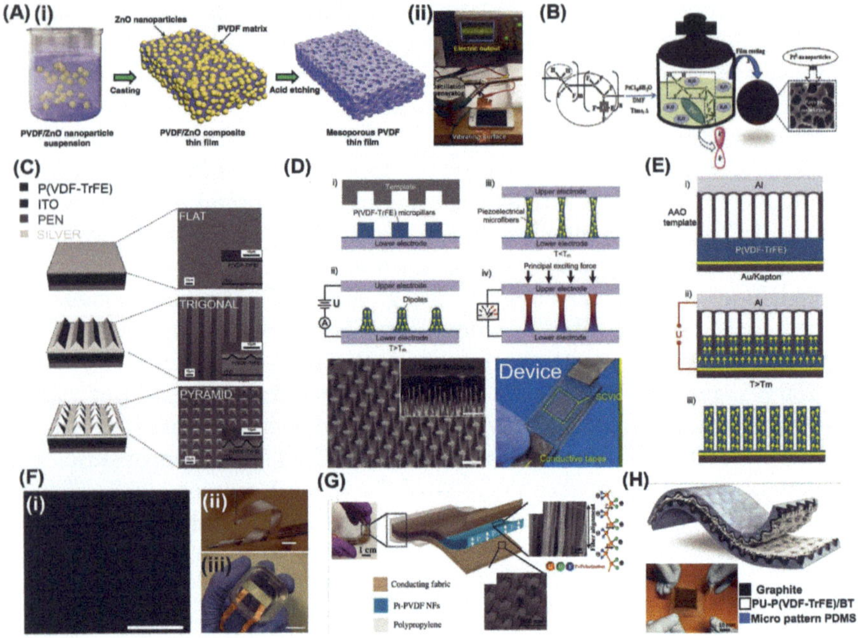

FIGURE 10.7 PVDF-based PENGs. (A) (i) Fabrication process of ZnO nanoparticle etched sponge-like mesoporous PVDF-based PENG which (ii) operates smart phone from the surface oscillation of an oscillator (Source: Reproduced with permission Y. Mao, P. Zhao, G. McConohy, H. Yang, Y. Tong, X. Wang, Adv. Energy Mater. 4 (2014) 1301624 [31]). (B) Mechanism of in-situ poled microporous PVDF film formation by using Pt-salt (Source: Reproduced with permission S.K. Ghosh, T.K. Sinha, B. Mahanty, S. Jana, D. Mandal, J. Appl. Phys. 120 (2016) 174501 [32]). (C) Schematic diagram and SEM images of planar, trigonal and pyramid structure P (VDF-TrFE) films (Source: Reproduced with permission J.-H. Lee, H.-J. Yoon, T. Y. Kim, M. K. Gupta, J. H. Lee, W. Seung, H. Ryu, S.-W. Kim, *Adv. Funct. Mater.* 25 (2015) 3203.). (D) Scheme of the experimental steps used for building and exciting self-connected vertically integrated PENG (top panel) consisting of microfiber arrays (SEM micrographs in the bottom left panel) and a digital photograph of the final device (bottom right panel) (Source: Reproduced with permission X. Chen, H. Tian, X. Li, J. Shao, Y. Ding, N. An, Y. Zhou, *Nanoscale* 7 (2015) 11536). (E) Scheme of the process used for the fabrication of nanoporous AAO template-based P (VDF-TrFE) nanowire (Source: Reproduced with permission X. Chen, J. Shao, N. An, X. Li, H. Tian, C. Xu, Y. Ding, *J. Mater. Chem. C* 3 (2015) 11806). (F) (i) SEM image of electrospun aligned P (VDF-TrFE) nanofibers which (ii) show excellent flexibility and (iii) can be used to fabricate an accelerometer (Source: Reproduced with permission L. Persano, C. Dagdeviren, Y. Su, Y. Zhang, S. Girardo, D. Pisignano, Y. Huang, J. A. Rogers, *Nat. Commun.* 4 (2013), 1633). (G) Schematic diagram of Pt-NPs doped PVDF aligned nanofiber-based all-fiber PENG (Source: Reproduced with permission S. K. Ghosh, D. Mandal, *Nano Energy* 53 (2018), 245). (H) Schematic and original photograph of omnidirectional stretchable nanofiber-based PENG. (Source: Reproduced with permission S. Siddiqui, H. B. Lee, D.-I. Kim, L. T. Duy, A. Hanif, N.-E. Lee, *Adv. Energy Mater.* 8 (2018) 1701520).

dimethylformamide (DMF) solvent. The solvent/nonsolvent phase separation followed by the interfacial interaction of Pt-nanoparticles with the $-CF_2$ bond of P (VDF-HFP) is demonstrated to enhance the piezoelectric properties, through the formation of micropores behaving as microdipoles, and that synergically interact with the induced nanodipoles, which belongs to the β-phase form of PVDF (Fig. 10.7B) [32,40]. The resulting soft material behave as a ferroelectret and the measured piezoelectric coefficient was ~ -686 pC/N, comparable with PZT. No electrical poling was used. Mahanty et al. reported on ferroelectret soft film of porous P (VDF-HFP) made by etching ZnO nanoparticles in a single step process. Electrical charges were trapped within the pores which effectively enhance the ferroelectric as well as the piezoelectric properties. The fabricated energy harvester using this pores film was further used for efficient wireless energy-harvesting applications [41]. In order to fabricate porous soft film, P (VDF-HFP) is an ideal material in comparison to PVDF as it has lower crystallinity than PVDF [42]. Additionally, a range of self-polarized ferroelectret-based energy harvesters were reported by using rare earth salts, such as, Ce^{3+} [43], Er^{3+} [44], Eu^{3+}[45] and Yb^{3+}[46] modified PVDF. In these cases, a resonance structure was formed between the rare earth salts and solvents of PVDF. This resonance structure, by forming hydrogen bonding with the $-CF_2$ units of the PVDF chain, induces both the formation of the $\beta-$phase and a localized alignment of the PVDF chain. Also in this work, no further electrical poling was required for energy generation. In addition it has been demonstrated that, more generally, porous films experienced enhanced displacement compared to smooth films under same applied pressure and therefore enabled improved performances in piezoelectric energy generation.

The use of organic nano- and microstructure is especially appealing due to the apparent strain/stress confinement effect and improved strain/stress tolerance capability of nano/microstructures. It has been demonstrated that a mechanical stress applied along one axis can cause remarkable deformations along orthogonal directions [33]. In piezoelectric materials, such geometrical strain confinement induces better dipole alignment that increased the piezoelectric potential and energy conversion efficiency. With this aim several nano/microstructures were fabricated. For instance, trigonal line-shaped and pyramid shaped micropatterned P (VDF-TrFE) film-based NGs generate nearly five and seven times larger voltage output, respectively, compared to piezoelectric generators based on flat film. This effect is explained by different strain levels induced by the same applied force depending on the shape and size of the geometrical features (Fig. 10.7C) [47]. A bioinspired design was effective in enhancing the power generation and sensitivity. The skin of the human finger is indeed composed of epidermal and dermal layers which are interlocked through microstructural array. This interlocked microstructure is related to the enhanced sensitivity of human skin towards external stimuli such as, vibration, strain, pressure, temperature. Inspired from this design, PVDF/reduced graphene oxide (rGO)-based microdomes patterned with an interlocked geometry were fabricated [34]. Under the application of an external pressure, the stress was mostly concentrated at the contact points of the microdome patterns thus enabling superior sensitivity and power generation capability when compared with planar structure film. In another approach, P (VDF-TrFE) film consisting of micropillar array (Fig. 10.7D), was manufactured by electrohydrodynamic (EHD) pulling method [48] and P (VDF-TrFE)/$BaTiO_3$ composite film based on micropillar arrays was prepared by nanoimprinting. [35] These films were further used for fabricating energy-harvesting devices which operated several consumer electronic gadgets and were employed in healthcare monitoring applications.

In-situ poled P (VDF-TrFE) nanowires were prepared by anodized aluminum oxide (AAO) template (Fig. 10.7E) [49]. The nanoconfinement within the nanoporous template yields a preferential alignment of both nanowires and the polymer chains leading to superior sensitivity in one step [49,50]. The resulting NGs show high energy conversion efficiency and were used for monitoring vital signs. An easy and effective tool for the fabrication of in-situ poled one dimensional (1D) nanofiber is represented by the electrospinning technology [51]. It has been observed that the direction of dipoles in the molecular chain of electrospun PVDF nanofibers is almost perpendicular to the fiber main axis [36,52] and, highly aligned nanofibers show superior dipolar alignment and piezoelectric performances when compared to random nanofibers [37]. This is mainly related to use of rotating collectors. Indeed, in such case, during the fabrication process, in addition to the stretching forces exerted on the electrified solution jets by the high electric field applied (up to 1 MV/m in far field electrospinning), mechanical stretching force acts during the fibers collection. As a result, in-situ poled-stretched nanowires of ultrahigh aspect ratio (i.e., several centimeter length and micro/nanosize diameter) are produced. Persano et al. prepared a large area, flexible, free-standing mat of highly aligned arrays piezoelectric P (VDF-TrFE) fibers by electrospinning onto a fast rotating collector (Fig. 10.7F (i)) [37]. This fiber mat exhibited excellent flexibility and mechanical robustness and can be repeatedly bent and folded without any damage (Fig. 10.7F (ii)). At a bending frequency of 2 Hz, the device exhibits 40 nA of output current and 1.5 V of output voltage. Owing to its outstanding piezoelectric properties, a fiber mat-based sensor was capable to detect extremely small pressure as low as 0.1 Pa. Using the fiber array, various types of sensors, such as accelerometers, orientational, and vibrometer sensors can also be easily fabricated (Fig. 10.7F (iii)).

The inclusion of nanofillers into PVDF nanofibers further enhances the orientation of the dipoles through interfacial interactions between the nanofillers and the PVDF chain [53]. The main challenge in the fabrication of energy harvesters based on electrospun nanofibers, is represented by the fabrication of proper electrodes given the intrinsic porosity of the electrospun mats which easily leads to the formation of short circuits. In most cases, fiber-based energy harvesters were designed by embedding the nanofiber mat between two metallic plates or foils in a sandwich configuration. However, this conventional design limits the piezoelectric output performance due to a large mismatch between the mechanical properties (such as, Young's modulus and Poisson's ratio) of the two dissimilar materials (polymer and metal). It has been indeed demonstrated that fiber-based conductive materials, when used as electrodes, enable superior output performances.

For instance, a novel approach has been proposed where pure PVDF nanofibers were sandwiched between poly (3,4-ethylenedioxythiophene) (PEDOT) coated PVDF fiber used as electrodes [54]. This high performance and mechanically robust device was further used for weight measurement by harvesting human movements such as foot strikes and walking. Furthermore, an all-fiber energy harvesters was reported where highly flexible Pt-NPs doped aligned PVDF nanofibers were prepared by electrospinning and then sandwiched between microfiber-based conducting fabric electrode (Fig. 10.7G) [55].

By following a similar strategy, an all-fiber acoustoelectric nanogenerator was fabricated by using hydrated metal salt ($MgCl_2 \cdot 6H_2O$) (Mg-salt) reinforced PVDF–Mg nanofibers as an active layer and interlocked conducting microfiber-based electrode [38]. An alternative strategy for the realization of high performance, omnidirectionally stretchable NGs was proposed by Siddiqui et al. (Fig. 10.7H) [56]. The device, consisting of electrodes of graphite deposited onto a 3D micropatterned stretchable substrate and of a stacked mat of piezoelectric nanofibers alternatively composed of $BaTiO_3$ nanoparticles embedded in polyurethane and poly (vinylidene fluoride-trifluoroethylene) (PVDF-TrFE) nanofibers, was successfully used to harvest the energy from human kinematics by placing it over the knee cap of a human being. In another strategy, a PVDF nanofiber membrane was sandwiched between conductive nanofiber membranes and CNT electrodes [57]. The conductive nanofiber membrane was made by mixing PVP, PEDOT:PSS, ethyl alcohol, ethylene glycol and ionic liquid. The flexibility of the nanogenerator was enhanced by the use of electrospun thermoplastic polyurethane nanofiber membrane as a substrate. This device was capable to harvest several human motions such as, walking, running, arm movement and wind flow movements. Furthermore, the piezoelectric membrane device was also designed by a three-layer structure in which the PVDF nanofiber membrane was sandwiched between two PVDF–rGO nanofibers-based electrode membranes. The whole device was assembled in single electrospinning process [58]. Finally, by using a thermal fiber drawing process, thermally stable single PVDF micro- and nanoribbons with high piezoelectric coefficient of −58.5 pm/V were demonstrated [59].

10.2.3 Biodegradable materials-based piezoelectric nanogenerators

Biodegradable piezoelectric materials are specifically appealing for next generation self-powered technologies. Indeed, if one considers properties such as biocompatibility, biodegradability, or at least recyclability, conventional inorganic and organic piezoelectric materials rarely meet these requirements. Hence, the development of eco-friendly and sustainable technologies is highly important. Most of the existing electronic devices are nondecomposable thus leading to the generation of large volumes of electronic waste (e-waste) (53.6 million tons in 2019 and an expected 74.7 million tons by 2030 worldwide) [60], and threatening the health of ecological systems. Since e-waste contains hazardous components, even the recycling of these materials contaminates ecosystems with toxins that can enter in the living bodies by inhalation, skin exposure, and oral intake of contaminated foods [61]. Hence, transient and ingestible electronics made of completely biodegradable and edible materials are particularly attractive. Such devices can be completely degraded by the decomposition and digestion processes after a certain period of use, thus reducing the amount of e-waste produced. Apart from this, biodegradable piezoelectric materials are the ideal candidate for next generation implantable biomedical devices for in-vivo operation [62]. In contrast to the traditional organic and inorganic materials, the biodegradable natural piezoelectric materials possess reduced toxicity and can be further absorbed by the body after a certain period of use. In these regards, only few approaches have been reported on biodegradable piezoelectric materials-based PENGs. Several natural biological waste materials such as, fish swim bladder ($d_{33} \sim 22$pC/N) [63], fish skin ($d_{33} \sim -3$pC/N) [64], fish scale ($d_{33} \sim -5$pC/N) [65], prawn shell ($d_{33} \sim -2$pC/N) [66] were used as a "biopiezoelectric nanogenerator (BPNG)" and applied towards self-powered technology. For example, fish swim bladder (Fig. 10.8A) and fish scale (Fig. 10.8B)-based BPNGs were demonstrated to operate consumer electronics, such as, LEDs and capacitors (Fig. 10.8C). The fish skin and prawn shell-based BPNG were explored as a self-powered healthcare monitoring sensors. The piezoelectricity from the fish swim bladder, fish scale and fish skin was mainly related to the presence collagen fibers. On the other hand, prawn shell was composed of chitin fiber which was also used as piezoelectric material. In these ways the biowaste materials were transformed from waste product into added-value product. In addition, commercially available gelatin [69] and poly-L-lactic

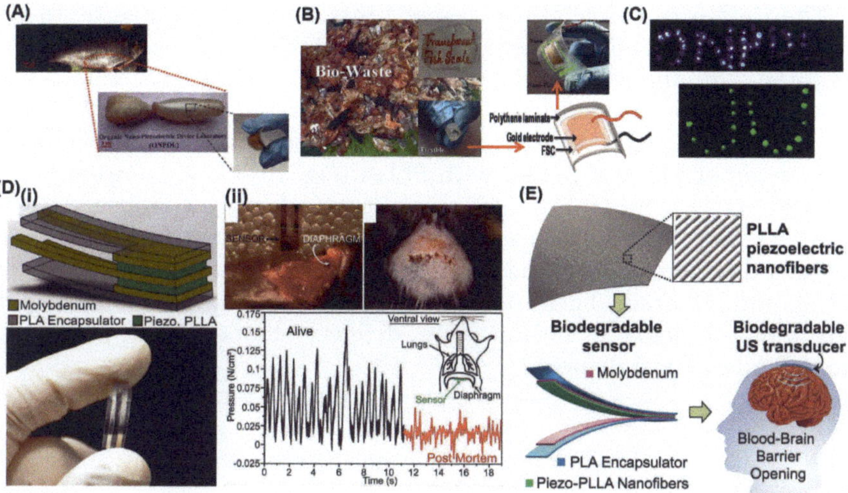

FIGURE 10.8 Biodegradable PENGs. (A) Designing nanogenerator from fish swim bladder collected from sweet water fish (*Catla catla*) (Source: Reproduced with permission S. K. Ghosh, D. Mandal, *Nano Energy* 28 (2016) 356) and (B) fabrication steps of nanogenerator from biowaste fish scales collected from common market which was further used for (C) operation of LEDs (Source: Reproduced with permission S. K. Ghosh, D. Mandal, *Appl. Phys. Lett.* 109 (2016) 103701). (D) (i) Schematic (top panel) and photograph (bottom panel) of biodegradable piezoelectric PLLA pressure sensor which (ii) its implanted into a mouse abdominal cavity with diaphragmatic membrane (top panel). Measured force signals (bottom panel) generated by the implanted sensor when the mouse was alive and under anesthesia (*black* lines), and when the mouse was euthanized by overdose of anesthetics (*red* lines) (Source: Reproduced with permission E. J. Curry, K. Ke, M. T. Chorsi, K. S. Wrobel, A. N. Miller, A. Patel, I. Kim, J. Feng, L. Yue, Q. Wu, C.-L. Kuo, K. W.-H. Lo, C. T. Laurencin, H. Ilies, P. K. Purohit, T. D. Nguyen, *PNAS* 115 (2018) 909). (E) Schematic of a biodegradable piezoelectric device based on PLLA nanofibers used as ultrasonic transducer and implanted into brain. (Source: Reproduced with permission E. J. Curry, T. T. Le, R. Das, K. Ke, E. M. Santorella, D. Paul, M. T. Chorsi, K. T. M. Tran, J. Baroody, E. R. Borges, B. Ko, A. Golabchi, X. Xin, D. Rowe, L. Yue, J. Feng, M. D. Morales-Acosta, Q. Wu, I.-P. Chen, X. T. Cui, J. Pachter, T. D. Nguyen, *PNAS* 117 (2020) 214).

acid (PLLA) [67] were used to prepare nanofiber mats exhibiting $d_{33} \sim 20pC/N$ and $d_{33} \sim 3 \pm 1pC/N$ respectively. These nanofibers were also used to fabricate all-fiber PENG in order to demonstrate their applicability from skin conformable healthcare monitoring sensors to portable electronic gadgets.

On the other hand, very limited approach has been observed to implement the biodegradable PENGs towards implantable devices. Recently, a biodegradable piezoelectric PLLA film-based device (Fig. 10.8D (i)) was implanted into the abdominal cavity of a mouse to convert the mechanical energy of diaphragmatic motion into electricity (Fig. 10.8D (ii)) [68].

The device can precisely measure pressures in the range 0–18 kPa and has the potential to replace biodegradable electronic devices. In addition, an aligned PLLA nanofiber-based device (5 × 5 mm) was implanted into the abdominal cavity of a mouse and connected to a small printed circuit board (PCB) via a subcutaneous biodegradable wire made of Mo and coated with PLLA (Fig. 10.8E) [70]. Furthermore, the device was implanted into craniotomy defect in a mouse skull in order to demonstrate its use as a biodegradable ultrasonic transducer for blood–brain barrier opening, in order to facilitate the delivery of drugs into the brain. These demonstrations can strongly impact on various medical fields including drug delivery, tissue engineering, and implanted medical devices.

10.3 Pyroelectric nanogenerator

Pyroelectric nanogenerator (PyNG) is an energy-harvesting device which converts thermal energy into electric energy by means of nanostructured pyroelectric materials. In the pyroelectric effect, the power is generated from time-dependent temperature fluctuation (dT/dt) in contrast to the thermoelectric effect where spatial temperature gradients (dT/dx) were used to generate electricity. Pyroelectric materials are of interest since under the right conditions can potentially operate with high thermodynamic efficiency and, compared to thermoelectric materials, do not require bulky heat sinks to maintain a temperature gradient [71]. In recent years, the pyroelectric energy-harvesting systems have come out as promising field of research towards energy harvesting.

The short circuit pyroelectric output current is given by $i = \frac{dQ}{dt} = pA\frac{dT}{dt}$ where, Q, p, A and t are pyroelectric charge, pyroelectric coefficient, effective surface area and time respectively. The pyroelectric coefficient ($p^{\sigma,E}$) of the material at constant electric field (E) and constant stress (σ) is given by the following equation, $p^{\sigma,E} = \left(\frac{dP}{dT}\right)_{\sigma,E}$, where P is the spontaneous polarization

The pyroelectric mechanism is based on thermally induced random oscillation of electric dipoles around their equilibrium axis as shown in Fig. 10.9. At room temperature, dipoles will randomly oscillate within a

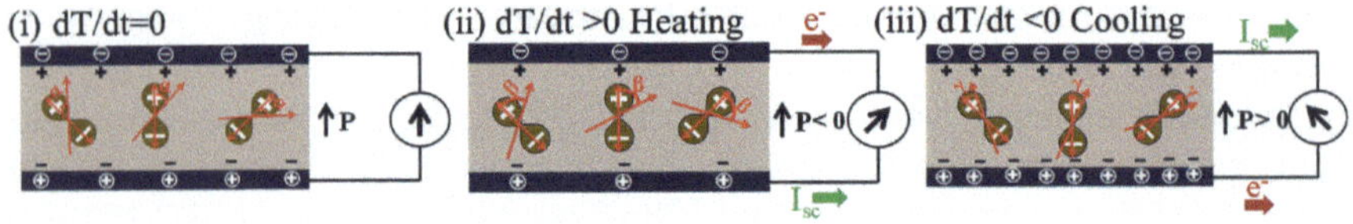

FIGURE 10.9 Schematic representation of the operating principle of pyroelectric current generation in PyNG.

certain degree (such as, α) from their respective polar axes (Fig. 10.9 (i)). The total average strength of the spontaneous polarization (P) from the dipoles is constant at a fixed temperature ($dT/dt = 0$); thus, there is no output current from the PyNG (Fig. 10.9 (i)). When ΔT increases ($dT/dt > 0$), the dipoles oscillate within a larger degree of wiggling angle (such as, β) around their respective polar axes (Fig. 10.9 (ii)). As a result, the total average spontaneous polarization is decreased ($P < 0$), which results in a flow of electrons/current from one electrode to other. When ΔT decreases ($dT/dt < 0$), the spontaneous polarization is enhanced ($P > 0$) since the electric dipoles oscillate within a smaller degree of wiggling angles (such as, γ) due to lower thermal activity (Fig. 10.9 (iii)). In such a case, electrons/current will flow in the reverse direction. In this overall phenomena, the wiggling angles were found as, $\beta > \alpha > \gamma$. The temperature change across the material mainly controls the pyroelectric coefficient of the materials and lead to change of the output current of the PyNGs. In addition, in a recent publication, the importance of the resistance and capacitance of the voltmeter used during measurements has been highlighted by Li et al. by demonstrating an increase of 2296 times of the voltage output when the voltmeter resistance increases from 10 MΩ to 100 TΩ [72].

10.3.1 The progress of pyroelectric nanogenerator

The PyNG was first developed in 2012 and reported by Yang et al. The device was based on ZnO NWs and exhibit pyroelectric voltage $\sim 1.2-1.5$ nC/cm^2 K and current coefficient $\sim 2.5-4 \times 10^4$ V/mK (Fig. 10.10A) [73]. The practical application of the PyNG was demonstrated by using a PZT film-based PyNG showing a pyroelectric coefficient of about -80 nC/cm^2 K, output voltage ~ 22 V and current density ~ 171 nA/cm^2 at a heating/cooling rate ~ 0.2 K/s which was further used for driving a liquid crystal display (LCD), LED, Li-ion battery and a wireless temperature sensor (Fig. 10.10B) [78]. Furthermore, ferroelectric $0.7Pb(Mg_{1/3}Nb_{2/3})O_3-0.3PbTiO_3$ (PMN-PT) single crystal was used as active material in PyNG. The corresponding open circuit output voltage is 1.1 V and the current is 10 nA at room temperature [79]. Apart from operating portable electronic gadgets, the pyroelectric effect was also used in different research fields. For example, light-induced pyroelectric effect in ferroelectric $BaTiO_3$-based device was used to fabricate a self-powered 365 nm UV light photodetector [80]. A new approach was introduced where the PyNG was used for water splitting thus representing a novel alternative hydrogen source. The PVDF and PZT-based PyNG generated sufficiently large electric potential between two electrodes to split water into hydrogen and oxygen gas [74].

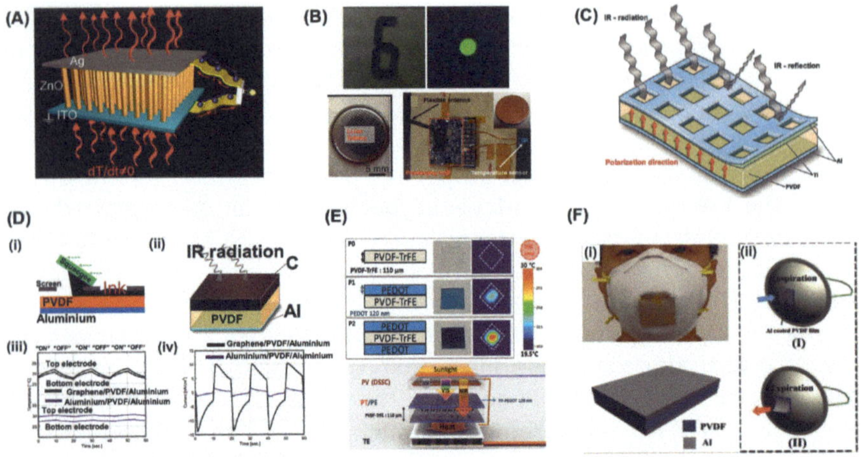

FIGURE 10.10 (A) Schematic diagram showing the structure of the ZnO NWs-based PyNG (Source: Reproduced with permission Y. Yang, W. Guo, K. C. Pradel, G. Zhu, Y. Zhou, Y. Zhang, Y. Hu, L. Lin, Z. L. Wang, *Nano Lett.* 12 (2012) 2833). (B) Photographs of the consumer electronics operated by using PZT film-based PyNG (Source: Reproduced with permission Y. Yang, S. Wang, Y. Zhang, Z. L. Wang, *Nano Lett.* 12 (2012) 6408). (C) Schematic of micropatterned top electrode on a film of PVDF for IR light harvesting (Source: Reproduced with permission D. Zabek, J. Taylor, E. L. Boulbar, C. R. Bowen, *Adv. Energy Mater.* 5 (2015) 1401891). (D) (i) Schematic of the screen printing process to deposit graphene ink on a film of PVDF, showing (ii) the final device structure, which generated (iii) higher temperature fluctuation and (iv) output current (Source: Reproduced with permission D. Zabek, K. Seunarine, C. Spacie, C. Bowen, *ACS Appl. Mater. Interfaces* 9 (2017) 9161). (E) PEDOT-coated P (VDF–TrFE) film showing higher temperature generation capability, which was further hybridized with a solar cell (Source: Reproduced with permission T. Park, J. Na, B. Kim, Y. Kim, H. Shin, E. Kim, *ACS Nano* 9 (2015) 11830). (F) (i) PVDF-based wearable PyNG integrated into N95 mask for (ii) harvesting temperature fluctuations generated by regular human breathing (Source: Reproduced with permission H. Xue, Q. Yang, D. Wang, W. Luo, W. Wang, M. Lin, D. Liang, Q. Luo, *Nano Energy* 38 (2017) 147).

According to the pyroelectric equation ($i = pA\frac{dT}{dt}$), there are two strategies that can be pursued to increase the pyroelectric output current in devices with a specific surface area. One strategy is to increase dT/dt, which can be achieved by improving the heat transfer of the PyNG. The other one is to use materials with high pyroelectric coefficient. In organic pyroelectric materials which show low pyroelectric coefficients (PVDF) ~ $27\,\mu C/m^2/K$, P (VDF-TrFE) ~ $40\,\mu C/m^2/K$ when compared to ceramic-based materials, such as PZT ~$380\,\mu C/m^2/K$ and PMN − 0.25PT ~$746\,\mu C/m^2/K$ [71] the first strategy has been pursued. Zabek et al. deposited partially covered Al electrodes on a film of PVDF by micropatterning to enhance heat transfer and achieve larger temperature fluctuations under IR bulb (Fig. 10.10C). It has been demonstrated that the electrode micropatterning can significantly improve the open circuit voltage by 380% and the closed circuit current by 420% when compared with flat electrodes [75]. Electrodes based on interconnected graphene nanoplatelets (GNPs) have also been demonstrated to improve both the electrical conductivity and the thermal radiation absorbance capability of the device thus enhancing the pyroelectric effect in PVDF films (Fig. 10.10D (i)) [76]. More in general, metal electrodes including Al and gold (Au), have high thermal and electrical conductivity but low radiation absorbance ($\sim 0.1 - 0.3$). Accordingly, upon illumination, most of the thermal energy is reflected or transmitted at the electrode surface thus leading to small changes of the temperature across the pyroelectric active material. On the contrary, electrodes based on graphene inks absorb most of the available radiation heat energy (Fig. 10.10D (ii)), thereby maximizing the rate of temperature change dT/dt and improving pyroelectric energy-harvesting device performance and effectiveness (Fig. 10.10D (iii)). In graphene ink/PVDF/Al systems the closed circuit pyroelectric current improves by 7.5 times, the open circuit voltage by 3.4 times, and the harvested energy by 25 times when compared to a standard Al/PVDF/Al system electrode design (Fig. 10.10D (iv)). Furthermore, photothermal effects in poly (3,4-ethylenedioxythiophene)s (PEDOTs) were explored for pyroelectric conversion by coating them onto both sides of a film of P (VDF-TrFE) (Fig. 10.10E) [77]. The photothermally driven pyroelectric harvesting film provided a very fast electric output with a high voltage output of 15 V. Benefitting from the transparency of the PEDOT electrodes, the device was further hybridized with a solar cell and a thermoelectric device. Under sunlight irradiation (AM 1.5G) the photoconversion efficiency was increased by 20% using the light transmitted through the photovoltaic cell as a heat source that was converted into pyroelectric and thermoelectric output simultaneously from the high photothermal PEDOT electrodes. Another strategy so far proposed consists in the combination of inorganic and organic materials in order to enhance the pyroelectric coefficient. For example, PZT/PVDF-HFP (50/50 vol.%) composite materials show pyroelectric coefficients of 450 $\mu C/m^2 K$ at 70°C [81].

Furthermore, PyNG has also been used to harvest energy from our regularly wasted heat energy. For example, a wearable PyNG was manufactured using PVDF thin film integrated in a N95 respirator for scavenging the energy of human breathing (Fig. 10.10F (i)) [82]. Using the dynamic process of human breathing, this system creates $dT/dt \sim 12°C$. The measured open-circuit voltage is 42 V and the short-circuit current is 2.5 μA. The maximum power reached up to 8.31 μW with an external load of 50 $M\Omega$, and was used to directly drive electronics, such as a liquid crystal display and LEDs. The temperature fluctuation generated by a cup of hot coffee was also used to drive low-power electronic device (such as a digital watch and LED) by using a PVDF-based PyNG [83].

Therefore, the PyNG is proven useful in recovering a significant amount of waste heat that would otherwise be lost as by-product of industry and even during our daily life.

10.4 Coupled piezo- and pyroelectric nanogenerator

The performance of the nanogenerator can be improved by using several coupled energy sources. Essentially, the hybrid energy harvesters are fabricated by employing two or more energy conversion mechanisms, such as piezoelectric, pyroelectric, triboelectric, photovoltaic and thermoelectric making sure that each coupled harvesting method does not limit the others [84]. Here, we will be focusing only on the hybrid energy harvesters consisting of pyroelectric and piezoelectric systems. Since all the pyroelectric materials are also piezoelectric, there was a great deal of interest in the scientific community to attempt to combine both methods of energy harvesting. An innovative approach to detect temperature and pressure simultaneously was demonstrated by Song et. al. in ferroelectric $BaTiO_3$-based pyropiezoelectric sensor [85]. A flexible 4×4 array sensor was developed to sense temperature and pressure variations induced by a human finger in real-time (Fig. 10.11A (i)) and for mapping the instantaneous pyro- and piezoelectric voltage of each unit of the array (Fig. 10.11A (ii)).

Furthermore, a PZT film-based hybrid nanogenerator has been demonstrated and placed in a car exhaust line to collect mechanical and thermal energies simultaneously [86]. Under normal operation (i.e., 500 − 3000 rpm), the temperature of the PZT film reach up to 80°C and the speed of the exhaust gas is 8 m/s. The PZT film-based nanogenerator stably generates very high piezoelectric-pyroelectric current, $J_{sc} = 3\,\mu A/cm^2$ at 3000 rpm, due to the large temperature fluctuations and mechanical vibrations. Stretchable hybrid nanogenerator for powering wearable electronic gadgets has also been investigated. A micropatterned P (VDF-TrFE) film, in combination with a PDMS-CNT composite and graphene nanosheets used as electrodes, was demonstrated and successfully used under 30% stretching (Fig. 10.11B) [87]. In addition, a biomimetic all-organic stretchable energy harvester is designed by using a xylitol-added poly (3,4-ethylenedioxythiophene):poly (4-styrenesulfonate) (PEDOT:PSS/Xyl) film as electrode and a polyaniline-reinforced 1D aligned PVDF nanowires as active material (Fig. 10.11C) [88]. The device performs simultaneous energy scavenging of temperature and pressure changes under pressure and stretch conditions (~ 20%). Also as an epidermal sensor, the device simultaneously detected and discriminated both the subtle pressure and the thermal deviations exposed to a surface of the epidermis. Real-time recording and wireless transfer of physiological signals to a smartphone indicate an effective way to accomplish remote health monitoring for early intervention.

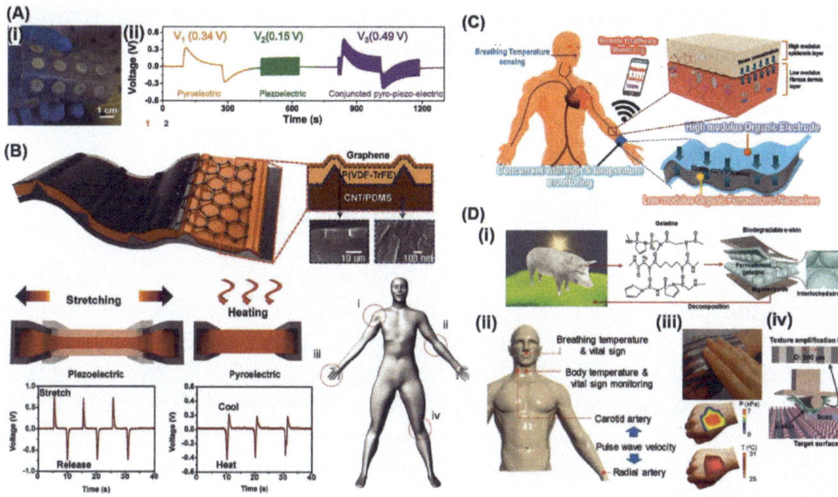

FIGURE 10.11 (A) (i) $BaTiO_3$-based 4×4 piezopyroelectric dual sensing matrix which (ii) can simultaneously harvest piezo- and pyroelectric output voltage (Source: Reproduced with permission K. Song, R. Zhao, Z. L. Wang, Y. Yang, *Adv. Mater.* 31 (2019) 1902831). (B) P (VDF-TrFE) film-based stretchable piezo-/pyroelectric hybrid device (Source: Reproduced with permission J.-H. Lee, K. Y. Lee, M. K. Gupta, T. Y. Kim, D.-Y. Lee, J. Oh, C. Ryu, W. J. Yoo, C.-Y. Kang, S.-J. Yoon, J.-B. Yoo, S.-W. Kim, *Adv. Mater.* 26 (2014) 765). (C) PVDF nanofibers-based temperature–pressure hybrid sensor (Source: Reproduced with permission S. K. Ghosh, T. K. Sinha, M. Xie, C. R. Bowen, S. Garain, B. Mahanty, K. Roy, K. Henkel, D. Schmeißer, J. K. Kim, D. Mandal, *ACS Appl. Electron. Mater.* 3 (2021) 248). (D) (i) Schematic of the fabrication process of biodegradable porcine skin gelatin-based piezopyroelectric sensor which (ii) simultaneously detect and discriminated the human breathing temperature and vital sign, (iii) can be used as a wearable (5×5 pixel) tactile device for pressure–temperature mapping and (iv) can recognize the surface texture of an unknown objects (Source: Reproduced with permission S. K. Ghosh, J. Park, S. Na, M. P. Kim, H. Ko, *Adv. Sci.* 8 (2021) 2005010, DOI: 10.1002/advs.202005010).

In addition, an ecological skin sensor based on biodegradable porcine skin gelatin (Fig. 10.11D (i)) has been developed that can simultaneously detect the space-resolved pressure and temperature (Fig. 10.11D (ii)) [89].

For practical applications, an array (5×5 pixel) was fabricated. When in contact with an object, the sensor demonstrated an excellent mapping of contact pressure (0–1.2 kPa) and temperature (25°C–26°C, $\Delta T = 1$°C). Thanks to its conformability and flexibility, the sensor array worn on the human hand is able to map both pressure (0–7 kPa) and temperature (25°C–31°C) distribution of another hand in contact with it (Fig. 10.11D (ii)). In addition, the skin sensor was further used for the perception of the surface smoothness of commercial knitted textile fabrics, such as polyester, cotton, silk, and nylon upon sliding over them (Fig. 10.11D (iii)). These demonstrations represented the excellent capability of the biodegradable skin sensor to discriminate the spatial resolution of simultaneous temperature and pressure measurements and surface texture of an unknown object, demonstrating potential for robotic skins, prosthetic limb and wearable electronics with zero waste footprints.

10.5 Conclusion and future outlook

In summary, several multifaceted approaches have been adopted to improve the performance and efficiency of energy-harvesting devices and particularly, in piezoelectric and pyroelectric hybrid NGs. The energy-harvesting ability is not limited to the enhancement of the figures of merit of the material used but also to the engineering of both the device and material structure. In this chapter we have discussed the performance of the different materials and device structures developed so far. Although a huge amount of research has been done on the piezoelectric and pyroelectric nanogenerator, there are still ample opportunities for development. For example, multiple energy-harvesting approaches, such as combining mechanical, thermal, solar, and magnetic energies scavenging by a single device have not been studied in depth so far. In addition, flexible and rollable magnetoelectric IoT devices are still in the early stage of research [90]. Furthermore, thin two-dimensional (2D) materials having a wide range of unique optical, electrical, and mechanical properties are still basically unexplored for flexible, stretchable, and adaptive electronic skin implementation [91]. Furthermore, the development of bioabsorbable energy harvesters is still at an early stage. So far, several efforts have been made for the realization of implantable self-powered devices by using synthetic materials. However, the implementation of bioabsorbable self-powered devices based on biomaterials is still ongoing and hopefully, upcoming innovations in piezoelectric/pyroelectric materials can solve the current hurdle, which is mainly represented by their lower piezoelectric coefficient and the poorly explored study of the origin of piezoelectricity in biomaterials, since they do not follow the classic models of piezoelectricity based on idealized, crystalline structures.

Acknowledgment

Dr. S. K. Ghosh would like to acknowledge the financial support from the European Commission's Marie Skłodowska-Curie Actions (MSCA), through the Marie Skłodowska-Curie Individual Fellowship (896811—BIOIMD—H2020-MSCA-IF-2019).

Conflicts of interest

The author does not have any conflicts of interest.

References

[1] S.R. Steinhubl, E.D. Muse, E.J. Topol, Sci. Transl. Med. 7 (2015) 283rv3.
[2] S.K. Ghosh, D. Mandal, J. Mater. Chem. A 9 (2021) 1887.
[3] B. Stadlober, M. Zirkl, M. Irimia-Vladu, Chem. Soc. Rev. 48 (2019) 1787.
[4] X. Liu, J. He, S. Zhao, Y. Liu, Z. Zhao, J. Luo, et al., Nat. Commun. 9 (2018) 4365.
[5] J. Lu, Y. Song, F. Lei, X. Du, Y. Huo, S. Xu, et al., Sci. Rep. 10 (2020) 5269.
[6] H. You, Y. Jia, Z. Wu, F. Wang, H. Huang, Y. Wang, Nat. Commun. 9 (2018) 2889.
[7] Z. Liang, C.-F. Yan, S. Rtimi, J. Bandara, Appl. Catal. B 241 (2019) 256.
[8] S. Kumar, M. Sharma, A. Kumar, S. Powar, R. Vaish, J Ind Eng Chem 77 (2019) 355.
[9] Q. Zheng, Q. Tang, Z.L. Wang, Z. Li, Nat. Rev. Cardiol. 18 (2021) 7.
[10] F. Mushtaq, H. Torlakcik, M. Hoop, B. Jang, F. Carlson, T. Grunow, et al., Adv. Funct. Mater. 29 (2019) 1808135.
[11] C.R. Bowen, H.A. Kim, P.M. Weaver, S. Dunn, Energy Environ. Sci. 7 (2014) 25.
[12] Z.L. Wang, J. Song, Science 312 (2006) 242.
[13] S. Xu, Y. Qin, C. Xu, Y. Wei, R. Yang, Z.L. Wang, Nat. Nanotechnol. 5 (2010) 366.
[14] G. Zhu, R. Yang, S. Wang, Z.L. Wang, Nano Lett 10 (2010) 3151.
[15] Y. Hu, Y. Zhang, C. Xu, G. Zhu, Z.L. Wang, Nano Lett 10 (2010) 5025.
[16] D. Choi, M.-Y. Choi, W.M. Choi, H.-J. Shin, H.-K. Park, J.-S. Seo, et al., Adv. Mater. 22 (2010) 2187.
[17] S. Lee, S.-H. Bae, L. Lin, Y. Yang, C. Park, S.-W. Kim, et al., Adv. Funct. Mater 23 (2013) 2445.
[18] Y. Hu, Y. Zhang, C. Xu, L. Lin, R.L. Snyder, Z.L. Wang, Nano Lett 11 (2011) 2572.
[19] R. Yang, Y. Qin, L. Dai, Z.L. Wang, Nat. Nanotechnol. 4 (2009) 34.
[20] Y. Qi, N.T. Jafferis, K. Lyons, C.M. Lee, H. Ahmad, M.C. McAlpine, Nano Lett 10 (2010) 524.
[21] Y. Qi, J. Kim, T.D. Nguyen, B. Lisko, P.K. Purohit, M.C. McAlpine, Nano Lett 11 (2011) 1331.
[22] K.-I. Park, J.H. Son, G.-T. Hwang, C.K. Jeong, J. Ryu, M. Koo, et al., Adv. Mater. 26 (2014) 2514.
[23] L. Gu, N. Cui, L. Cheng, Q. Xu, S. Bai, M. Yuan, et al., Nano Lett 13 (2013) 91.
[24] K.-I. Park, M. Lee, Y. Liu, S. Moon, G.-T. Hwang, G. Zhu, et al., Adv. Mater. 24 (2012) 2999.
[25] K.-I. Park, C.K. Jeong, J. Ryu, G.-T. Hwang, K.J. Lee, Adv. Energy Mater. 3 (2013) 1539.
[26] C.K. Jeong, J. Lee, S. Han, J. Ryu, G.-T. Hwang, D.Y. Park, et al., Adv. Mater. 27 (2015) 2866.
[27] R. Ganeshkumar, C.W. Cheah, R. Xu, S.-G. Kim, R. Zhao, Appl. Phys. Lett. 111 (2017) 013905.
[28] J.H. Jung, M. Lee, J.-I. Hong, Y. Ding, C.-Y. Chen, L.-J. Chou, et al., ACS Nano 5 (2011) 10041.
[29] M.M. Alam, S.K. Ghosh, A. Sultana, D. Mandal, Nanotechnology 26 (2015) 165403.
[30] P. Martins, A.C. Lopes, S. Lanceros-Mendez, Prog. Polym. Sci. 39 (2014) 683.
[31] Y. Mao, P. Zhao, G. McConohy, H. Yang, Y. Tong, X. Wang, Adv. Energy Mater. 4 (2014) 1301624.
[32] S.K. Ghosh, T.K. Sinha, B. Mahanty, S. Jana, D. Mandal, J. Appl. Phys. 120 (2016) 174501.
[33] L. Persano, A. Catellani, C. Dagdeviren, Y. Ma, X. Guo, Y. Huang, et al., Adv. Mater. 28 (2016) 7633.
[34] J. Park, M. Kim, Y. Lee, H.S. Lee, H. Ko, Sci. Adv. 1 (2015) e1500661.
[35] X. Chen, X. Li, J. Shao, N. An, H. Tian, C. Wang, et al., Small 13 (2017) 1604245.
[36] D. Mandal, S. Yoon, K.J. Kim, Macromol. Rapid Commun. 32 (2011) 831.

[37] L. Persano, C. Dagdeviren, Y. Su, Y. Zhang, S. Girardo, D. Pisignano, et al., Nat. Commun. 4 (2013) 1633.
[38] B. Mahanty, S.K. Ghosh, S. Jana, K. Roy, S. Sarkar, D. Mandal, Sustain. Energy Fuels 5 (2021) 1003.
[39] Y. Zhang, C.R. Bowen, S. Deville, Soft Matter 15 (2019) 825.
[40] S.K. Ghosh, T.K. Sinha, B. Mahanty, D. Mandal, Energy Technol 3 (2015) 1190.
[41] B. Mahanty, S.K. Ghosh, S. Garain, D. Mandal, Mater. Chem. Phys. 186 (2017) 327.
[42] S. Abbrent, J. Plestil, D. Hlavata, J. Lindgren, J. Tegenfeldt, Å. Wendsjö, Polymer 42 (2001) 1407.
[43] S. Garain, T.K. Sinha, P. Adhikary, K. Henkel, S. Sen, S. Ram, et al., ACS Appl. Mater. Interfaces 7 (2015) 1298.
[44] S.K. Ghosh, M. Xie, C.R. Bowen, P.R. Davies, D.J. Morgan, D. Mandal, Sci. Rep. 7 (2017) 16703.
[45] P. Adhikary, S. Garain, S. Ram, D. Mandal, J Polym Sci B Polym Phys 54 (2016) 2335.
[46] S.K. Ghosh, A. Biswas, S. Sen, C. Das, K. Henkel, D. Schmeisser, et al., Nano Energy 30 (2016) 621.
[47] J.-H. Lee, H.-J. Yoon, T.Y. Kim, M.K. Gupta, J.H. Lee, W. Seung, et al., Adv. Funct. Mater. 25 (2015) 3203.
[48] X. Chen, H. Tian, X. Li, J. Shao, Y. Ding, N. An, et al., Nanoscale 7 (2015) 11536.
[49] X. Chen, J. Shao, N. An, X. Li, H. Tian, C. Xu, et al., J. Mater. Chem. C 3 (2015) 11806.
[50] R.A. Whiter, V. Narayan, S. Kar-Narayan, Adv. Energy Mater. 4 (2014) 1400519.
[51] L. Persano, A. Camposeo, C. Tekmen, D. Pisignano, Macromol Mater Eng 298 (2013) 504.
[52] X. Liu, J. Ma, X. Wu, L. Lin, X. Wang, ACS Nano 11 (2017) 1901.
[53] S.K. Ghosh, D. Mandal, Nano Energy 53 (2018) 245.
[54] K. Maity, D. Mandal, ACS Appl. Mater. Interfaces 10 (2018) 18257.
[55] S.K. Ghosh, D. Mandal, in: H. Song, R. Venkatachalam, T.A. Nguyen, H.B. Wu, P. Nguyen-Tri (Eds.), Nanobatteries and Nanogenerators, Elsevier, 2021, p. 555.
[56] S. Siddiqui, H.B. Lee, D.-I. Kim, L.T. Duy, A. Hanif, N.-E. Lee, Adv. Energy Mater. 8 (2018) 1701520.
[57] M.-H. You, X.-X. Wang, X. Yan, J. Zhang, W.-Z. Song, M. Yu, et al., J. Mater. Chem. A 6 (2018) 3500.
[58] B. Li, F. Zhang, S. Guan, J. Zheng, C. Xu, J. Mater. Chem. C 4 (2016) 6988.
[59] M. Kanik, O. Aktas, H.S. Sen, E. Durgun, M. Bayindir, ACS Nano 8 (2014) 9311.
[60] V. Forti, C.P. Balde, R. Kuehr, G. Bel, The Global E-waste Monitor 2020: Quantities, Flows and the Circular Economy Potential, United Nations University/United Nations Institute for Training and Research, International Telecommunication Union, and International Solid Waste Association, Bonn/Geneva/Rotterdam, 2020.
[61] A. Zabala, Nat. Sustain. 2 (2019) 353.
[62] D. Jiang, B. Shi, H. Ouyang, Y. Fan, Z.L. Wang, Z. Li, ACS Nano 14 (2020) 6436.
[63] S.K. Ghosh, D. Mandal, Nano Energy 28 (2016) 356.
[64] S.K. Ghosh, D. Mandal, ACS Sustain. Chem. Eng. 5 (2017) 8836.
[65] S.K. Ghosh, D. Mandal, Appl. Phys. Lett. 109 (2016) 103701.
[66] S.K. Ghosh, D. Mandal, Appl. Phys. Lett. 110 (2017) 123701.
[67] A. Sultana, S.K. Ghosh, V. Sencadas, T. Zheng, M.J. Higgins, T.R. Middya, et al., J. Mater. Chem. B 5 (2017) 7352.
[68] E.J. Curry, K. Ke, M.T. Chorsi, K.S. Wrobel, A.N. Miller, A. Patel, et al., PNAS 115 (2018) 909.
[69] S.K. Ghosh, P. Adhikary, S. Jana, A. Biswas, V. Sencadas, S.D. Gupta, et al., Nano Energy 36 (2017) 166.
[70] E.J. Curry, T.T. Le, R. Das, K. Ke, E.M. Santorella, D. Paul, et al., PNAS 117 (2020) 214.
[71] C.R. Bowen, J. Taylor, E. LeBoulbar, D. Zabek, A. Chauhan, R. Vaish, Energy Environ. Sci. 7 (2014) 3836.

[72] Q. Li, S. Li, D. Pisignano, L. Persano, Y. Yang, Y. Su, Nano Energy 86 (2021) 106045.
[73] Y. Yang, W. Guo, K.C. Pradel, G. Zhu, Y. Zhou, Y. Zhang, et al., Nano Lett. 12 (2012) 2833.
[74] M. Xie, S. Dunn, E.L. Boulbar, C.R. Bowen, Int. J. Hydrog. Energy 42 (2017) 23437.
[75] D. Zabek, J. Taylor, E.L. Boulbar, C.R. Bowen, Adv. Energy Mater. 5 (2015) 1401891.
[76] D. Zabek, K. Seunarine, C. Spacie, C. Bowen, ACS Appl. Mater. Interfaces 9 (2017) 9161.
[77] T. Park, J. Na, B. Kim, Y. Kim, H. Shin, E. Kim, ACS Nano 9 (2015) 11830.
[78] Y. Yang, S. Wang, Y. Zhang, Z.L. Wang, Nano Lett. 12 (2012) 6408.
[79] Y.J. Ko, Y.K. Park, B.K. Yun, M. Lee, J.H. Jung, Curr Appl Phys 14 (2014) 1486.
[80] N. Ma, Y. Yang, Nano Energy 40 (2017) 352.
[81] L.F. Malmonge, J.A. Malmonge, W.K. Sakamoto, Materials Research 6 (2003) 469–473.
[82] H. Xue, Q. Yang, D. Wang, W. Luo, W. Wang, M. Lin, et al., Nano Energy 38 (2017) 147.
[83] F. Gao, W. Li, X. Wang, X. Fang, M. Ma, Nano Energy 22 (2016) 19.
[84] H. Ryu, H.-J. Yoon, S.-W. Kim, Adv. Mater. 31 (2019) 1802898.
[85] K. Song, R. Zhao, Z.L. Wang, Y. Yang, Adv. Mater. 31 (2019) 1902831.
[86] Y.J. Ko, D.Y. Kim, S.S. Won, C.W. Ahn, I.W. Kim, A.I. Kingon, et al., ACS Appl. Mater. Interfaces 8 (2016) 6504.
[87] J.-H. Lee, K.Y. Lee, M.K. Gupta, T.Y. Kim, D.-Y. Lee, J. Oh, et al., Adv. Mater. 26 (2014) 765.
[88] S.K. Ghosh, T.K. Sinha, M. Xie, C.R. Bowen, S. Garain, B. Mahanty, et al., ACS Appl. Electron. Mater. 3 (2021) 248.
[89] S.K. Ghosh, J. Park, S. Na, M.P. Kim, H. Ko, Adv. Sci. (2021) 2005010. Available from: https://doi.org/10.1002/advs.202005010.
[90] S.K. Ghosh, K. Roy, H.K. Mishra, M.R. Sahoo, B. Mahanty, P.N. Vishwakarma, et al., ACS Sustain. Chem. Eng. 8 (2020) 864.
[91] S.K. Ghosh, D. Mandal, in: S. Zafeiratos (Ed.), 2D Nanomaterials for Energy Applications, Elsevier, 2020, p. 1.

Index

Note: Page numbers followed by "*f*" and "*t*" refer to figures and tables, respectively.

F

FSC
www.fsc.org
MIX
Papier | Fördert
gute Waldnutzung
FSC® C083411

Kolibri 360 GmbH
Ferdinand-Jühlke-Straße 7
99095 Erfurt, Deutschland
produktsicherheit@kolibri360.de